Wild Rangelands

Conservation Science and Practice Series

Published in association with the Zoological Society of London

Wiley-Blackwell and the Zoological Society of London are proud to present our *Conservation Science and Practice* series. Each book in the series reviews a key issue in conservation today. We are particularly keen to publish books that address the multidisciplinary aspects of conservation, looking at how biological scientists and ecologists are interacting with social scientists to effect long-term, sustainable conservation measures.

Books in the series can be single or multi-authored and proposals should be sent to:

Ward Cooper, Senior Commissioning Editor, Wiley-Blackwell, John Wiley & Sons, 9600 Garsington Road, Oxford OX4 2DQ, UK
Email: wcooper@wiley.com

Each book proposal will be assessed by independent academic referees, as well as our Series Editorial Panel. Members of the Panel include:

Richard Cowling, Nelson Mandela Metropolitan University, Port Elizabeth, South Africa
John Gittleman, Institute of Ecology, University of Georgia, USA
Andrew Knight, University of Stellenbosch, South Africa
Georgina Mace, Imperial College London, Silwood Park, UK
Daniel Pauly, University of British Columbia, Canada
Stuart Pimm, Duke University, USA
Hugh Possingham, University of Queensland, Australia
Peter Raven, Missouri Botanical Gardens, USA
Helen Regan, University of California, Riverside, USA
Alex Rogers, Institute of Zoology, London, UK
Michael Samways, University of Stellenbosch, South Africa
Nigel Stork, University of Melbourne, Australia

Previously published

Reintroduction of Top-Order Predators
Edited by Matt W. Hayward and Michael J. Somers
ISBN: 978-1-4051-7680-4 Paperback; ISBN: 978-1-4051-9273-6 Hardcover; 480 pages; April 2009

Recreational Hunting, Conservation and Rural Livelihoods: Science and Practice
Edited by Barney Dickson, Jonathan Hutton and Bill Adams
ISBN: 978-1-4051-6785-7 Paperback; ISBN: 978-1-4051-9142-5 Hardcover; 384 pages; March 2009

Participatory Research in Conservation and Rural Livelihoods: Doing Science Together
Edited by Louise Fortmann
ISBN: 978-1-4051-7679-8 Paperback; 316 pages; October 2008

Bushmeat and Livelihoods: Wildlife Management and Poverty Reduction
Edited by Glyn Davies and David Brown
ISBN: 978-1-4051-6779-6 Paperback; 288 pages; December 2007

Managing and Designing Landscapes for Conservation: Moving from Perspectives to Principles
Edited by David Lindenmayer and Richard Hobbs
ISBN: 978-1-4051-5914-2 Paperback; 608 pages; December 2007

Conservation Science and Practice Series

Wild Rangelands: Conserving Wildlife While Maintaining Livestock in Semi-Arid Ecosystems

Edited by

Johan T. du Toit, Richard Kock and James C. Deutsch

A John Wiley & Sons, Inc., Publication

This edition first published 2010, © 2010 by Blackwell Publishing Ltd

Blackwell Publishing was acquired by John Wiley & Sons in February 2007. Blackwell's publishing program has been merged with Wiley's global Scientific, Technical and Medical business to form Wiley-Blackwell.

Registered office: John Wiley & Sons Ltd, The Atrium, Southern Gate, Chichester, West Sussex, PO19 8SQ, UK

Editorial offices: 9600 Garsington Road, Oxford, OX4 2DQ, UK
The Atrium, Southern Gate, Chichester, West Sussex, PO19 8SQ, UK
111 River Street, Hoboken, NJ 07030-5774, USA

For details of our global editorial offices, for customer services and for information about how to apply for permission to reuse the copyright material in this book please see our website at www.wiley.com/wiley-blackwell

Library of Congress Cataloguing-in-Publication Data
Wild rangelands : conserving wildlife while maintaining livestock in semi-arid ecosystems / edited by Johan du Toit, Richard Kock, and James Deutsch.
    p. cm.
  Includes index.
  ISBN 978-1-4051-7785-6 (pbk. : alk. paper) – ISBN 978-1-4051-9488-4 (hardcover : alk. paper)
  1. Grassland animals – Conservation.   2. Range management.   3. Livestock – Environmental aspects.   I. Du Toit, Johan T.   II. Kock, Richard.   III. Deutsch, James.
  QL115.3W55 2010
  636.08′45 – dc22

2009038740

ISBN: 978-1-4051-7785-6 (paperback) and 978-1-4051-9488-4 (hardback)

A catalogue record for this book is available from the British Library.

Set in 10.5 on 12.5 pt Minion by Laserwords Private Limited, Chennai, India
Printed and bound in Malaysia by Vivar Printing Sdn Bhd

1   2010

Contents

Contributors

Steven R. Archer is a Professor at the University of Arizona. Trained as a plant ecologist and ecosystem scientist, his research focuses on the factors driving shifts in the abundance of grasses and woody plants in drylands, and how ecosystem processes are impacted when trees or shrubs replace grasses. sarcher@ag.arizona.edu

Sumanta Bagchi is a Ph.D. student studying rangeland function and dynamics in the Trans-Himalayas. He is based at India's Nature Conservation Foundation and the Department of Biology at Syracuse University. sumanta@ncf-india.org

Nick Baker is Lecturer in Higher Education and Ecology at the University of Queensland. His wildlife research focuses on the ecology of mammalian carnivores and their prey, especially in high-use national parks where human–carnivore interactions are inevitable. n.baker@uq.edu.au

Ricardo Baldi is a researcher at the Consejo Nacional de Investigaciones Científicas y Técnicas of Argentina, and coordinates the Wildlife Conservation Society's Patagonian and Andean Steppe Program. He investigates the ecological and human-driven processes affecting the functionality of populations of large native herbivores, particularly the guanaco, in Argentine Patagonia. rbaldi@cenpat.edu.ar

Batjav Batbuyan heads the social and economic section of the Mongolian Institute of Geography and directs the Mongolian NGO Center for Nomadic Pastoralism Studies. He focuses on Mongolian economic and regional development and internal migration, working on issues related to pastoral development including herder cooperatives, pastoral land tenure and community-based management. B_batbuyan@yahoo.com

Yash Veer Bhatnagar is Senior Scientist at the India-based Nature Conservation Foundation and Director of the India Program of the Snow Leopard Trust. He is interested in carnivore and mountain ungulate ecology, and is involved in policy development and capacity enhancement for snow leopard conservation. yash@ncf-india.org

Pablo Carmanchahi is a researcher at the Consejo Nacional de Investigaciones Científicas y Técnicas of Argentina, and a technical advisor to a wild guanaco management and conservation programme in La Payunia Provincial Reserve. Pablo researches the eco-physiological and population consequences of guanaco management, and develops sustainable alternatives for arid-lands production. pablocarman@gmail.com

Sarah Cleaveland is Reader of Comparative Epidemiology at the University of Glasgow. A veterinary epidemiologist, her research focuses on the epidemiology of infectious disease – including rabies, canine distemper, bovine tuberculosis, brucellosis, and echinococcosis – at the human–wildlife–domestic animal interface in northern Tanzania. s.cleaveland@bio.gla.ac.uk

Gabriel M. Crowley is biodiversity information officer with the Cooperative Research Centre for Tropical Savannas Management. She has worked on the ecology of Australia's tropical savanna region, and assists its managers by supporting information networks and producing user-driven publications and internet resources. Gabriel.crowley@cdu.edu.au

Charles Curtin is a landscape ecologist interested in how pastoralism sustains arid-land economy, ecology and culture. He directs research for the Malpai Borderlands Group, and works on cross-border conservation in East Africa and the Middle East. Charles is a lecturer and research associate at the Massachusetts Institute of Technology, and a faculty member at Antioch University. ccurtin@earthlink.net

James C. Deutsch directs the Africa Program of the Wildlife Conservation Society, helping protect and manage seventeen globally important conservation landscapes. He formerly taught conservation biology and behavioral ecology at Cambridge University and Imperial College and studied Uganda kob and cichlid fishes. jdeutsch@wcs.org

Sumiya Enkhbold is an ecologist whose research and field activities focus on community-based wildlife management and conservation in Mongolia. Ceo_taij@yahoo.com

Mauricio Failla is a wildlife manager focused on the conservation and sustainable use of Patagonian wildlife, particularly guanacos and birds. He is the former Director of the Provincial Wildlife Office of Río Negro, in Argentina's Patagonia. patagoning@argentina.com

María E. Fernández-Giménez is Associate Professor in the Warner College of Natural Resources at Colorado State University and Program Leader for the Rangeland Ecology degree. Her interests include community-based rangeland management institutions, and integration of scientific and local knowledge in collaborative ecosystem management, especially in Mongolia. Maria.Fernandez-Gimenez@ColoState.edu.

Pablo Ferrando focuses on the economic valuation of natural resources. He is an independent consultant for the United Nations Development Programme-Argentina and the Inter-American Development Bank. He has also worked for the Argentine government on conservation and sustainable use of wild South American camelids. pabloferrando@hotmail.com

Thomas L. Fleischner is Professor of Environmental Studies at Prescott College, where he teaches natural history, conservation biology and environmental policy. Co-founder of the North Cascades Institute, he has served on the Board of Governors of the Society for Conservation Biology and as President of its Colorado Plateau Chapter. tfleischner@prescott.edu

Martín Funes is Coordinator of the Wildlife Conservation Society's Patagonian and Andean Steppe Program in Argentina. His research focuses on carnivore and raptor ecology, exotic species ecology, monitoring techniques and threats to guanaco populations in northern Patagonia. mfunes@wcs.org

Stephen T. Garnett is Director of the School for Environmental Research at Charles Darwin University. His research covers many facets of sustainability in northern Australia and Southeast Asia, including society, culture, economics and the environment. stephen.garnett@cdu.edu.au

Katherine Homewood is Professor in Anthropology at University College London. She works at the interface between conservation and development. Her Human Ecology Research Group focuses on the implications of less-developed countries' rural resource use for environment and wildlife, and of environmental conservation and management for human welfare and livelihoods. k.homewood@ucl.ac.uk

Chloe Inskip is a Ph.D. student at the Durrell Institute of Conservation and Ecology, University of Kent. Her research focus is human–carnivore conflicts in tropical forest systems. Previously, she worked for the North of England Zoological Society (Chester Zoo), assisting to develop and implement their *in situ* felid conservation programmes. ci32@kent.ac.uk

Mike Kock is a senior field veterinarian for the Wildlife Conservation Society's Global Health Program. He works on rangeland and forest wildlife in Africa, the United States and Middle East. His location in South Africa facilitates his involvement in the AHEAD (Animal & Human Health for the Environment and Development) programme on mixed-species rangelands in trans-boundary southern Africa. mdkock@kingsley.co.za

Richard Kock is Programme Manager of Wildlife Health at the Zoological Society of London. He focuses on wildlife health capacity-building and economically important infectious diseases at the interface among humans, livestock and wildlife in rangeland systems in East, Central and North Africa, and in desert and grassland systems in the Middle East and South Asia. richard.kock@zsl.org

Alex S. Kutt is a research scientist at Australia's Commonwealth Scientific and Industrial Research Organisation (CSIRO) Sustainable Ecosystems. He studies the relationship among cattle grazing, fire, land condition and wildlife pattern, and how this information can be applied to biodiversity conservation in tropical savanna rangelands. alex.kutt@csiro.au

John D.C. Linnell is a senior researcher at the Norwegian Institute for Nature Research. He focuses on wildlife ecology, human–wildlife conflict and the science–policy interface, especially regarding large carnivores in human-dominated landscapes. He coordinates interdisciplinary projects in this field in Norway, Europe and India. john.linnell@nina.no

Silvio Marchini coordinates the Brazil-based Escola da Amazônia and People and Jaguars Coexistence Project. He is working on his Ph.D. thesis, entitled *Human Dimensions of the Conflicts between People and Jaguars in Brazil*, at the Wildlife Conservation Research Unit of the University of Oxford. silvio@escoladaamazonia.org

Charudutt Mishra is Science and Conservation Director of the U.S.-based Snow Leopard Trust and Trustee of India's Nature Conservation Foundation. He studies carnivores, mountain ungulates, rangelands, human–wildlife conflict and human ecology, and oversees research and conservation programmes in the Himalayas and Central Asia. charu@conservation.in

Tsewang Namgail is a Ph.D. student studying niche relationships of mountain ungulates in the Trans-Himalayas. He is based at the Nature Conservation Foundation in India and the Resource Ecology Group, Wageningen University, the Netherlands. namgail@ncf-india.org

Michael Norton-Griffiths has been a resident of East Africa since 1969. He conducts independent research into the economics of land use and land tenure in agricultural and pastoral systems. His website is www.mng.com. mng5939@gmail.com

Andrés Novaro directs the Patagonian and Andean Steppe Program of the Wildlife Conservation Society, and is a researcher at the Consejo Nacional de Investigaciones Científicas y Técnicas of Argentina. He focuses on the conservation of guanaco migrations, and studies the effects of poaching and predation on the recovery of guanacos and other wildlife. anovaro@wcs.org

John Odden is a researcher at the Norwegian Institute for Nature Research. He focuses on the conservation biology of large carnivores, including their ecology, behaviour, population dynamics and conflicts with livestock. john.odden@nina.no

Gregory Rasmussen is a researcher with the University of Oxford's Wildlife Conservation Research Unit. He is Director of Painted Dog Conservation in Zimbabwe, a research and conservation organization he founded in 1993 focusing on African wild dog population dynamics, with an interest in Allee effects, predator–prey interactions, home-range utilization and predator–farmer conflict. gregory.rasmussen@zoo.ox.ac.uk

Kent H. Redford is Director of the WCS Institute and Vice-president of Conservation Strategy at the Wildlife Conservation Society, where he has worked since 1997. His expertise includes park-based conservation, traditional resource use, poverty alleviation and conservation, and South American mammals. kredford@wcs.org

Mohammed Y. Said is Research Scientist at the International Livestock Research Institute in Kenya. He assesses trade-offs between poverty alleviation and wildlife conservation to improve policy and management options for pastoral lands. His background is in ecological monitoring specialized in aerial counts, remote sensing, mapping, spatial analysis and modeling. m.said@cgiar.org

George B. Schaller is Senior Conservationist with the Wildlife Conservation Society and Vice-president of Panthera, an NGO working to conserve the world's wild cats. For 50 years he has worked in the wilds of Asia, Africa and South America, studying and helping protect species as diverse as the mountain gorilla, lion, jaguar, tiger, giant panda and wild sheep and goats of the Himalayas. gschaller@wcs.org

Katie M. Scharf is a joint J.D.-Ph.D. (History) student at Yale University, with research interests in environmental law and environmental history. She began studying wildlife management in Mongolia in 2000, when she was a Henry Luce Scholar with the United Nations Development Programme in Ulaanbaatar. katherine.scharf@aya.yale.edu

Anthony R.E. Sinclair is Professor of Ecology at the Centre for Biodiversity Research, University of British Columbia, where he has been since 1975. He conducts research in Tanzania, Canada, Australia and

New Zealand, focusing on how ecosystems work and their conservation. sinclair@zoology.ubc.ca

D. Michael Thompson is Funding Manager with the Environment Agency for England and Wales, where he has worked on environmental protection projects in Kenya, Madagascar and Mozambique. Michael's recent work has focused on water provision and waste-water treatment. His research interests are the drivers of livelihood choices and their impact on land-use change. dmichaelthompson@hotmail.co.uk

Gavin Thomson is a virologist at TAD Scientific in Pretoria, South Africa. He extensively researches the diagnosis and control of a wide range of diseases, with a specialization in foot and mouth disease and African swine fever. gavin@tadscientific.co.za

Johan T. du Toit is Head of the Department of Wildland Resources and Co-director of the Jack H. Berryman Institute at Utah State University, where he is a professor. His research interests focus on large mammals as drivers of terrestrial ecosystem processes, with most of his experience being in African savannas. johan.dutoit@usu.edu

Adrian Treves founded the Carnivore Coexistence Lab at the Nelson Institute for Environmental Studies, University of Wisconsin-Madison. He studies balancing human needs and wildlife conservation, exploring people's conflicts with large carnivores. He measures problem-carnivore behaviour using spatial predictive models, and people's responses to and perceptions of conflict. atreves@wisc.edu

Ray Victurine is Director of Conservation Finance at the Wildlife Conservation Society. His work focuses on developing incentives, new financial opportunities and market mechanisms to value ecosystem services and generate revenue for conservation of wildlife and wild places. His experience in development and conservation financing in Africa and Latin America spans over 20 years. rvicturine@wcs.org

Brian Walker is Research Fellow at Commonwealth Scientific and Industrial Research Organisation (CSIRO) Sustainable Ecosystems in Australia, and Chair of the Resilience Alliance (www.resalliance.org). His interests are in the resilience of social–ecological systems. brian.walker@csiro.au

Susan Walker is a wildlife biologist in the Patagonian and Andean Steppe Programme of the Wildlife Conservation Society. Her research interests are conservation of the Andean cat, and landscape connectivity for Patagonian wildlife. swalker@wcs.org

John C.Z. Woinarski is Principal Scientist with the Northern Territory's Department of Natural Resources Environment The Arts and Sport, and an Adjunct Professorial Fellow with the School for Environmental Research, Charles Darwin University, Darwin, Australia. His research interests relate to biodiversity conservation in northern Australia. john.woinarski@nt.gov.au

Monica L. Wrobel has managed the Wildlife Conservation Society's Africa Program since 2000, during a rapid expansion to its present scope of activities in sub-Saharan Africa and Madagascar. Previously, Monica has focused on land-use planning and on species recovery projects for endangered birds and mammals, preceded by a 10-year background at various zoological institutions. mwrobel@wcs.org

Alexandra Zimmermann's interest is human–wildlife conflict. Having developed community-based conflict mitigation projects for jaguars in Brazil and elephants in India, she researches spatial analysis and decision-making models for best practice in conflict mitigation. She works at the Chester Zoo and is reading for a D.Phil at the University of Oxford. a.zimmermann@chesterzoo.org

Preface

The subject of this book is a dramatic rolling landscape of grass, trees and shrubs that stretches beyond the far horizons, for which there is increasing concern and debate about its appropriate or inevitable uses: livestock, wildlife, both or neither? It is a debate that is unlikely to be settled because value judgements are involved, but for the debate to be rational it requires a synthesis of current ideas and knowledge. We have thus invited a diverse set of experts to discuss the challenges of conserving wildlife while maintaining livestock communities in ecosystems that could be referred to as *wild rangelands*. We assume that anything 'wild' exists without complete human control – neither tamed nor cultivated. We use the term *wildlife* in its most common application, referring to indigenous free-ranging mammals and birds with an emphasis on large mammals (>5 kg). The term *rangeland* can be widely applied to any terrestrial ecosystem in which livestock production is practised with animals feeding on wild plants. This classification includes desert and tundra and most of what lies between, but the emphasis of this book is on semi-arid ecosystems. It was in such ecosystems that humans first arose, in which the genome we carry with us was selected, and in which wild ungulates were first domesticated to begin the livestock breeds of today. Wild rangeland is our natural habitat. Let any adequately nourished and unthreatened person sit on a rock with a vista of wild rangeland and you will have a contented person. But wild rangelands are rapidly disappearing, and it is in response to that unsettling fact this book has been written.

By definition, semi-arid ecosystems are moisture-limited, but humans have devised ways to artificially provide drinking water to livestock and irrigated water to crops. As a result, rangelands have undergone dramatic anthropogenic changes in the distribution, timing, and intensity of herbivory, coupled with conversion to croplands. Wildlife resources are decimated in most rangelands outside of protected areas because of hunting by humans for meat and to eliminate competition with, and predation on, livestock. Habitat change and loss then follow, closing options for wildlife populations to return to their former numbers and natural ranges. Mounting pressure to tap new energy resources is causing some rangelands to be impacted by oil and gas extraction, while others are being planted with biofuel crops to satisfy the ever-hungry machinery of modern life. Above everything, global warming is forcing change across all social–ecological systems that currently represent rangelands, with advancing aridity in most rangelands and increasing climatic variability across all.

As researchers, managers, politicians and administrators, we can only strive for a better understanding of the processes and predictions of global change, while devising and implementing action to protect the resilience of ecosystems and conserve the environmental services that so many people depend on for their livelihoods. This entails conserving biodiversity, of which, in rangelands, wildlife represents a fundamentally important part. Its ultimate importance lies in the plethora of co-evolved effects that wildlife species exert as top-down agents of ecosystem processes. Its proximate importance lies in the benefits people can derive from its natural abundance and accessibility for a multitude of sustainable uses, both consumptive and non-consumptive, in those areas of rangeland that can still be called 'wild'.

The seed of this book germinated in January 2006 when the conservation programmes of the Zoological Society of London and the Wildlife Conservation Society (New York) jointly organized a symposium in Regent's Park, London. That symposium, which was the brainchild of Peter Coppolillo and Richard Kock, was organized with the valuable assistance of Deborah Body. Key supporters were Georgina Mace from the Institute of Zoology, London, and John Robinson from the Wildlife Conservation Society, who approved funding for a diverse group of rangeland specialists from all continents to travel to London for a symposium entitled 'Wild Rangelands: Conservation in the World's Grazing Ecosystems'. From the outset, it was envisioned that a book would result from the symposium but it was only during and after the symposium that the book project really gained momentum. We invited some additional key contributors who had not participated in the symposium but we felt were needed to fill important gaps. Georgina Mace arranged and funded a meeting in London in December 2006, at which she and the editors met with Ward Cooper of Wiley–Blackwell Publishing and we finalized the design of this volume. Linda DaVolls played a very important part in the arrangement of that meeting and then subsequently, together with Lucinda Haines, in the circulation, editing and final approval of the publishing contract and authors' agreements. From then on, the book took a long and winding path as authors wrote their drafts, reviewers commented on them, editors asked for changes, authors sent back revisions, some authors dropped out, others replaced them and so on. Throughout this time, we worked closely with the editorial team at Wiley–Blackwell, where Delia Sandford encouraged us along and then passed on the baton to Rosie Hayden who patiently and tactfully kept us on task through to the end.

Drafts of each chapter were commented on by at least two reviewers, some of whom were authors of other chapters in this volume and others were independent experts from whom we specifically requested advice and comments. These included Peter Adler, Andrew Ash, Roy Behnke, Joel Berger, Luigi Boitani, Ivan Bond, Chris Carbone, Paul Cross, Eric Dinerstein, Paul Ferraro, Stephen T. Garnett, Doug Jackson-Smith, Urs Kreuter, Steve Osofsky, Fred Wagner, Brian Walker, Jane Wheeler, Kent H. Redford, Robin Reid and Ray Victurine. The contributions of each and every reviewer were extremely valuable for improving the quality of the published work.

Several authors have special acknowledgements to people and institutions for helping them individually. Steven R. Archer is grateful to Katie Predick, N. Pierce, S. Woods and Rob Wu for reviewing drafts and/or assisting with graphics, while his work was supported in part by NSF grants DEB-9981723 and DEB-0618210 and USDA-NRI grant 2005-35101-15408. Thomas L. Fleischner dedicates his chapter to the memory of Joy Belsky, and is grateful for assistance from Jayne Belnap, Lisa Floyd-Hanna, Ed Grumbine, Andy Holycross, Beth Painter, Dirk Van Vuren, Conrad Bahre and Dave Krueper. Katherine Homewood and D. Michael Thompson are grateful for research permission from the governments of Kenya and Tanzania, for past funding from DFID contracts R6828 and R7638 and EU contract ERBIC18*CT 960070 and for ongoing funding from the government of Belgium through DGIC for work in collaboration with ILRI. Charudutt Mishra and his co-authors acknowledge the Whitley Fund for Nature and the Ford Foundation for providing core support to their academic and conservation programmes, and they thank R. Raghunath for help in preparing the map. Alexandra Zimmermann is grateful for helpful comments from Claudio Sillero-Zubiri.

Throughout the project, Monica L. Wrobel was in the middle of the e-mail traffic among the editors, contributing her time without complaint and going far beyond the call of duty to help move the process along. One of her brain waves was to involve Jocelyn Ziemian as a copyeditor and manager of the final revisions, tasks that Jocelyn performed with utmost skill and efficiency.

Finally, our sincere thanks go to all the wise and patient authors who contributed their time and thought to this drawn-out project. There are no 'silver bullet' solutions but we feel confident this book will serve its purpose by supplying a volume of ideas and information to those with influence over the alternative futures of the world's rangelands.

Johan T. du Toit
Richard Kock
James C. Deutsch

Foreword

Rangelands or Wildlands? Livestock and Wildlife in Extensive Ecosystems

We realized as we wrote this foreword that much of our professional life has been spent in a study of rangelands, including Tanzania's Serengeti with its throbbing stream of migrating wildebeest, the uplands of Tibet, and the great steppes of Mongolia. Rangelands always lie in our thoughts or at the border of memory. To understand the ecology of savannas and grasslands – biomes that cover about 40% of the earth's land surface – is an urgent task, given the human assault on the environment through greed, ignorance and negligence. Science is ample justification for our research; it is the only tool we have for obtaining the kind of information that can lead to effective long-term conservation. Much of the biological world remains unknown, even the functioning of a seemingly simple ecosystem such as grassland. With ever-increasing rangelands being converted to agriculture, the fragile soils given to the wind, or degraded by too much livestock, conservation becomes an imperative, both for the survival of species and for the pastoral cultures. Mankind turns rangelands into deserts and calls it development.

For us to spend years of fieldwork in an area, often under harsh conditions, requires a passionate involvement that goes beyond science. On the steppes the sky seems wider and deeper, and the horizon gives way to horizon. In this austere vastness everything is silence and solitude and light yet vibrant with life. Grasslands and savannas somehow provide a sense of place. All of us who work in such open spaces may have this fusion of feeling with fact. That is perhaps not surprising. Humans had their roots in the savannas of Africa, and many hunter – gatherer societies were until recent times associated with such an environment. Human society, in all its diversity, evolved some 13,000 years ago by cultivating plants and domesticating wild ungulates. We thus have empathy not only with that kind of landscape but also with the human societies whose livelihoods depend on it.

Over the millennia, a distinctive way of life has evolved on rangelands in which pastoralists with their livestock shift seasonally to better pastures. Almost all of the world's rangelands now support livestock for all or part of the year. But traditional herding practices are changing rapidly as the human population increases and governments prefer that nomads be settled as a way of exercising better control. Many grasslands are being converted to agriculture, often in arid zones basically unsuited for this, but new genetic strains of crops now make it possible. Intensive agriculture provides more food per capita than does pastoralism – a trend we see in the Ngorongoro Conservation Area of Tanzania, one of the first areas set aside specifically to support pastoralists.

Communal pastures are giving way to fenced private plots, which, as so often happens, then become degraded because of overstocking with livestock. This is true whether we speak of the large ranches in the United States and Australia or the small ones in Kenya. Even the vast uplands of the Tibetan Plateau in China, areas too high for agriculture, have in recent years been divided into household plots. In times of drought or deep snow, livestock cannot now shift to better pastures. Fences, their construction subsidized by the government, further reduce the flexibility of any grazing strategy. Only the Mongolian government has so far resisted the temptation to fence its great eastern steppes, the finest temperate grassland surviving in the world. Whatever the land tenure system, development focuses on increasing livestock numbers, not on the sustainability of rangelands.

No ecosystem remains static. Grasslands can turn to shrub or desert, a trend certain to increase as global warming shifts vegetation zones. Similarly, cultures also adapt and change, sometimes rapidly and often in a way that is detrimental to the land. This makes conservation a process rather than a goal, and all we can do is devise new ideas, concepts, policies, and approaches to address the changing economic, social and ecological conditions. A basic goal is to conserve ecosystem function for the pastoralists who have no alternative livelihood, as well as for the plants and animals which have no alternative either.

The rangelands represent biological treasures in their beauty and variety. Some species are endemic, such as the wild tulips of Central Asia and the burrowing owl of North America, and some represent unsurpassed wildlife spectacles, among them the great migrations of Tibetan antelope, Mongolian gazelles, wildebeest and once, long ago, the American bison. All large ungulates are threatened by fences. Fences in Botswana caused mass death when wildlife was prevented access to drinking water, and the critically endangered Przewalski gazelle in China dies when it tries to squeeze through fences too high to jump. Tibetan wild asses run into and break fences, something for which they are disliked. And should a household tolerate a large herd of asses on its private land eating grass that has been saved as winter feed for domestic yak, sheep and goats?

Government policies, based on ignorance, have an impact on rangelands as well. Prairie dogs, once abundant and now uncommon on the prairies of North America, were treated as vermin because they were thought to compete with livestock for forage even as evidence showed that their activities benefited rangelands. The black-footed ferret, a specialist predator of prairie dogs, became almost extinct as a consequence. In a replay of a disastrous policy, China is poisoning the plateau pika, the ecological equivalent of the prairie dog, on the Tibetan Plateau, and this will ultimately have a wide-ranging effect on the ecosystem, on everything from snow finches that nest in pika burrows to the nutrient content of plants. Must the same mistakes be made again and again? The time and kind of interventions, rather than haphazard and impulsive action, are critical to successful management.

Competition and conflict between humans, livestock, and wildlife will increase as species receive better protection in many areas. Wolves in Tajikistan, lions in Tanzania, snow leopards in Mongolia – everywhere large predators come into conflict when they prey on livestock, particularly where natural prey has been eliminated or greatly reduced. It is a difficult issue, yet one that needs at least partial resolution. Wildlife management is actually people management. Local concerns must be addressed if wildlife and pastoralists are to coexist under increasingly crowded conditions. Local people have a basic stake in the land, their livelihood is not expendable, and any management solution must also consider their interests, knowledge, culture, and cooperation. Solutions are always complex and require an integrated approach for long-term success. Management must also confront the issue of limits – on number of people, domestic animals, and wildlife, as well as on fences, roads and other development. Ideally, communities should take the initiative in protecting and managing their own resources: conservation ultimately depends on their goodwill and participation. Such initiatives are slowly evolving and need to be widely emulated. For example, several Tibetan communities in China have formed conservation committees; they protect and monitor wildlife, and they have set aside areas for wild yak, blue sheep and other species where livestock grazing is prohibited.

The complex and conflicting demands of biodiversity conservation, economic development and survival of local cultures must attain some level of dynamic stability. Yet, every country also wants to save fragments of its natural heritage, the beauty of its landscape and all it contains as a part of its past and a gift to the future. Ecological wholeness, aesthetics, spiritual values and ethics are an integral part of conservation too. Protected areas must be established and maintained without compromise, unaffected by greed and as a witness to our moral obligation assuring the survival of species. They are also essential repositories of species, especially vulnerable ones that cannot exist with humans; places where the structure and function of an ecosystem can be studied and monitored and sites against which changes elsewhere can be measured.

However, most protected areas cannot endure as islands of habitat in a sea of humanity; basic ecological processes cannot persist in isolation. What we know of species loss in such habitats makes it imperative to protect whole landscapes. In spite of the conflicting demands of society and wilderness in which development usually has priority over conservation, rangelands can provide 'soft boundaries' between strictly protected areas that exclude most human use and rural communities in which livestock grazing is consistent with the maintenance of natural habitats and coexistence with wild species. Rangelands thus provide a buffer between two opposing land uses thereby allowing community-based conservation. Such a plan conforms to the 'Malawi Principles', formulated for the Convention on Biological Diversity in 1998. Underlying these are two basic tenets. One is that all stakeholders must be involved in conservation planning and the other is that such planning should consider the ecosystem rather than an individual species as the unit of management. These principles are particularly applicable to rangelands because there is often considerable information on the dynamics of ecosystems from both stakeholders and scientists, and pastoralism lends itself well to long-term use and conservation of the indigenous biota.

We have made these comments to emphasize that the aim of rangelands conservation is to maintain healthy, productive and diverse ecosystems with vigorous populations of all animal and plant species coexisting in most areas with pastoralists and their livestock. There is a dynamic relationship between these and we must assure that it is maintained. Rangelands are resilient. We are fortunate that large expanses still endure, even if all have been modified by humans to some extent. Good management options persist and solutions need only to be applied based on knowledge, wisdom and persistence. The contributions to this volume range from Mongolia and Argentina to the United States, and they cover many issues and details including grassland management, disease, predation, history and policy. The facts and ideas can, with slight modification, be applied to any of the world's rangelands in our search for ecological harmony based on love, respect and responsibility for the land.

Anthony R.E. Sinclair
George B. Schaller

$$\textbf{1}$$

Introduction: A Review of Rangeland Conservation Issues in an Uncertain Future

Monica L. Wrobel and Kent H. Redford

Wildlife Conservation Society, NY, USA

Rangelands are typically those areas used by people to graze their livestock, converting plants into products of use to humans, like meat, milk, blood, leather and wool (Brown 2001). In ecological terms, there is no one habitat type for which the term *rangelands* applies; instead, these areas are typically classified as grassland, savanna or scrublands, or some combination of these. In fact, livestock can be grazed in any type of ecosystem where there is sufficient accessible food and may include dry forests, coniferous forests, broadleaf riparian forests, deserts (Chapter 9) and even agricultural lands after crops have been harvested.

Around 40% of the earth's land surface supports the world's grazing domestic animals. An estimated 200 million people are pastoralists tending cattle, sheep or goats (Brown 2008), and with increasing demand for animal products in human diets, the number of livestock is expected to grow substantially. Additionally, a large number of wild animals that also graze or browse share these rangelands. Millions of wild ungulates, such as Tibetan antelope, guar, guanacos, bison, antelope, elk and zebra uneasily share the plant resources with cattle, camels, yaks, sheep and goats. To wild animals, these are not rangelands, but their grassland homes. This book addresses the opportunity to conserve wildlife in places where people manage livestock and to safeguard these places as wild rangelands.

Wild Rangelands: Conserving Wildlife While Maintaining Livestock in Semi-Arid Ecosystems,
1st edition. Edited by J.T. du Toit, R. Kock, and J.C. Deutsch.
© 2010 Blackwell Publishing

As human populations shift geographically, so also do rangelands; organizations such as the Food and Agriculture Organization of the United Nations keep data on the evolving importance of livestock grazing to national economies. The most expansive areas of semi-arid grassland, savanna and shrub land supporting domestic livestock across several countries are on the African continent: across the band of countries in the Sahel region south of the Sahara and in east and southern Africa (EOE 2009), where populations depend heavily on livestock economies for food and employment. Similarly, the Middle East, Central Asia, Mongolia, Northwest China and the Indian subcontinent depend on domestic livestock economies and on open access to rangeland to support local livelihoods (Brown 2008; EOE 2009). Rangeland is dominant in Australia, and grass-based livestock economies predominate in several countries in South America. The Great Plains of North America support grazing cattle as a form of land use in semi-arid land that is not otherwise suited to growing grain (Brown 2008).

Rangelands are therefore globally important and are some of the last, great wild lands. They are important for conservation as many of them continue to sustain wildlife outside of national parks but are under no formal protection. Amongst the millions of people also supported there, many are very poor and dependent on access to land for their livestock. Despite concern by researchers, conservationists and development practitioners, these rangelands and their inhabitants have not been adequately addressed in national or international policies.

The low and variable productivity characteristic of semi-arid ecosystems and the resulting need for extensive areas means that large, connected areas are essential for both domestic animals and wildlife. However, the new century finds that patterns of habitat fragmentation, typical of the world's forests, now extends to rangelands and threatens the existence of all who rely on them. Similarly, the drivers of deforestation – unclear land tenure, short-term profit-taking, underrated values of natural systems and poor resource management – are poised to further wreak havoc worldwide on rangelands. It is vital to better understand these systems and their residents and develop approaches that ensure the resilience of these coupled socio-ecological systems in the face of economic, social, environmental and climate changes.

To address the issues, in early 2006, the Wildlife Conservation Society joined the Zoological Society of London in the lecture hall at Regent's Park, London, convening speakers from around the world for the symposium

'Wild Rangelands: Conservation in the World's Grazing Ecosystems'. The symposium set out to explore a variety of strategies for conserving wildlife on rangelands centred on: rangeland ecology; the interface of people, wildlife and livestock; and policy, planning and economics. Over 2 days, the discussions involved the challenges in maintaining livestock and wildlife in rangelands and ways to assist and engage the human population dependent on them. Since that symposium, the world has witnessed an increase in food and biofuel prices and land conversion, an unprecedented economic collapse and predictions of increasing stressors from climate change. These recent changes have made it even more urgent for conservationists, cultural advocacy groups, development planners and landholders to unite in the common agenda of ensuring the integrity of rangelands and the value of wild lands. This book seeks to share the outputs of the symposium and to highlight the challenges and opportunities for averting the loss of functioning ecosystems where historically humans, domestic animals and wildlife have coexisted.

Global drivers threatening rangelands

At the beginning of this century, the Millennium Ecosystem Assessment (MA) employed a scenario-building technique as a holistic, integrated and participatory approach to help in the understanding of intrinsic heterogeneity and uncertainty in ecosystem management (Lebel et al. 2005). The key driving forces of the scenarios in the MA included population, income, technological development, changes in consumption patterns, land use change and climate change (Alcamo et al. 2005).

The scenarios identified three regions as susceptible to rapid changes in ecosystems services. The first region was Central Africa because of increased population, expansion of agriculture, increase in food and water consumption, intensified use and contamination of ground waters. The second region was the Middle East, predicted to experience an increased population and income and therefore increased demands for food, meat and a higher dependence on food imports. In South Asia, the third region, the MA predicted intensified agriculture and a breakdown in ecosystem services (Alcamo et al. 2005). Across all the MA scenarios, a resounding message is that 10–20% of current grassland and forestland will be lost, mainly because of the expansion of agriculture and secondarily because of the expansion of cities and infrastructure (Alcamo et al. 2005).

Land degradation

The majority of the world's 3.3 billion cattle, sheep and goats graze on 40% of the earth's land surface, almost half of which is lightly to moderately degraded, with 5% being severely degraded (Brown 2008). Desertification, where soil cover is lost through wind and rain erosion driven by overgrazing and disappearance of vegetation, is increasing and affecting about a third of the world's dry lands (Montgomery 2007). In some rangelands like those of Argentina's Patagonian steppe, desertification has affected nearly 30% of the habitat (del Valle 1998). Dust storms appear to be increasing and moving great distances, for instance, depositing dust from Sahelian Africa westward across the Atlantic Ocean and landing in Caribbean reef systems (Brown 2001, 2008).

Better management, recovery and planning for use of grassland systems are rare, but possible. If intervention against land degradation comes early enough, at least for croplands, experience in Kazakhstan has shown that some recovery to grassland can occur (Brown 2001). In the United States, the U.S. Congress created the Conservation Reserve Program (CRP) in 1985 to reduce soil erosion and control overproduction of basic commodities. By 1990, 10-year contracts existed for 14 million hectares of highly erodible land with permanent vegetative cover. Farmers received payment to plant fragile cropland with grass or trees. The soil erosion in the United States decreased from 3.1 billion tons to 1.9 billion tons over 15 years (Zinn 2001; Brown 2008).

Internationally, the United Nations Convention to Combat Desertification (UNCCD) was created in 1994 with an increase in development aid targeted at this serious threat. The Conference of Parties (COP) met for its eighth session in September 2007 and adopted a 10-year strategic plan and framework for 2008–18. The four strategic objectives are to improve the living conditions of affected populations, improve the condition of affected ecosystems, generate global benefits through effective implementation of the UNCCD and mobilize resources to support implementation of the Convention through building effective partnerships between national and international actors. The primary form of the implementation is through National Action Programs (NAP) complemented by sub-regional and regional action programs where appropriate (UNCCD 2009).

The United Nations Environment Program (UNEP) estimates that an effective 20-year global effort to combat desertification would cost $10–22 billion per year. UNEP estimates that desertification currently costs countries

$42 billion per year in lost revenue (UNCCD 2009). This foregone income does not include estimates of ecosystem services yet to be quantified (Brown 2008). Since 2003, the Global Environmental Facility adopted land degradation as a focal area, increasing available financial resources (UNCCD 2009). In addition, synergistic approaches are being explored with the other international conventions, the UN Framework Convention on Climate Change and the Convention on Biodiversity (Mouat et al. 2006). Incorporating considerations about wildlife within rangelands to protect ecosystem functions and combat land degradation warrants much further attention.

Food demand, food prices and biofuel

Threats of over-exploitation and conversion of rangelands were once primarily driven by local livelihood issues. Now threats driven by increasing global markets for food and biofuels are intensifying the challenges to sustaining rangelands. All of the four scenarios of global change outlined in the MA predicted 50% increases in the world's total production of grain (Alcamo et al. 2005). Some of the gains in agricultural land will be at the expense of yet uncultivated natural land – much of which may be rangelands. Examples have been emerging in 2008, such as a reported deal by Abu Dhabi to develop 30,000 hectares of land in Sudan to ensure food security in the emirate. Other Middle Eastern countries are securing land deals from Brazil to Pakistan to guarantee supplies of cereals, meat and vegetables (Rice 2008).

The labelling of rangelands as 'marginal' lands or 'wastelands'– regarded as marginal for traditional agriculture – is commonplace in national and international policies and makes such areas prime targets for the kinds of predicted growth in some form of production system. Much of the increasing crop and biofuel production involves bringing such lands under agricultural production worldwide. A story in 2008 in the *New York Times* reported that thousands of farmers in the United States were taking their fields out of the government's biggest conservation program that otherwise paid farmers not to cultivate, to realize more profitable land use under wheat, soybean, corn or other crop production in response to price increases for those commodities (Streitfeld 2008). Elsewhere, rangelands are threatened by the rise in the planting of *Jatropha*, an arid-adapted succulent plant with high potential as a source of biofuel. Often, the areas cited to be of potential suitability for its production are marginal lands, a term that includes rangelands.

Increased income has resulted in growing demand in developing countries for animal products. Meat demand in developing countries rose from 11 to 24 kg/capita/year during the period 1967–97, achieving an annual growth rate of more than 5% by the end of that period (FAO 2006; Smith et al. 2007). By 2020, Rosegrant et al. (2001) forecast a further increase of 57% in global meat demand, mostly in South and Southeast Asia and sub-Saharan Africa (Smith et al. 2007). Grazing livestock is not the source of all of the increased demand, which includes that from fish and poultry. Nonetheless, processed feed from cereals for raising meat or poultry adds to an overall demand for bringing land under cultivation.

There are several reasons for including wildlife as grazers in rangeland systems and keeping stocking levels of domestic livestock low, despite pressures to use the land otherwise. Wild herbivores represent a suite of species that have evolved in the wider ecosystem that graze in less concentrated ways, avoid the risk of soil erosion and soil compaction and have a reduced need for water. Wild grazing species, converting vegetation for their energy and growth (i.e. meat) and depositing seeds and nutrients, altogether distribute biomass across the landscape. Wild species maintain environmental services over great lengths of time, in ecosystem processes that have co-evolved. Removing or displacing wild grazers from rangelands and creating the mixed crop–livestock food production systems recommended by development and agriculture sectors result in the transfer of nutrients in dung from the grasslands to pens. This interrupts the critical nutrient cycle by not returning nitrogen and phosphate in the form of manure to the rangeland itself, essentially 'nutrient mining' the landscape.

Increasing incentives for biofuel production, increasing demand for food with a concomitant intensification and expansion in area for crops and mixed crop/livestock systems will undoubtedly be at the expense of wildlife, pastoral peoples and native ecosystems. This multi-pronged threat highlights the need for the sustainable management of rangelands as rangelands and indeed as wild rangelands.

Poverty

The first of the UN Millennium Development Goals (MDG) is to halve extreme poverty and hunger by 2015. The World Bank states that this goal is within reach for all regions of the developing world, except for sub-Saharan Africa, where 800 million residents are in countries sliding deeper into poverty (Brown 2008). Rangelands play a key and under-appreciated role in development strategies for addressing global poverty.

When measuring the wildest areas on a spectrum from the completely wild to the extremely transformed (Sanderson et al. 2002), a spatial analysis of global poverty undertaken within that spectrum showed that the greatest number of the world's most poor, 7.2 million people, live in tropical and subtropical grasslands, savannas and shrub lands. Additionally, 3.3 million people live in desert and xeric shrub lands and 1.2 million live in montane grasslands and shrub lands. These areas, defined as the most wild places by Sanderson et al. (2002) are collectively sustaining the largest number of the world's most poor people (Redford et al. 2008). Especially in the areas where people are striving to live off the land, conservation groups are well positioned to develop new partnerships for the delivery of benefits and to ensure sustainable environments for some of the least accessible poor people in remote places in the world. There are opportunities and imperatives to combine conservation, donor and development attention, to improve the chances of sustaining rangelands, wild rangelands in particular. Redford et al. (2008) recognize that a new socially responsible, long-term approach to conservation of the world's wildlife and wild places may present a circumstance of an unusual synergy of poverty alleviation efforts for approximately 16 million people, along with conservation of the wildest places.

Climate change effects

Though only limited modelling work has been done, rangelands are predicted to be affected both directly and indirectly by climate change. Direct changes will affect species, ecosystems and livelihood systems. It will also change the area and distribution of the rangelands themselves. Indirect changes will include those resulting from mitigation efforts to diminish the causes of climate change, such as greenhouse gas emissions generated by agricultural or land conversion practices.

The Intergovernmental Panel on Climate Change (IPCC) produced an updated report in 2007, citing a prediction that a 1 to 2°C increase in temperature over that of the 20-year period preceding the year 2000 would place up to 30% of species at a higher risk of extinction. Effects might include a shift in species' ranges with the possibility of disrupting ecosystem functions (Bernstein et al. 2007). Rangelands are still generally regarded as relatively unimportant to global CO_2 levels compared to forest, but changes in their woody vegetation and/or release of carbon stored in soils warrant further evaluation. Neither rangeland nor semi-arid habitats are a focus *per se* in the IPCC report and therefore extrapolations from the agriculture data must be made. At mid to high latitudes, cereal productivity may increase and this could be through existing land under cultivation or by bringing new land (such as wild land) under cultivation. The predictions include tendencies for cereal productivity to decrease in low latitudes; in warmer regions, yields of most crops would be lower; crop damage and failure are expected; and drought and livestock deaths are expected to increase (Bernstein et al. 2007). Increases in the extent of arid and semi-arid land are predicted for Africa, and semi-arid vegetation is likely to be replaced by arid vegetation in Latin America. Worldwide decreases in freshwater availability are expected (Bernstein et al. 2007).

Livestock, predominantly ruminants such as cattle and sheep, are notable sources of methane gas (CH_4) from eructation and manure production, accounting for about one-third of global anthropogenic emissions of methane (EPA 2006; Smith et al. 2007). The IPCC notes that the intensive production of beef, poultry and pork is increasingly common with concomitant green house gas (GHG) emissions such as methane and nitrous oxide (Barker et al. 2007). Technological advances with the use of intensive systems – such as providing higher quality feed and producing lower emissions from better gut fermentation – are not sufficiently developed to counteract the effects to a great degree (Barker et al. 2007). Even so, the conversion of land to produce such modified, higher quality feeds could likely be at the expense of rangelands.

An increasing use of 'marginal lands' is expected to further contribute to climate change through degradation and loss of carbon storage in soils and vegetation. Loss of these natural rangelands, with their greater resilience to climate change, will further compound this problem. The risks associated with failures of traditional crops and vulnerability of livestock will deem such practices unjustified in degrading what wild rangelands remain. Despite these predictions, there is a poor understanding about the future of rangelands in a climate-changed world.

Seven global case studies

To understand the complexity of the issues and illustrate the lack of globally coordinated efforts towards rangeland conservation, this book presents seven case studies from around the world. These cases exemplify the challenges not only for sustaining grasslands, savannas and shrub lands for livestock production but also for restoring and maintaining the ecological integrity of those same landscapes as wild lands and as homes for their native fauna. These cases document examples in which livestock grazing or introduced plant species have led to degraded systems and economic collapse, ecological damage and extensive fires. They address issues such as access to benefits at individual and state levels, as well as inadequacies of conservation measures for either. They discuss how global market demands can drive stocking rates beyond sustainable limits, the effects of trans-boundary strife and the uncertainty of the effects of climate change on vegetation and grazing practices. Collectively, these case studies provide a glimpse into the global future of rangelands and their multiple values.

Global themes for rangeland conservation

The next section of the book presents a thematic approach to the problem of conserving the resilience of rangelands as socio-ecological systems. The emphasis in these chapters is on identifying global challenges and suggesting solutions. The authors do not shy away from recognizing the challenges posed by the dual objectives of wildlife conservation and livestock production. The chapters in this section span all levels of ecological and institutional organization from the micro-realm of soil decomposer communities to disease and health dynamics, the functional properties of top predators, management institutions, global economics and climatic effects. The authors map out the complexity of the issues and the spectrum of expertise and dialogue that would be available for conserving wild rangelands.

Lessons from the rangelands

Finally, 'lessons' are extracted from each chapter and presented not only as a synthesis of the book but and perhaps more importantly, also as a call for concerted action from the conservation community and policymakers at large to exploit every opportunity for conserving wild rangelands on a global scale. While the world attempts to monitor and regulate forestry and reduced deforestation, rangeland conservation needs much greater attention than currently exists in the global policy arena.

This multifaceted and practical book is itself an illustration of the scope of expertise that has come of age for tackling what has been an inexorable and recently accelerated degradation of a unique suite of ecosystems, on which initially wildlife and then mankind has depended for thousands of years. The very existence of these habitats has become imperilled more quickly than their managers have had the opportunity to learn about them. This book takes vital, early steps in addressing significant challenges and articulating what is necessary to sustain wild rangelands.

To move from the lessons laid out in this volume and begin to address the challenges, rangeland scientists and conservationists will need to work with economists and policymakers to define how productive intact systems can be valued as wild lands – far beyond the view of fodder for livestock or marginal land ready for conversion. There has been too little consideration of the roles of wild rangelands as providers of carbon storage, erosion prevention, nutrient cycling and moisture retention, as gene banks for drought-resistant species and sources of future drugs. With 50% or more of the world's rangelands being overgrazed, an alternative valuation will level the playing field for conservation, wildlife and the people living there and remove other short-term and destructive overuse of these fragile soils. At appropriate stocking levels of domestic herds in rangelands or indeed in appropriately 'under-stocked' wild lands, rangelands can serve as a means for cultural retention and poverty alleviation in some of the remotest, poorest and most vast areas. Conserving the ecological integrity of wild rangelands and their suite of ecosystem services will increase their resilience in the face of climate and economic change and offer a more robust future for both people and wildlife.

References

Alcamo, J., van Vuuren, D., Cramer, W. et al. (2005) Changes in ecosystem services and their drivers across the scenarios. In: Carpenter, S.R., Pingali, P.L., Bennett, E.M. & Zurek, M.B. (eds.) *Ecosystems and Human Well-Being: Scenarios. Volume 2, Millenium Ecosystem Assessment.* Island Press, Washington, D.C., pp. 297–373.

Barker, T., Bashmakov, I., Alharthi, A. et al. (2007) Mitigation from a cross-sectoral perspective. In: Metz, B., Davidson, O.R., Bosch, P.R., Dave, R. & Meyer, L.A. (eds.) *Climate Change 2007: Mitigation. Contribution of Working Group III to the Fourth Assessment Report of the Intergovernmental Panel on Climate Change.* Cambridge University Press, Cambridge and New York, pp. 619–690.

Bernstein, L., Bosch, P., Canziani, O. et al. (2007) Climate change 2007: synthesis report, summary for policymakers. An assessment of the Intergovernmental Panel on Climate Change. http://www.ipcc.ch/pdf/assessment-report/ar4/syr/ar4_syr_spm.pdf (accessed March 2009).

Brown, L.R. (2001) *Eco-Economy: Building an Economy for the Earth.* W.W. Norton & Company, Inc., New York.

Brown, L.R. (2008) *Plan B 3.0: Mobilizing to Save Civilization.* W.W. Norton & Company, Inc., New York.

Encyclopedia of Earth (EOE) (2009) The Encyclopedia of Earth Anthropogenic Biomes. http://www.eoearth.org/eoe-maps/gm/Anthromes/gmviewer.html (accessed March 2009).

Environmental Protection Agency. (2006) *Global Anthropogenic Non-CO2 Greenhouse Gas Emissions: 1990–2020.* United States Environmental Protection Agency, Washington, D.C. http://www.epa.gov/nonco2/econ-inv/downloads/GlobalAnthroEmissionsReport.pdf (accessed 26 March 2007).

Food and Agricultural Organization of the United Nations (FAO) (2006) FAOSTAT Agricultural Data. http://faostat.fao.org (accessed March 2007).

Lebel, L., Thongbai, P., Kok, K. et al. (2005) Sub-global scenarios. In: Capistrano, D., Samper, C., Lee, M.J., Raudsepp-Hearne, C. (eds.) *Ecosystems and Human Well-Being: Multiscale Assessments. Volume 4, Millenium Ecosystem Assessment.* Island Press, Washington, D.C., pp. 229–259.

Montgomery, D.R. (2007) *Dirt: The Erosion of Civilizations.* University of California Press, Berkeley and Los Angeles.

Mouat, D., Lancaster, J., El-Bagouri, I. & Santibañez, F. (2006) (eds.) *Opportunities for Synergy Among the Environmental Conventions: Results of National and Local Level Workshops.* United Nations Convention to Combat Desertification, Bonn, Germany.

Redford, K.H., Levy, M.A., Sanderson, E.W. & de Sherbinin, A. (2008) What is the role for conservation organizations in poverty alleviation in the world's wild places? *Oryx* 42, 1–14.

Rice, X. (2008) Abu Dhabi develops food farms in Sudan. *The Guardian* [online]. http://www.guardian.co.uk/environment/2008/jul/02/food.sudan (accessed 2 July 2008).

Rosegrant, M., Paisner, M.S. & Meijer, S. (2001) *Long-Term Prospects for Agriculture and the Resource Base, Rural Development Strategy Background Paper #1.* The World Bank, Washington, D.C.

Sanderson, E., Jaiteh, M., Levy, M., Redford, K., Wannebo, A. & Woolmer, G. (2002) The human footprint and the last of the wild. *BioScience* 52, 891–904.

Smith, P., Martino, D., Cai, Z. et al. (2007) Agriculture. In: Metz, B., Davidson, O.R., Bosch, P.R., Dave, R. & Meyer, L.A. (eds.) *Climate Change 2007: Mitigation. Contribution of Working Group III to the Fourth Assessment Report of the Intergovernmental Panel on Climate Change.* Cambridge University Press, Cambridge and New York, pp. 497–540.

Streitfeld, D. (2008) As prices rise, farmers spurn conservation program. New York Times, [online], 9 April 2008. http://www.nytimes.com/2008/04/09/business/09conserve.html?scp=1&sq=farmers%20spurn%20conservation&st=cse

United Nations Convention to Combat Desertification (UNCCD) (2009) Financing to combat desertification, Fact Sheet 8. http://www.unccd.int/publicinfo/factsheets/pdf/Fact_Sheets/Fact_sheet_08eng.pdf (accessed April 2009).

del Valle, H.F. (1998) Patagonian soils: a regional synthesis. *Ecologia Austral* 8, 103–123.

Zinn, J. (2001) *Conservation Reserve Program: Status and Current Issues.* Congressional Research Service, Washington, D.C.

Part I
Thematic Reviews

Riding the Rangelands Piggyback: A Resilience Approach to Conservation Management

Brian Walker

Commonwealth Scientific and Industrial
Research Organization, Australia

Introduction

Aside from deliberate destruction for short-term profits, most unwanted management outcomes in rangelands stem from a combination of having the wrong mental model of how the system works and applying partial solutions to perceived problems. Recent findings from comparative studies of regional-scale social–ecological systems indicate that systems like 'wild rangelands' can have a number of alternate pairs of stability regimes (configurations of states) at different scales, each pair separated by a defined threshold. Loss of resilience in a desired regime leads to a 'flip' into an alternate, usually undesired, regime, from which it is either difficult or impossible to recover. The thresholds between regimes are marked by changes in feedbacks in the system. The interactions amongst the thresholds strongly determine the future trajectory of the system as a whole.

Because these linked social–ecological rangeland systems behave as non-linear, complex adaptive systems, goals of optimizing for particular products or states (a command-and-control approach to management) will likely fail. Goals aiming to enhance the resilience of desired system regimes (albeit

Wild Rangelands: Conserving Wildlife While Maintaining Livestock in Semi-Arid Ecosystems, 1st edition. Edited by J.T. du Toit, R. Kock, and J.C. Deutsch.
© 2010 Blackwell Publishing

not the 'best' state) are far more likely to be achieved. Learning how to ride the rangelands piggyback, nudging them away from trajectories that are likely to cross undesired thresholds and allowing them to self-organize within the set of acceptable trajectories is a much better option than top-down, command-and-control management.

The conventional approach to managing rangelands, for wildlife conservation or for livestock production, is based on one or the other variant of maximum sustainable yield (MSY). In a conservation context it usually amounts to trying to get, or keep, the ecosystem in some perceived optimal state that will deliver maximum benefits. There are some unfortunate problems in the application of this approach that cause unexpected, usually unwanted, outcomes. To start with, though they do take normal variation in climate and other variables into account, they tend to ignore extreme events – the events that tend to dominate ecosystem dynamics (Walker et al. 1986). The approach tends to treat the different sectors (ecological, economic, social) separately, assuming that problems from these different sectors do not interact – the issue of partial solutions. An all-too-common third assumption is that change in these systems is linear and incremental, when, in fact, it is mostly non-linear and often lurching. Perhaps the most problematic assumption is the one to do with achieving an optimal state. What I hope to convey is that there is no such thing as an optimal state, either of an ecosystem or a social–ecological system; it is akin to asking what might be the optimal state of biological evolution.

In contrast to the optimal state/MSY approach, the critical premise of resilience management and governance is that ecosystems and social–ecological systems have non-linear dynamics and can exist in alternate stability regimes (they have alternate attractors). An outcome of this premise is that sustainable use and development of rangeland systems – wild or domesticated – rests on three properties of these systems: resilience, adaptability and transformability (Walker et al. 2004). *Resilience* is the capacity of a system to absorb disturbance and reorganize so as to retain essentially the same function, structure and feedback – in other words, to have the same identity, remaining in the same system regime. *Adaptability* is the capacity of actors in the system (people, in the case of social–ecological systems) to manage resilience by (i) changing the positions of thresholds between alternate regimes or (ii) controlling the trajectory of the system – avoiding crossing a threshold (or engineering such a crossing). *Transformability* is the capacity to become a fundamentally different system when ecological, social and/or economic conditions make the existing system untenable. I will return to this at the end, but the focus of this paper is on resilience and adaptability.

Resilience

Resilience places an emphasis on thresholds between alternate attractors, or regimes, of a system. A shift from one regime to an alternate regime (*sensu* Scheffer & Carpenter 2003) occurs when such a threshold on a controlling variable is crossed. The threshold effect occurs at the point where a feedback in the system changes, leading to changes in function and structure and therefore identity (Walker et al. 2004). A system 'regime' is therefore, by definition, the set of system states that have essentially the same structure, function and feedbacks.

Considering the resilience of any particular system raises the question 'resilience of what, to what?' (Carpenter et al. 2001) and leads to an important distinction between general and specified resilience. Specified resilience is easier to grasp; an example is the resilience of the grassy regime of a rangeland to changes in rainfall and grazing pressure. General resilience does not specify any particular regime shift or shock to the system. General resilience of a system is enhanced by such attributes as diversity, tightness of feedback, modularity (fully connected systems are more susceptible to any negative inputs), openness (possibilities for immigration and emigration) and reserves (memory, seed banks) (cf. also Levin 1999). In addition, aspects of the social domain, such as overlapping governance structures (Ostrom 2005), also confer resilience.

One of the most important mistakes in landscape management is to underestimate or ignore the importance of scale and cross-scale effects. One cannot understand or manage a landscape by focusing on only the scale of interest or concern. It is necessary to consider the dynamics of the system at scales above and below. Resilience effects occur at all scales, and as an illustration we will consider three of them: patch, landscape and region.

Patch scale

1. Species composition and resilience of production (or biomass)

This example comes from southeast Australia, comparing a patch of rangeland in good condition with 21 perennial grass species (far from water), with a patch close to water that had been heavily grazed (Walker et al. 1999).

Rather than consider the species merely in terms of their phylogeny, we examined them in terms of the functional roles they performed in the ecosystem. Specifically, we measured attributes of the species that influenced their primary production, carbon sequestration and water use. Data on such attributes are not commonly available and the best that could be found, in relation to the attributes we wished to estimate, were

- maximum height (production and C storage);
- mature plant biomass (production and C storage);
- specific leaf area (water use efficiency);
- longevity (C storage and turnover);
- leaf litter quality (nutrient turnover times, C storage).

These five attributes were estimated for the 21 species and a matrix of functional similarity (or dissimilarity) calculated. The abundance of each species represents a typical rank–abundance curve for ecological communities (Figure 2.1), with a few very abundant species making up the bulk of the biomass and a long tail of species present in low abundance.

Two important aspects of the grass community emerged from this analysis: (i) The dominant species were significantly more dissimilar to each other than would be expected from a random sorting of attributes. In other words, they were functionally different and therefore complementary in terms of their roles in the ecosystem. (ii) Each of these dominant species had one or more minor species that was functionally very similar to it – a functional analogue.

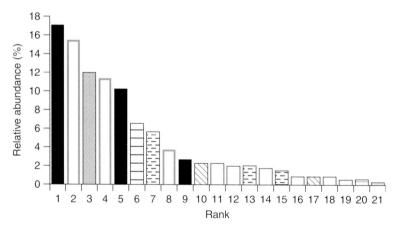

Figure 2.1 **Functional similarities between dominant and minor species. Species with the same bar pattern are functionally similar to each other (from Walker et al. 1999).**

On examining the community that had been heavily grazed, four of the species that were dominant in the 'good-condition' community had either disappeared or become very rare. We predicted that they would have been replaced by one of their functional analogues from amongst the minor species, and this turned out to be the case for three of them.

An important implication of this study is that, without the response diversity (the existence of the functional analogues), it is most likely that a regime shift would have occurred, resulting in a different functioning community with different controls and processes.

The conclusion is that ecosystem performance is promoted by high functional diversity (complementarity) and that resilience is promoted by high response diversity. This finding has been backed up by studies in three other kinds of ecosystems (rainforests, coral reefs and lakes) (Elmqvist et al. 2003). The message for ranchers and conservation managers is that the minor, apparently unimportant, species in the rangelands they manage likely confer resilience of the rangeland to droughts and heavy grazing spells. These species are not redundant, so try to retain them.

2. Community composition and resilience of the grass : woody balance

The response diversity example illustrates the kinds of ecological community attributes that influence resilience. An example of the kind of regime shift that can occur when resilience is lost is the shift from an open grassy rangeland to one dominated by a thicket of shrubs – a commonly observed change in Africa, Australia and the Americas. The dynamics are well described in the model developed by Anderies et al. (2002). Figure 2.2 depicts the structure of the model and controls on its dynamics. 'F' is a critical feedback in the system, responsible for the threshold effect.

Figure 2.3 describes how the threshold effect arises, illustrating the stable isoclines for grass shoots and grass crowns/roots under increasing amounts of shrubs. The shrub increase is induced by keeping fire out of the system. With no shrubs (t = 0 in Figure 2.3), there are two stable joint equilibria for grass shoots and crowns – a high one and a zero (in reality it would not be zero, but grass would be essentially absent). They are separated by an unstable equilibrium, or threshold. The resilience of the grass system (the amount of change in shoots and crowns that the system can undergo without a regime shift) is represented by the distance between the unstable equilibrium and the joint (high) stable equilibrium. As shrubs increase, so does the distance

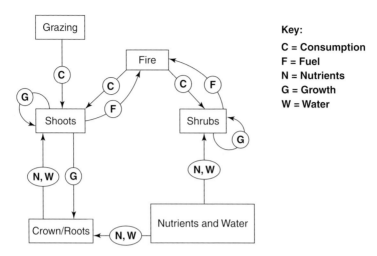

Figure 2.2 **Structure of a model of the dynamics of grass and shrubs in a rangeland (after Anderies et al. 2002).**

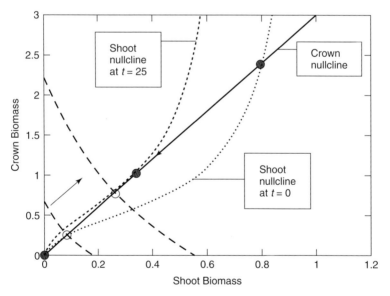

Figure 2.3 **Zero isoclines for grass shoot and crown dynamics in relation to increasing amounts of shrubs (After Anderies et al. 2002).**

between the two declines (resilience declines) and it becomes progressively easier for a bout of heavy grazing or a drought to push the system across the threshold into the 'zero' grass regime.

Grass dynamics are relatively fast, responding to seasonal rainfall effects. Managers tend to focus on the grass. The more insidious increase in shrubs occurs less noticeably and by the time they are seen as a problem, they have, all too often, resulted in a regime shift. Even if livestock or all other grazers are removed from the system, there is insufficient grass to carry a fire and the woody vegetation needs to re-structure and self-thin before the fire can again come back into the system and permit sufficient grass for a reverse regime shift. The following is the message to managers: Stop focusing on only the grass biomass dynamics and pay close attention to shrubs, especially the shrub : grass ratio. Keep shrub biomass well below the level where grass fuel cannot control it.

Landscape scale

Using the same basic model of shrub–grass dynamics at a patch scale, as described above, Janssen et al. (2002) examined the consequences of sheep grazing distribution on the net grass : shrub composition at the scale of a landscape.

Sheep exhibit contagion in grazing behaviour, remaining on a site for longer than optimal, in terms of access to grass. The phenomenon is commonly observed, leading to heavily grazed patches. Janssen et al. (2002) explored the consequences by dividing a large single paddock into a number of 'cells', according to the distance to water. The closer the cells are to water, the more attractive they are to sheep. In addition, the more vegetation there is in a cell, the more attractive it is, and the sheep move to a new cell when the vegetation drops below a critical level.

Running the model of linked cells with a single flock of sheep showed a sequential set of patch-scale regime shifts, resulting in the whole paddock flipping to a degraded (thicket) state at a lower mean grazing pressure than occurs without the spatially explicit grazing. This was not what was originally expected, but it concurs with the observations of graziers. The following is the message to graziers: The lower the stocking rate, the later the thresholds will be crossed, but the only way to achieve the same level of resilience, as if sheep behaved individually, is to induce them to move (by herding) to new 'cells' (areas that were not recently grazed in the paddock), before the flock is ready to move of its own accord.

Regional scale

At the regional level, several kinds of thresholds may apply, of ecological and/or social natures.

The first threshold relates to species persistence compared to percentage cover and connectivity of habitat. The relationship between biodiversity (species richness and community diversity) and the percentage cover of native habitat in a region appears to be non-linear. There is a regional-scale threshold of around 30% natural vegetation cover, claimed in a number of publications (e.g. Radford & Bennett 2004). Below this critical amount of natural vegetation cover, a suite of species types drop out of the system (cannot be sustained). The nature and generality of such a threshold needs more investigation; it is introduced here more as a question to be posed to rangeland managers: Is there evidence, or a suspicion, that there are lower limits to native habitat cover below which some species disappear? And how does the condition of the native habitat (e.g. grazed vs. not grazed) influence the level where this threshold occurs? Assuming a linear relationship is likely to lead to unexpected (and unwanted) losses of species.

The density of water points for herbivores presents another possible threshold. In the southern African drought of 1981–82, where water points were on average 2 km or less apart, massive mortality occurred, ascribed to the absence of reserve grazing areas (Walker et al. 1987). The proximate threshold effect is the level of reserve grazing, below which the community of herbivores suffers massive mortality in a drought. The ultimate threshold effect is the density of water points that determines the critical amount of reserve grazing. Questions facing managers (and scientists) are as follows: How sharp is the threshold? Is it merely a steeply changing relationship or do the dynamics of reserve grazing biomass, in relation to density of water points, reflect alternate stable state behaviour, as described by Scheffer and Carpenter (2003)? These are important questions to tackle since they have long-term implications for water-point planning in wildlife regions.

The long-term effects of such a drought event also need to be examined. Is the effect reversible? Does the system return to its original composition? It has been suggested that such a threshold, in regard to the effects of a drought, has irreversible effects and may result in the local extinction of species – such as occurred first with roan (*Hippotragus equinus*) and then with sable (*Hippotragus niger*) in the Tuli Block region of Botswana (B. Page, personal communication, 1982) – compared to regions where the threshold was not exceeded.

In rangelands, there may also be a social-attitude threshold along a slow variable of changing economic and social conditions. Up to some point in environmental awareness and preference, society values production above conservation in terms of perceived benefits. Under these conditions, government reflects social preferences and demands, and the feedbacks on regulations, taxes, subsidies, etc. tend to favour production.

Tipping points in social behaviour and preferences lead to sudden changes in governance, and should this occur in the production versus conservation attitudes, as preferences for conservation slowly increase, the result will be a change in the feedback, switching in favour of conservation. This is as yet only postulated, but it could explain the sudden negative shifts in the welfare of farmers in some parts of the world, or the change in their activities from being purely primary producers to partly conservation stewards. It is something to be considered in developing longer term strategies for rangeland regions.

Cross-scale

In addition to individual aspects of resilience at various scales, resilience can also be either enhanced or reduced through cross-scale interactions. Interested readers can refer to Peterson et al. (1998). Manipulating a system so as to maintain resilience at one scale can lead to changes (often reductions) in resilience at other scales, or in other places. A common example in the livestock industry is the attempt by governments to help individual farmers, through subsidies or floor prices, to survive in the face of declining terms of trade. While this works for a while at the scale of the individual farm, it leads to a decline in resilience at the scale of the industry and a much larger collapse for farmers when the assistance is eventually and inevitably withdrawn. This happened in Australia in the 1980s when the floor price for wool was eventually withdrawn, after the market price had diverged too far and the country had accumulated a so-called wool mountain.

Multiple interacting thresholds

Thus far, the examples of resilience have been of single-threshold effects. A comparison of several regional-scale social–ecological systems, however (Kinzig et al. 2006), has highlighted that in most systems there is invariably more than one threshold. In each of four quite different regions,

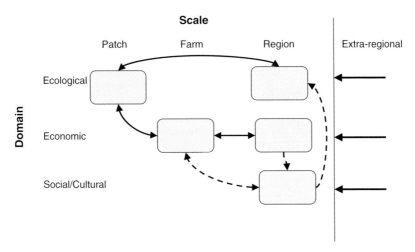

Figure 2.4 **Multiple interacting thresholds in four case studies (from Kinzig et al. 2006). Each box represents one or more thresholds.**

different kinds of thresholds with associated potential regime shifts were identified at local, farm and regional scales, in the ecological, economic and social domains. The distribution of thresholds is shown in Figure 2.4.

Figure 2.4 highlights three important points:

1. The threshold that is crossed first depends on how close the system is to that threshold AND on the nature of the disturbance. Even though an ecological threshold may be the closest of all, if the 'shock' that hits the system is an economic one, it may well be a farm scale economic threshold that is crossed first.
2. Crossing one threshold can lead to a cascade effect in crossing others, or it might preclude crossing others.
3. It is inadequate to consider individual thresholds and regime shifts. The 'system' consists of the whole patch, farm, region and ecological, economic and social domain complex. The future trajectory of this overall system depends on the interactions amongst the multiple thresholds that it contains and on which of them is crossed first.

The message to managers is fairly obvious (the above three points), but an overarching message is this: Stop focusing on only the scale of immediate interest and on the ecological, economic or social issue of most concern.

Start imagining your system as a version of Figure 2.4 and think about where thresholds in that space might occur (i.e. not only in the 'boxes' where they occurred in the four case studies).

Identifying thresholds

How does one identify the controlling (slow) variables that have thresholds on them and how does one discover where the thresholds are? Can one learn about a threshold before it is crossed? Most of the examples we have so far are thresholds that have been crossed; the others are proposed or hypothesized possible thresholds (see Walker & Meyers 2004 for a discussion on the database of thresholds). This is the topic of a separate publication: *The Resilience Workbook*, an ongoing development by members of the Resilience Alliance. There are two versions – one for scientists and one for practitioners (see http://www.resalliance.org/3871.php). Some pointers from this work-in-progress may offer some clues, all fairly commonsense:

- Do not consider the ecosystem in isolation. Begin with a stakeholder analysis and ensure that all stakeholder interests are included, including those at scales above the focal scale (above the wild rangeland itself).
- Determine what it is that needs to be resilient and in the face of what kinds of shocks (resilience of what, to what).
- Develop conceptual models of the system. Start with simple box-and-arrow diagrams, develop state-and-transition type models (cf. Westoby et al. 1989) and from these develop simple models of alternate regimes and the drivers and disturbances that can cause regime shifts. Achieving this requires incorporating what is known about the system into hypotheses about where threshold effects might lie and hence the controlling (often slow) variables involved.
- Use observations and experiments to test these hypotheses in an iterative way to develop a model of the system's resilience.

System identity and the significance of a regime shift

Consider a wild rangeland as a system depicted in Figure 2.4. Can we consider the overall system in terms of alternate regimes, as a whole? The individual thresholds in each 'box' separate alternate regimes at the scale and in the domain concerned. It is harder, however, to think about alternate regimes of the overall system. One such overall regime (a desired one) might be 'with self-maintaining, diverse plant and animal communities' and its alternate regime might be 'without them'.

This raises the issue of iconic species or species groups. How we define a significant change, or a regime shift, depends on our objectives. For example, a significant shift may be interpreted as one involving the presence or absence of large predators, seasonal migrants, riverine forests, or even a particular species. If this is what is considered significant and important, it defines the 'identity' of the system (Walker et al. 2004; Cumming & Collier 2005). All states within a regime of the system have, by definition, the same identity.

Each 'identity' variable has, potentially, a threshold between persistence and loss. The challenge for planners and managers is to identify the control variables on which the thresholds occur and at which levels of those variables they are likely to be. Do the thresholds (control variables) interact? The interactions amongst these multiple thresholds determine the trajectory of the system as a whole.

The further challenge that the 'identity' poses to planners and managers is forging an agreement on the identity of the system they are trying to manage. Different stakeholders will have different views on what defines the system and what constitutes an undesirable change. A careful and structured approach to stakeholder involvement is required; two good starting points are the companion modelling approach (Bousquet et al. 2005) and the work of Brown, Adger and colleagues (cf. Brown et al. 2001).

Dealing with resilience

As stated earlier, adaptability is the capacity of actors in the system (people) to manage resilience, either by changing the positions of thresholds or by controlling the trajectory of the system. It follows that it is just as important (if not more important) for managers of rangelands to understand and consider adaptability as it is to consider ecological resilience. Adaptability is an integral part of being resilient. Social capital (leadership, trust), networks, information flow and governance are all crucial attributes of adaptability and a wild rangeland or any other social–ecological system that has low adaptability has little capacity to avoid unwanted regime shifts.

The third attribute of a social–ecological system that determines its long-term welfare, mentioned at the beginning of this chapter, is transformability. Again, it is not the scope of this chapter to deal with it in detail. A good example, however, is the transformation, described elsewhere in this book, occurring in the lifestyle of the northern Tanzanian Maasai people. They are transforming from pure nomadic pastoralists to sedentary mixed agricultur-alists and cattle farmers (Chapter 13). They have little choice but to undertake this transformation if they are to persist as a social group. There is a potential tension between adaptability and transformability. Becoming better adapted to remaining in a regime that is no longer viable (usually due to changes at other scales) can reduce the transformative capacity of the system.

Some insights from a comparison of regional case studies

A comparison of 15 regional-scale social–ecological systems (Walker et al. 2006) led to a set of propositions about how such systems work. Though much is still to be learned, based on these propositions some initial observations relating to wild rangelands are worth noting:

- Most changes in social–ecological systems are not problematic. They do not lead to lasting negative effects. Yet, often, much time and money is spent in trying to prevent or reverse them.
- Such changes are, in fact, often *necessary*, self-organizing processes that enhance the system's resilience. The apparent goal of many resource man-agement agencies to prevent change and keep things constant in time and space works against what is needed to remain resilient.
- Identify changes that do matter – those related to threshold. An example of this has occurred in Kruger National Park, South Africa, where scientists and managers have developed a set of what they term *thresholds of potential concern* (Biggs & Rogers 2003).

Sustainable rangeland management amounts to

- identifying critical thresholds that lead to unacceptable system regimes;
- understanding the interactions amongst multiple thresholds;

- developing the adaptive capacity to avoid crossing these thresholds; and
- allowing the system to self-organize within the range of acceptable trajectories.

It does not consist of trying to pick the 'optimal' system state or trajectory.

Finally, for the long-term well-being of humans and ecosystems in rangelands, it is better to learn how the system changes and to know how to deflect it from unwanted trajectories than to confront it head on and force it into some perceived optimal state or trajectory. Learning to ride such systems piggyback (C.S. Holling, personal communication, 1978), nudging them away from undesired thresholds and allowing them to self-organize within the set of acceptable trajectories is a strategy far more likely to succeed than command-and-control management for optimal states delivering maximum benefits. Perhaps, when we know all there is to know about rangelands ecosystem dynamics, about the sociology of rangelands managers and about the relationships between the two across relevant scales, then an MSY approach might work. But until then, riding the rangelands piggyback is the better option.

References

Anderies, J.M., Janssen, M.A. & Walker, B.H. (2002) Grazing management, resilience and the dynamics of a fire driven rangeland. *Ecosystems* 5, 23–44.

Biggs, H.C. & Rogers, K.H. (2003) An adaptive system to link science, monitoring, and management in practice. In: du Toit, J.T., Rogers, K.H. & Biggs, H.C. (eds.) *The Kruger Experience: Ecology and Management of Savanna Heterogeneity.* Island Press, Washington, D.C., pp. 59–80.

Bousquet, F., Trebuil, G. & Hardy, B. (2005) *Companion Modelling and Multi-Agent Systems for Integrated Natural Resource Management in Asia.* International Rice Research Institute, Los Banos, Philippines.

Brown, K., Tomkins, E. & Adger, W.N. (2001) *Trade-Off Analysis for Participatory Coastal Zone Decision-Making.* Overseas Development Group, University of East Anglia, Norwich.

Carpenter, S.C., Walker, B.H., Anderies, J.M. & Abel, N. (2001) From metaphor to measurement: resilience of what to what? *Ecosystems* 4, 765–781.

Cumming, G.S. & Collier, J. (2005) Change and identity in complex systems. *Ecology and Society* 10(1), 29 [online]. http://www.ecologyandsociety.org/vol10/iss1/art29/

Elmqvist, T., Folke, C., Nystrom, M. et al. (2003) Response diversity, ecosystem change and resilience. *Frontiers in Ecology and Environment* 1, 488–494.

Janssen, M.A., Anderies, J.M., Stafford-Smith, M. & Walker, B.H. (2002) In: Janssen, M.A. (ed.) *Complexity and Ecosystem Management: The Theory and Practice of Multi-Agent Systems.* Edward Elgar Publishing, Cheltenham, pp. 103–123.

Kinzig, A.P., Ryan, P., Etienne, M., Allison, H., Elmqvist, T. & Walker, B.H. (2006) Resilience and regime shifts: assessing cascading effects. *Ecology and Society* 11(1), 20 [online]. http://www.ecologyandsociety.org/vol11/iss1/art20/

Levin, S.A. (1999) *Fragile Dominion: Complexity and the Commons.* Perseus Books Group, Cambridge.

Ostrom, E. (2005) *Understanding Institutional Diversity.* Princeton University Press, Princeton.

Peterson, G., Allen, C.R. & Holling, C.S. (1998) Ecological resilience, biodiversity, and scale. *Ecosystems* 1, 6–18.

Radford, J.Q. & Bennett, A.F. (2004) Thresholds in landscape parameters: occurrence of the white-browed treecreeper *Climacteris affinis* in Victoria, Australia. *Biological Conservation* 117, 375–391.

Scheffer, M. & Carpenter, S.R. (2003) Catastrophic regime shifts in ecosystems: linking theory to observation. *Trends in Ecology and Evolution* 18, 648–656.

Walker, B.H., Anderies, J.M., Kinzig A.P. & Ryan, P. (2006) Exploring resilience in social-ecological systems through comparative studies and theory development: introduction to the special issue. *Ecology and Society* 11(1), 12 [online]. http://www.ecologyandsociety.org/vol11/iss1/art12/

Walker, B.H., Emslie, T.H., Owen-Smith, N. & Scholes, R.J. (1987) To cull or not to cull: lessons from a southern African drought. *Journal of Applied Ecology* 24, 381–401.

Walker, B.H., Holling, C.S., Carpenter, S.C. & Kinzig A.P. (2004) Resilience, adaptability and transformability. *Ecology and Society* 9(2), 3 [online]. http://www.ecologyandsociety.org/vol9.iss2/art5

Walker, B.H., Kinzig, A.P. & Langridge, J. (1999) Plant attribute diversity, resilience, and ecosystem function: the nature and significance of dominant and minor species. *Ecosystems* 2, 1–20.

Walker, B.H., Matthews, D.A. & Dye, P.J. (1986) Management of grazing systems – existing versus an event-orientated approach. *South African Journal of Science* 82, 172.

Walker, B.H. & Meyers, J.A. (2004) Thresholds in ecological and social-ecological systems: a developing database. *Ecology and Society* 9(2), 3 [online]. http://www. ecologyandsociety.org/vol9/iss2/art3

Westoby, M., Walker, B.H. & Noy-Meir, I. (1989) Opportunistic management for rangelands not at equilibrium. *Journal of Range Management* 42, 266–274.

(3)

Addressing the Mismatches between Livestock Production and Wildlife Conservation across Spatio-temporal Scales and Institutional Levels

Johan T. du Toit

Department of Wildland Resources,
Utah State University, UT, USA

The focus in this chapter is the management gap that can arise on rangelands because livestock production and wildlife conservation typically operate at different spatio-temporal scales and institutional levels. The livestock produced on rangelands is, in the vast majority of cases, managed as the private property of individuals, families or companies, each operating within short time horizons (seldom exceeding a decade) and prescribed geographical areas. Indigenous wildlife resources, however, are comprised of co-evolved species adapted to extensive ecosystems and managed as common property for multiple objectives – as local short-term protein sources, pests, targets for trophy hunters, attractions for ecotourists, or components of global biodiversity to be conserved for future generations. These mismatches between livestock and wildlife management have to be at worst recognized, and at best overcome, in working towards the goal of integrated livestock production and ecosystem conservation (du Toit & Cumming 1999).

Rangeland ecosystems, which might or might not include wild lands embedded within a human-transformed matrix, are ideally managed as

Wild Rangelands: Conserving Wildlife While Maintaining Livestock in Semi-Arid Ecosystems,
1st edition. Edited by J.T. du Toit, R. Kock, and J.C. Deutsch.
© 2010 Blackwell Publishing

social—ecological systems of which people, livestock and wildlife are integral components. But the dilemma faced by conservation practitioners is how to deploy limited financial and human resources to respond effectively to perceived threats to conservation within these social—ecological systems. This is the 'implementation crisis', which requires an effective operational model for implementing conservation action (Knight et al. 2006). In the case of rangelands, this dilemma carries a particularly heavy burden of responsibility because so many people depend on rangelands for their livelihoods, and rangelands represent such vast expanses of the terrestrial biosphere (Chapter 1). If conservation practitioners 'get it wrong', they waste time and money and damage whatever trust may have existed between land users and conservation agencies. To get it right, each intervention should be applied at the spatio-temporal scale(s) and institutional level(s) at which the problem is generated, even if the immediate problem is perceived at a different scale or level. By way of analogy, consider how a microscope is used – the investigator racks up and down using lenses of different magnification before selecting a setting that provides the clearest view – but the final setting could not be arrived at without first racking up and down. In contrast, rangeland conservation interventions are often directed at 'pet projects' that cannot possibly lead to long-term solutions because the causes of the problem occur at different (usually wider or higher) spatio-temporal scales or levels of institutional organization. In this chapter, I will use rangeland case studies to illustrate these issues. My purpose is neither to criticize nor preach, but to focus some attention on the need for conservation organizations to evaluate interventions at multiple spatio-temporal scales and institutional levels before formulating the most effective strategy for allocating effort and resources.

The problem of cross-scale interactions

According to the Millennium Ecosystem Assessment (MA) (2003), which was a 4-year worldwide project to compile information on the links between ecosystem change and human well-being, there are two main reasons as to why scale matters when assessing ecosystem services. First, social—ecological systems and their processes operate across multiple scales and can be influenced differently by the same driving forces operating at different scales, so that what is measured at one scale does not necessarily apply at another. Second, social—ecological issues (e.g. hunting of wildlife) can have cross-scale interactions and investigating the issue from the bottom up can lead to a different outcome than if it were investigated from the top down. The MA's conceptual working group thus recommended that where cross-scale interactions apply (and they almost always do), a single level or scale of appropriate response is unlikely to be found and so integrated multi-scale responses should be developed.

For extensive rangelands, it follows that wildlife conservation issues should be expected to have ramifications that extend beyond the scope of any immediate problem as initially perceived. By the same logic, the *causes* of a conservation problem might well be rooted outside the spatio-temporal scope of the problem as it appears to conservation practitioners on the ground (Sinclair & Byrom 2006). Indeed, with the progression of globalization, we can increasingly expect that some local rangeland conservation problems (or opportunities) are driven by market forces operating at a global scale. I present below examples from India and Africa to illustrate this point.

Subsistence pastoralists in southern India commonly herd their cattle into wildlife reserves in an unregulated practice that, on face value, represents the multiple use of natural resources in reserves and the distribution of benefits beyond reserve boundaries. By analysing the practice in some detail, however, Madhusudan (2005) found that 'subsistence' pastoralists herding their cattle into the Bandipur National Park and Tiger Reserve are driven by financial profits from cow dung collection and this highly organized practice effectively mines nutrients from the wildlife reserve. Villagers along Bandipur's northern boundary maintain cattle densities of approximately 236 animals/km^2 by herding them in and out of the reserve. All cow dung is collected, dried and exported to the highlands where there is a lucrative market in manure (organic fertilizer) for coffee plantations. Driven by the international demand for organically grown coffee, which is ironically consumed mainly by educated people who are typically discerning and environmentally conscious, the income from dung sales allows villagers to buy industrial fertilizers for their own crops and still make a net profit. This unsustainable practice could also emerge, if it has not already, in rangelands adjacent to the coffee-producing regions of East Africa.

Another example of a local conservation problem driven by global-scale forces can be found in the bush-meat problem of Ghana. In this case, however, the problem has emerged partly because of a lack of organization in the production of livestock on West African rangelands. From a comparative analysis of trends in biomass and harvest of (i) indigenous large mammals in terrestrial reserves in Ghana and (ii) marine fish in the Gulf of Guinea,

Brashares et al. (2004) have revealed some fascinating cross-scale interactions. Most of Ghana's 20 million people live less than 100 km from the coastline and marine fishing has traditionally been the primary source of protein and employment. Marine fish stocks have, however, declined steeply over the past 40 years in an inverse relationship with the fishing efforts of commercial fleets in these waters. The foreign fishing fleet is now dominated by subsidized European Union vessels, which have increased their fish harvests from the Gulf of Guinea by a factor of 20 since 1950. These catches are mostly consumed in Europe and it is only the surplus that becomes available in West Africa, but when local fish supplies decline, the resulting protein demand drives up the consumption of bush meat. Attempts to reduce bush meat consumption to sustainable levels are futile without international agreements to regulate the marine fishing activities of foreign fleets, especially those receiving perverse incentives from the European Union. But then, alternative protein sources also need to be found and this is problematic because livestock production systems on West African rangelands and meat markets across the region are still insufficiently developed. Conserving terrestrial wildlife in West Africa thus requires an integrated programme covering marine, forest and rangeland ecosystems, operating across institutional levels from the European Commission to the Ghanian household and across temporal scales to address the immediate nutritional demand for protein, as well as the long-term requirements for sustainable livestock production and meat marketing systems across West African rangelands.

Achieving integrated multi-scale responses to all conservation problems, but particularly those in extensive rangelands, is difficult when conservation organizations base their project funding decisions on responses to a request-for-proposals (RFP) process. For proposals to achieve success through the RFP process, they have to meet standardized criteria that confine the budget, duration and scope of the project. Furthermore, the conservation community depends heavily on the charity of donors who wish to see their funds used for tangible achievements within a short timeframe and typically with an emphasis on charismatic species and landscapes. The result is a plethora of multiple interventions limited in geographical and temporal scope, each of which is apparently 'doable', at best cost-effective and at worst a low financial risk. Many (but admittedly not all) such interventions cannot produce sustainable solutions because they address symptoms in a particular place and time, but the causes, symptoms and interventions interact *across* spatial and temporal scales.

Working across spatial scales

Over the past century or so the world's flagship terrestrial national parks (Yellowstone, Kruger, Serengeti, etc.) have all emerged, to greater or lesser degrees, from management regimes that emphasized the command-and-control approach inside the park and the 'good fences make good neighbours' approach to human activities outside the park (e.g. see Mabunda et al. 2003; Wagner 2006). More recently, with the development of private game reserves, conservancies and controlled hunting areas adjacent to large parks, together with transfrontier conservation areas (Chapter 5), the spatial scale of wildland management is steadily widening. This has been greatly assisted by research tools developed through advances in remote sensing and geospatial technology. Nevertheless, the spatial scale of relevance to the management of a grassland or savanna ecosystem can be far wider than its geographical extent, especially if the rivers that flow through the ecosystem have their tributaries and catchments in distant highlands. To illustrate this point, I will use the Serengeti–Mara (Tanzania and Kenya) and Kruger (South Africa) ecosystems as two contrasting examples.

The Serengeti–Mara Ecosystem (SME) covers about 25,000 km^2 and is defined by the seasonal migration pattern of approximately 1.3 million wildebeest, *Connochaetes taurinus*. Most of the migration occurs within the Serengeti National Park and Ngorongoro Conservation Area in Tanzania, but in the dry season (August–November) it moves northward, with hundreds of thousands of wildebeest congregating in Kenya's Masai Mara National Reserve (Maddock 1979; Thirgood et al. 2004). This spectacular migration represents a UNESCO (United Nations Educational, Scientific and Cultural Organization) world heritage site and a major earner of foreign currency from tourist traffic through Tanzania and Kenya; the migration is sustained by adequate habitat in the core protected areas of Serengeti and Mara as well as various surrounding buffer areas (smaller game reserves, conservation areas, controlled hunting areas, etc.). There is, however, increasing concern that the less buffered western region of the ecosystem is undergoing rapid land transformation and human settlement, which presents a mounting threat of poaching when the migration is moving through this area on its northbound route (Thirgood et al. 2004). Poaching can nevertheless be controlled with patrolling, and a recent study has concluded that the current anti-poaching effort in Serengeti is more than sufficient to prevent any significant population-level impacts on the common large mammal species there (Hilborn et al. 2006). So if there is sufficient habitat, if the main migration route is buffered around most of the ecosystem's periphery and if there are adequate financial and human resources for anti-poaching operations, can we assume the SME will remain safe for future generations?

Unfortunately, if we adjust our microscope's resolution and expand the spatial scale to look for other factors crucial to the migration, a more worrisome problem comes into focus. The northern range of the wildebeest migration provides not only the food resources needed by the population during the dry season but also perennial drinking water in the Mara River. The Mara River Basin extends in a northeasterly direction into Kenya, far beyond the boundaries of the SME, where land use changes and a hydropower project threaten the only supply of drinking water in the migration's dry season range. Predictions of an eco-hydrological model point to developments in the upper catchments of the Mara River in Kenya, compounded by expected climatic changes, as the greatest threat to the survival of the wildebeest migration (Gereta et al. 2002). The model predicts that during future drought conditions, which have historically occurred with approximately 7-year intervals for moderate drought and approximately 20-year intervals for severe drought, anywhere between 20 and 80% of the migrating wildebeest will die. With anticipated rates of water extraction upstream for both irrigation and diversion to a hydropower project, combined with reduced watershed yield due to land transformation, the flow into the SME will be insufficient to meet the water consumption needs of the wildebeest after evaporation has been accounted for. There are now various efforts underway to implement an integrated catchment management plan, such as the World Wildlife Fund (WWF) Mara River Basin Initiative, that began field operations in 2004. Nevertheless, the management of the Mara River and its rapidly changing catchments had not previously been considered a priority for the conservation of the SME (see Sinclair & Norton-Griffiths 1979; Sinclair & Arcese 1995; Sinclair et al. 2008), apparently because this impending problem occurs at a spatial scale above that of the more immediate problems like poaching.

By contrast, managers and scientists working in the Kruger National Park identified integrated catchment management as a priority several decades ago; although in fairness to their Serengeti counterparts the Kruger situation is less complex because the catchments are at least in the same country as the park. Kruger's major rivers flow from west to east with their catchments in the South African Highveld, where mining, agriculture, and commercial forestry are the main water users. After flowing out of the park into Mozambique, they eventually discharge into the Indian Ocean. Intensive water sampling began in Kruger in 1983, and concerns over the flow rate, flow regime and water quality in these rivers led to the initiation of the Kruger Rivers Research Program in 1990. This far-reaching interdisciplinary programme catalysed change in the overall management of Kruger by ushering in a new era of scientist–manager–stakeholder interaction, which has subsequently expanded to other issues including the management of elephants, *Loxodonta africana* (Biggs & Rogers 2003). While Kruger's rivers are by no means fully protected from future developments in their catchments, some key remedial actions have been taken to reduce the anthropogenic modification of flow regimes (O'Keeffe & Rogers 2003). And, very importantly, there are institutional structures in place now whereby the managers of Kruger can be represented within catchment management agencies and can serve the park's objectives by working on an ecologically appropriate spatial scale, which transcends the park's geographical boundaries.

Advances in information technology, remote sensing, and geospatial analysis should make it possible for terrestrial ecosystems to be managed at multiple scales, but there is still the inertia of entrenched management systems that were developed at inappropriately small scales. An example is the attempt to control brucellosis in elk, *Cervus elaphus*, and bison, *Bison bison*, in the Greater Yellowstone Ecosystem (GYE) of the western United States. Brucellosis, caused by the bacterium *Brucella abortus*, spread to the United States through livestock from Europe and spread into the GYE bison and elk populations in the early part of the twentieth century. Transmission among ungulates is primarily through ingestion of body fluids associated with birth, and although morbidity is rare in most ungulates, for which brucellosis is a chronic illness that is difficult to cure, it can be transferred to humans (e.g. in unpasteurized milk). Through vaccination and test-and-slaughter campaigns brucellosis has been effectively eradicated as a public health risk in developed countries, but if it is identified in cattle in the United States then the entire herd is usually slaughtered. As a consequence, there are efforts underway to control the disease in elk and bison to prevent it spreading to cattle. However, as pointed out by Bienen & Tabor (2006), the brucellosis test-and-slaughter protocol is designed for cattle managed on a ranch scale, which is not suitable for the much larger scale at which bison and elk range within the GYE. Removing animals that test positive for brucellosis and reducing a local population by 10–25% does not control the disease at the ecosystem level and, in fact, probably exacerbates it. This is because all infected animals cannot be removed, while some recovered animals are removed because they test positive but have actually become resistant and would reduce the spread of the disease if they were left in the population. The problem here is that the control of brucellosis in GYE elk and bison falls largely under the authority of state livestock agencies in Montana, Wyoming and Idaho, which follow protocols designed for ranches.

These examples illustrate that meeting the long-term interests of wildlife conservation and agricultural production on rangelands requires the collaborative efforts of appropriate agencies with the willingness and ability to work at multiple spatial scales. This is not an unattainable goal, as has been demonstrated with the management of migratory geese in North America. The mid-continent population of lesser snow geese *Chen caerulescens caerulescens* has increased about fourfold since the 1970s, imposing a powerful top-down effect on coastal vegetation and soils at their Arctic summer breeding sites (Jefferies et al. 2004). However, attempts at vegetation and soil restoration at heavily impacted sites, such as along the Hudson Bay coast of Canada, would be futile because the reason for the snow goose population explosion is the bottom-up effect of greatly increased food availability in the migration's winter range, which is up to 5000 km away in mid-western and southern United States. The expansion of grain crops in the United States, combined with the intensification of production with nitrogenous fertilizer, provides an allochthonous nutritional subsidy to the population in the form of spilt grain, sprouting seed, green stubble, etc. The quickest solution is to reduce the goose population dramatically, and through collaboration between the Canadian Wildlife Service and the U.S. Fish and Wildlife Service, a spring goose hunt was successfully introduced in both countries for this specific purpose in 1999. If the desired 50% reduction of the population can be achieved and if the required annual offtake (approximately 1 million birds/yr) can be maintained, then the degraded Arctic habitat might eventually return to its former state. Recovery is unpredictable because the breeding grounds could have settled into an alternative stable state characterized by hypersaline, compacted and denuded soils. Nevertheless, the success story is about government agencies in Canada and the United States collaboratively addressing an ecological problem at spatial scales ranging from Arctic salt-marshes to transcontinental migratory flyways.

Working across temporal scales

The history of rangeland management is unfortunately replete with examples of practices and policies directed at short-term gains from livestock production without adequate consideration of the longer term consequences. In range-lands across all continents there have been comparatively rapid reductions in the grass : wood ratio in association with pastoralism and the provision of arti-ficial water sources for livestock (Chapter 4). There have also been dramatic invasions of weeds that outcompete indigenous herbaceous species, partly because of well-intended but short-sighted attempts to reclaim overgrazed rangelands by seeding with exotic species and partly because of accidental introductions of weeds with imported grains and fodder. A prime example of the latter is the invasion of cheatgrass (*Bromus tectorum*) throughout the United States. In less than one century, this annual grass from Eurasia has occupied a niche from which it is unlikely to be eradicated and in which it has greatly reduced the animal production potential of rangelands, especially in the Intermountain West. Because it seeds and dies off before the end of the summer, it provides a vast and highly combustible supply of fine fuel for wildfires (Brooks et al. 2004), which are now occurring at intervals of only 3–5 years on rangelands that previously burned every 30–110 years (McKnight 2008). An important result is that the indigenous shrub-steppe ecosystem, dominated by sagebrush (*Artemesia tridentata*), is steadily declin-ing in extent because big sagebrush does not re-sprout after burning. Intense recurrent fires kill sagebrush in expanses that exceed the plant's wind-borne seed dispersal range (approximately 30 m) from unburnt patches (Knick & Rotenberry 1997). Linked to this spatial reduction of the shrub-steppe are worrying population declines of shrub-steppe obligate wildlife species such as sage grouse (*Centrocercus* spp.) and pygmy rabbits (*Sylvilagus idahoensis*). Indeed, western American rangelands (and probably Australian rangelands too; see Chapter 8) have been transformed by exotic plant invasions over the past century to such an extent that scientists and managers are at a loss to prescribe effective strategies to contain the spread of these plant pests, let alone eradicate them and restore the indigenous communities. The problem has arisen essentially because managers were unprepared for the short time scale and wide spatial scale over which species can invade an ecosystem before the severity of their effects are fully recognized.

Keeping the focus on the western United States, another vexing problem in rangelands at higher elevations appears, somewhat ironically, to be caused by artificially *extended* fire intervals, opposite to the problem in the shrub-steppe ecosystem. Here, suppression of forest fires since Euro-American settlement has resulted in dramatic changes to woodlands and savannas. On the rocky, mid-elevation slopes, pinyon (*Pinus* spp.) and juniper (*Juniperus* spp.) trees have proliferated throughout the Great Basin in pinyon–juniper associations that are now enormously greater in extent, density, and crown cover than when those landscapes were managed with patch-burning by Native Americans. The outcome is reduced forage production for livestock and vastly increased woody fuels for intense fires that are getting more and more difficult to control (Gruell 1999; Tausch 1999). On higher elevation slopes the transition from Native American to Euro-American management has removed the fire-driven disturbance regime to which aspen trees (*Populus tremuloides*) are adapted. Aspen stands now occupy only approximately 40% of their estimated historic range in the western United States, where they are being replaced by conifers (mainly *Picea* and *Abies* spp.) that suppress understorey vegetation at even low levels (10–20%) of coniferous canopy cover (Stam et al. 2008). This reduced production of understorey vegetation obviously reduces the carrying capacity of the stand for livestock and wildlife, thus potentially causing public lands to become overstocked because of the insidious but steady erosion of grazing acreage in fixed allotments on which ranchers maintain prescribed stocking rates. Furthermore, the increase in the conifer : aspen ratio is associated with impaired watershed yields (LaMalfa & Ryel 2008), with potentially significant implications for communities dependent on the rivers flowing from these conifer-encroached catchments.

The above western American examples illustrate how in the very short time, in ecological terms, since human settlement (with Native Americans and Euro-Americans imposing contrasting effects) there have been dramatic, undesired, and effectively irreversible changes to rangelands in and around the Great Basin. Paradoxically, the amplitude of these changes is increased through the efforts of livestock managers to *prevent* change by suppressing fire, introducing water, erecting fences and maintaining relatively stable stocking rates within ecosystems that are naturally dynamic. And then ecosystem dynamics present an additional challenge when managers strive to restore disturbed ecosystems to their 'natural' states, because the benchmarks of 'natural' states tend to be arbitrary. It all depends on which episode in history is chosen to represent the desired state.

Along the banks of the Chobe River in northern Botswana, the managers of lucrative tourism companies have chosen an ecological state prevailing in the mid-1970s as their benchmark, since that included abundant wildlife under a spectacular canopy of tall riparian trees. Since then, however, the riparian woodland has all but disappeared (Mosugelo et al. 2002) and been replaced by shrub land occupied by a high density of large mammals, especially elephants, which occur at local densities of more than 4 animals/km^2 in the dry season when their only water supply is the Chobe River (Gibson et al. 1998). Local politicians and influential stakeholders in the tourism industry claim that the elephants are overpopulating the area and that they should be culled to return the system to its former state. However, an intensive 5-year study funded by the Governments of Botswana and Norway and involving an international team of scientists found no ecological basis for the culling argument (Skarpe et al. 2004). The 1970s 'Garden of Eden' was an unstable state and an artefact of two major anthropogenic disturbances near the end of the nineteenth century. Intensive commercial hunting for ivory had virtually eradicated the formerly abundant elephant population by 1880; then in 1896 the exotic rinderpest virus swept through Botswana and decimated the artiodactyls (Walker 1989; Vandewalle 2003). The result was a window of opportunity for a pulse of seedling growth that enabled the riparian woodland to become established in the decades while the populations of browsing mammals were recovering. By the 1970s, the system was aesthetically appealing with abundant wildlife again and spectacular trees along the grassy riverfront, but was poised to collapse and re-organize back into the mega-herbivore-driven system it had been when it attracted the ivory hunters of the nineteenth century. Seeing and understanding this process requires that it be analysed on a temporal scale that far exceeds the lifetimes of the present managers and stakeholders in the system. The challenge for ecologists is to educate those managers and stakeholders to view an ecosystem while ranging back and forth across the time axis, like an observer racking up and down on a microscope, and to appreciate that any 'benchmark' condition is temporary.

Working across levels of institutional organization

In addition to spatio-temporal scales, we need to consider the organizational levels at which conservation interventions are directed. In recent decades, most large conservation non-governmental organizations (NGOs) have embraced community-based natural resource management (CBNRM) as the desired process for implementing conservation in developing countries and this entails 'grass-roots' level of institutional organization. This is the most impoverished and politically disenfranchised level and so CBNRM requires the devolution of rights, responsibility and authority down through the hierarchy from national governments to local communities. Yet, this seldom happens in countries where the devolution of authority would compromise the vested interests of the ruling elite. Denial of such devolution can account for most failures of potentially promising CBNRM programmes, including the pioneering CAMPFIRE programme in Zimbabwe (Jones & Murphree 2004). Like it or not, quality of national governance has to be considered for larger conservation interventions since the richest concentrations of biodiversity occur in developing countries with the worst levels of political corruption (Smith et al. 2003). In such countries, injections of foreign revenue for diffuse, national-level integrated conservation and development programmes have proved to be virtually futile.

An alternative approach, which makes financial sense but can be controversial in some cases, is for conservation funds to be provided as direct payments to the specific individuals living within and managing the biodiversity that needs to be conserved (Ferraro & Kiss 2002). In developing countries, however, where biodiversity conservation depends on external funds, it is essential that political concessions are secured at all levels. This is to avoid the crippling stigma of 'conservation imperialism', by which a country's rural poor serve the wishes of the more privileged citizens of the world. Nevertheless, this approach can work well in countries such as the United States where, for example, private landowners are eligible to receive conservation incentives directly from the federal government in the form of conservation easements.

Direct payments have indeed been suggested for a highly contentious issue on North American rangelands – controlling predators for the interests of livestock production. The federal government of the United States has been engaged in predator control in the western states since World War I, with a focus on killing coyotes (*Canis latrans*). Various methods have been used including poisons such as Compound 1080, or sodium monofluoroacetate, until it was banned in 1972 after which aerial gunning was intensified (reviewed by Wagner 1988). Despite a significant investment from the federal government (about $1.6 billion over 1939–98), and an impressive offtake of predators (80,000–130,000 animals per year, of which 75–90% were coyotes), a retrospective analysis of a 60-year data set found that government-administered predator control efforts were ineffective overall in preventing the steady decline of the sheep industry since the 1940s (Murray Berger 2006). Most of that decline can be accounted for by unfavourable market conditions. Therefore, based on the results of her study and assuming that the government's objective is to support the sheep industry and not simply kill predators, Murray Berger suggests that the tax dollars could be better spent as direct payments to individual sheep producers. They could each use the subsidy to control predators independently and/or compensate themselves for stock losses while offsetting rising input costs (e.g. of hay and fuel), depending on circumstances. The point is that a 'big-picture' analysis has revealed the success of a major rangeland management intervention such as predator control to be dependent on the institutional level(s) at which it is implemented.

The appropriate institutional level(s) from which ecosystem management can be most effective and those levels from which funds should be *paid* to meet the costs of such management must be identified. National governments or states are typically responsible for the year-to-year budgets of large-scale conservation programmes involving land procurement, ecosystem restoration, law enforcement, etc. and in developing countries those budgets are heavily subsidized by international NGOs and the aid programmes of industrialized countries. But the main costs of conserving terrestrial ecosystems are borne by the people inhabiting them. These include the opportunity costs of forgoing alternative land-use endeavours, transaction costs arising from the time and resources invested in negotiating conservation plans but which could be used for other profitable purposes, and direct costs such as losses of livestock and crops to wildlife. Such costs would not be problematic if they were offset by sufficient benefits from biodiversity for local custodians to dedicate their lives to conserving it. The cost : benefit ratio is, however, skewed in favour of the international community deriving most of the benefits of biodiversity while local communities are burdened with the costs, at least in the tropics.

In an attempt to quantify the benefits of conserving tropical biodiversity and identifying who derives those benefits, Balmford & Whitten (2003) identified five classes of benefits:

1. Consumption of indigenous resources such as food, timber, fibre and medicines
2. Nature-based tourism
3. Localized ecosystem services such as water supply and control of flooding, erosion and sedimentation
4. Dispersed ecosystem services such as nutrient cycling, climate regulation and carbon storage
5. Option, existence and bequest values

When monetary estimates are applied to each, it turns out that the values of localized benefits (classes 1 to 3) are considerably lower than the dispersed benefits accruing to the global community (classes 4 and 5). The reason is that there are so many people in the global community who depend, wittingly or not, on the dispersed ecosystem services provided by vast expanses of nature. There is also a rapidly growing segment of the global community that is comparatively wealthy and sufficiently educated to appreciate the non-use values of biodiversity. These include the possibility of using it in future (option value), just knowing it is there (existence value) and knowing it will be there for the benefit of future generations (bequest values). The new frontier of biodiversity conservation is thus the macroeconomic manipulation of global revenue streams to direct appropriate payments to the local custodians of biodiversity in return for the services they (could) provide to the global community. In the context of this book, an advance on this frontier might, for example, be a politically acceptable and sustainable system of direct payments of sufficient value to Maasai herders that they are prepared to embrace wildlife conservation and eschew agriculture (Chapter 14). The fundamental requirement is that donors are prepared to commit themselves to abandoning time horizons (typically less than 5 yr) when entering into agreements with unconventional and risky partners (Wunder et al. 2008). Of course, the obstacles are huge and time is running out, but the fact remains that ecosystems such as rangelands can be expected to lose most of their biodiversity unless the necessary connections are made across institutional levels to even out the imbalances of costs and benefits on a global scale (du Toit et al. 2004).

Potentially helpful suggestions

My target audience is conservation practitioners and planners within the livestock industry who are interested in closing the gaps the preceding sections have hopefully identified. The conservation literature is brimming with ideas for planning specific conservation actions and even for implementing them (reviewed by Knight et al. 2006), but this is a very complex business. Indeed, from the plethora of localized, short-term, species-specific conservation projects showcased in the annual reports of conservation agencies and NGOs it is apparent that sophisticated protocols for planning and implementing conservation action are seldom followed, at least in rangeland ecosystems. I will, thus, make some simple suggestions that I believe could be helpful in a stage *before* the detailed planning and implementation.

The starting point of any conservation intervention is a perceived problem or threat, and with wildlife on rangelands there are a great many problems that could be selected as starting points. The first basic step in deciding if any perceived problem is worth an intervention is to explore all of its associated patterns and trends (habitat types, wildlife populations and distributions, land use, settlement, infrastructure, hydrology, etc.) within the social–ecological system at smaller and larger spatio-temporal scales than those at which the 'problem' was originally perceived (Figure 3.1). This is where the microscope analogy comes in, zooming the magnification in and out to better understand the system in terms of its various alternate states, agents and controllers of change and the set of responders to such change (Pickett et al. 2003). This

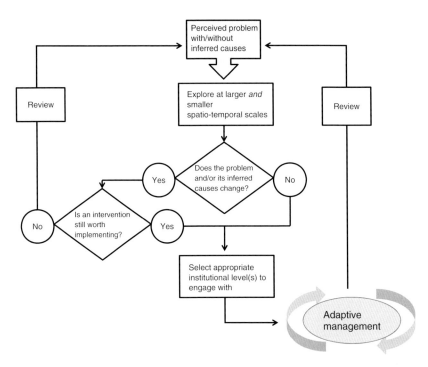

Figure 3.1 **A simple mechanistic process of decision-making to assist in searching for the 'right' focus, across spatio-temporal scales and institutional levels,** *before* **an adaptive management plan is designed and implemented for any perceived conservation problem in a rangeland ecosystem.**

exercise might reveal that a perceived problem is not actually a problem at all and is perhaps even an opportunity, such as the elephant-driven changes to the Chobe ecosystem in northern Botswana (Skarpe et al. 2004), in which case an intervention is not called for. On the other hand, it might be confirmed that a problem really does exist as originally perceived, or else a new focus is identified, and in either case a targeted intervention is still worth considering. Now the challenge is to identify the appropriate institutional level(s) to engage with.

A lesson emerging from the literature on integrated conservation and development projects (ICDPs) is that effectiveness is enhanced if interventions are sequenced across multiple levels of governance (Garnett et al. 2007). Depending on the spatial scale of the conservation problem the appropriate institutional level has to be engaged at the right time, typically securing concessions at higher levels before moving to lower levels. If a conservation intervention requires action at a local scale then it is to be expected that multiple levels of governance will be involved, potentially across all levels from international agreements to rural households (Figure 3.2). An example

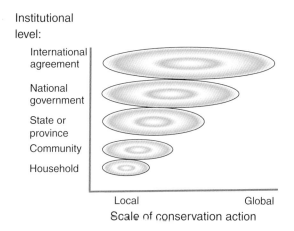

Figure 3.2 **Selecting institutional levels to engage with depends on the spatial scale of the required conservation action. Conservation efforts involving local action require the engagement of multiple institutional levels, with the appropriate sequence of engagement usually being from higher to lower levels. As the scale of required action broadens, the costs of engaging with lower institutional levels increases, both for the conservation project and for the local people and institutions having to bear opportunity and transaction costs.**

is the conservation of semi-arid African savannas, where the viability of wildlife-based CBNRM at the household level is dependent on cash from commercial trophy hunting. Whether elephant hunting may or may not be permitted can make or break the project, and that depends on concessions all the way down from international agreements (e.g. CITES) through national legislation to district governors, traditional leaders, etc. On the other hand, if conservation action is required at a global scale then it might be pointless to engage with any levels below national governments. An extreme example would be any attempt to reduce the component of desertification in tropical rangelands that is attributable to global warming.

The best outcome that can be expected from any conservation action is a self-organizing social–ecological system in which biodiversity and ecosystem services are managed adaptively for long-term sustainability (Figure 3.1, bottom right). Adaptive management does, however, require the best possible starting point and deciding when and where to begin is an entire process on its own. Past failures have demonstrated that an evaluation of opportunity costs is a good place to start, and opportunity costs in rangelands are largely determined by mean annual precipitation.

Rangelands occur where interactions of temperature, precipitation, altitude, latitude and potential evapotranspiration preclude dense stands of trees, generally resulting in tropical, subtropical and warm temperate rangelands in regions with less than 1000 mm annual precipitation. There is, thus, considerable variation in rangeland types, depending largely on the precipitation gradient. Across this arid-mesic gradient, the opportunity costs of wildlife conservation increase as the land's production potential increases under alternative uses such as commercial livestock production and cultivation, ultimately tailing off somewhat in regions of high rainfall with leached soils (Figure 3.3). Consequently, the most successful examples of CBNRM on rangelands occur in arid countries like Namibia, where the opportunity costs are low (du Toit 2004). In the rangelands of Kenya, the percentage of rangeland converted to cropland ranges from less than 10% in districts with less than 550 mm of annual rainfall to greater than 50% in districts receiving approximately 1000 mm (Norton-Griffiths 2007). Simply put, any project to conserve wildlife on rangelands with fertile soils and reliable rainfall will be a non-starter unless wildlife can generate more revenue than agriculture so as to cover the opportunity costs and meet the direct costs of human–wildlife conflict. That requirement is unlikely to be met without external subsidies provided at the household level, such as through direct payments.

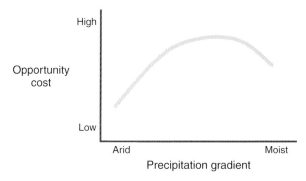

Figure 3.3 **Since the productivity of rangeland ecosystems is largely moisture-limited, the opportunity costs of wildlife conservation vary across rangelands in response to the gradient of mean annual precipitation. People living in the more arid rangelands (semi-deserts) incur the lowest opportunity costs since there are few profitable alternatives to community-based wildlife management involving ecotourism. As precipitation increases, opportunities for converting rangeland to cropland increase and then tail off somewhat in higher rainfall areas with leached soils (e.g. moist-dystrophic savannas).**

Another useful screening exercise is to consider the cost-efficiency of any planned conservation action and compare it against alternatives. In the context of biodiversity conservation, cost-efficiency is measured in units of biodiversity conserved per unit of financial expenditure. As with opportunity cost, the cost-efficiency of conservation is expected to vary across social–ecological systems according to a non-linear relationship (Figure 3.4). Rangelands, because they are generally flatter and more arable and inhabitable than mountains and forests, for example, tend to attract human settlement and infrastructure, or built capital, in the form of roads, railways, ploughed fields, towns, etc. Natural capital, of which wildlife is a component, is obviously low where built capital is high and in such areas the cost-efficiency of conservation is low because of high opportunity costs compounded by restoration costs. As the ratio of natural : built capital increases, the cost-efficiency of conservation increases to an optimum, but then decreases somewhat where ecosystems are already pristine and hence only few additional units of biodiversity can be conserved even with high expenditures. Cost-efficiency of conservation action is thus highest in social–ecological systems where there is some infrastructure. This was demonstrated for endangered African wild dogs (*Lycaon pictus*) in

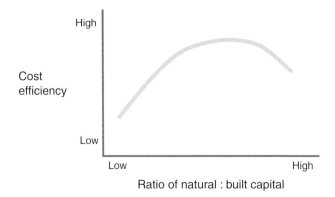

Figure 3.4   **Cost-efficiency of conservation effort, which can be measured as the gain in biodiversity units per financial unit expended, depends on the ratio of natural : built capital in the social–ecological system. Where the ratio is low, the cost-efficiency of conservation action is low because of the opportunity costs of foregoing existing land-use endeavours (e.g. agriculture), compounded by the direct costs of ecological restoration. Cost-efficiency increases to some optimal ratio of natural : built capital and then declines in pristine ecosystems where it is simply not possible to achieve immediate gains in biodiversity units, even with high expenditures on conservation action. The hypothetical curve does not return to a low level, however, since current expenditures on securing the conservation of pristine ecosystems should reduce the risk of biodiversity losses in the future.**

South Africa, where cost-efficiency was found to be higher if wild dogs are conserved on private ranches with the owners compensated for depredation, as compared with translocating those animals to wildlife reserves that require predator-proof fencing and continual monitoring (Lindsey et al. 2005). It is for this type of conservation with a focus on charismatic species, or biodiversity 'hotspots', that cost-efficiency estimates are particularly useful for screening plans and comparing options before moving to the implementation phase (Odling-Smee 2005), as well as for reviewing ongoing projects within an adaptive management framework (Figure 3.1).

Conclusion

The objective of conserving wildlife on rangelands is compatible with a growing acceptance that the social–ecological system is best served by management with an emphasis on ecosystem services rather than on livestock commodity production. Shifting between these management objectives requires a shift in mindset to accommodate issues of scale. The spatio-temporal scales by which rangelands are traditionally managed are those scales applicable to the life history parameters and spatial ecology of one species in the system: *Homo sapiens*. Managing domesticated animals across space and time in ways that conform to human ecology is a perfectly sustainable way to proceed, as the past few thousand years have demonstrated, as long as the necessary ecosystem services are in free supply. But when rangeland degradation and resource consumption expand to the extent that essential ecosystem services begin to fail, then the management emphasis must shift to restoring ecological heterogeneity. This requires that the biotic and abiotic components of the system be allowed to interact naturally within a space–time mosaic. Yet the benefits of restoring ecological resilience and securing ecosystem services remain subjects of academic discussion until conservation practitioners can implement pragmatic plans to achieve effective action. Such big challenges and woefully deficient resources mean that potential conservation projects have to be 'triaged' on the basis of those most likely to succeed. Identifying real problems and threats at the relevant scales, engaging at the appropriate institutional levels, and assessing opportunity costs and cost-efficiencies of alternative interventions are all fundamental to the success of this process.

References

Balmford, A. & Whitten, T. (2003) Who should pay for tropical conservation, and how could the costs be met? *Oryx* 37, 238–250.

Bienen, L. & Tabor, G. (2006) Applying an ecosystem approach to brucellosis control: can an old conflict between wildlife and agriculture be successfully managed? *Frontiers in Ecology and the Environment* 4, 319–327.

Biggs, H.C. & Rogers, K.H. (2003) An adaptive system to link science, monitoring, and management in practice. In: du Toit, J.T., Rogers, K.H. & Biggs, H.C. (eds.) *The Kruger Experience: Ecology and Management of Savanna Heterogeneity*. Island Press, Washington, D.C., pp. 59–80.

Brashares, J.S., Arcese, P., Sam, M.K. et al. (2004) Bushmeat hunting, wildlife declines, and fish supply in West Africa. *Science* 306, 1180–1183.

Brooks, M.L., D'Antonio, C.M., Richardson, D.M. et al. (2004) Effects of invasive alien plants on fire regimes. *BioScience* 54, 677–688.

Ferraro, P.J. & Kiss, A. (2002) Direct payments to conserve biodiversity. *Science* 298, 1718–1719.

Garnett, S.T., Sayer, J. & du Toit, J.T. (2007) Improving the effectiveness of interventions to balance conservation and development: a conceptual framework. *Ecology and Society* 12(1), 2 [online]. http://www.ecologyandsociety.org/vol12/iss1/art2/

Gereta, E., Wolanski, E., Borner, M. et al. (2002) Use of an ecohydrological model to predict the impact on the Serengeti ecosystem of deforestation, irrigation and the proposed Amala weir water diversion project in Kenya. *Ecohydrology and Hydrobiology* 2, 127–134.

Gibson, D., Craig, G.C. & Masogo, R. (1998) Trends of the elephant population in northern Botswana from aerial survey data. *Pachyderm* 25, 14–27.

Gruell, G.E. (1999) *Proceedings: Ecology and Management of Pinyon-Juniper Communities within the Interior West.* United States Department of Agriculture Forest Service, Rocky Mountain Research Station, Ogden, pp. 24–28, Proceedings RMRS-P-9.

Hilborn, R., Arcese, P., Borner, M. et al. (2006) Effective enforcement in a conservation area. *Science* 314, 1266.

Jefferies, R.L., Rockwell, R.F. & Abraham, K.F. (2004) Agricultural food subsidies, migratory connectivity and large-scale disturbance in Arctic coastal systems: a case study. *Integrative and Comparative Biology* 44, 130–139.

Jones, B.T.B. & Murphree, M.W. (2004) Community-based natural resource management as a conservation mechanism: lessons and directions. In: Child, B. (ed.) *Parks in Transition: Biodiversity, Rural Development and the Bottom Line.* Earthscan, London, pp. 63–103.

Knick, S.T. & Rotenberry, J.T. (1997) Landscape characteristics of disturbed shrub-steppe habitats in southwestern Idaho (USA). *Landscape Ecology* 12, 287–297.

Knight, A.T., Cowling, R.M. & Campbell, B.M. (2006) An operational model for implementing conservation action. *Conservation Biology* 20, 408–419.

LaMalfa, E.M. & Ryel, R. (2008) Differential snowpack accumulation and water dynamics in aspen and conifer communities: implications for water yield and ecosystem function. *Ecosystems* 11, 569–581.

Lindsey, P.A., Alexander, R., du Toit, J.T. et al. (2005) The cost efficiency of wild dog conservation in South Africa. *Conservation Biology* 19, 1205–1214.

Mabunda, D., Pienaar, D.J. & Verhoef, J. (2003) The Kruger National Park: a century of management and research. In: du Toit, J.T., Rogers, K.H. & Biggs, H.C. (eds.) *The Kruger Experience: Ecology and Management of Savanna Heterogeneity.* Island Press, Washington, D.C., pp. 3–21.

Maddock, L. (1979) The "migration" and grazing succession. In: Sinclair, A.R.E. & Norton-Griffiths, M. (eds.) *Serengeti: Dynamics of an Ecosystem.* University of Chicago Press, Chicago, pp. 104–129.

Madhusudan, M.D. (2005) The global village: linkages between international coffee markets and grazing by livestock in a South Indian wildlife reserve. *Conservation Biology* 19, 411–420.

McKnight, L. (2008) Challenging cheatgrass: can tools like the 'black fingers of death' fight this formidable invasive species? *RMRScience.* United States Department of Agriculture Forest Service, Rocky Mountain Research Station, Fort Collins.

Millennium Ecosystem Assessment. (2003) *Ecosystems and Human Well-being: A Framework for Assessment. A Report of the Conceptual Framework Working Group of the Millennium Ecosystem Assessment.* Island Press, Washington, D.C.

Mosugelo, D.K., Moe, S.R., Ringrose, S. et al. (2002) Vegetation changes during a 36-year period in northern Chobe National Park, Botswana. *African Journal of Ecology* 40, 232–240.

Murray Berger, K. (2006) Carnivore-livestock conflicts: effects of subsidized predator control and economic correlates on the sheep industry. *Conservation Biology* 20, 751–761.

Norton-Griffiths, M. (2007) How many wildebeest do you need? *World Economics* 8, 41–64.

Odling-Smee, L. (2005) Dollars and sense. *Nature* 437, 614–616.

O'Keeffe, J. & Rogers, K.H. (2003) Heterogeneity and management of the Lowveld rivers. In: du Toit, J.T., Rogers, K.H. & Biggs, H.C. (eds.) *The Kruger Experience: Ecology and Management of Savanna Heterogeneity.* Island Press, Washington, D.C., pp. 447–468.

Pickett, S.T.A., Cadenasso, M.L. & Benning, T.L. (2003) Biotic and abiotic variability as key determinants of savanna heterogeneity at multiple spatiotemporal scales. In: du Toit, J.T., Rogers, K.H. & Biggs, H.C. (eds.) *The Kruger Experience: Ecology and Management of Savanna Heterogeneity.* Island Press, Washington, D.C., pp. 22–40.

Sinclair, A.R.E. & Arcese, P. (eds.) (1995) *Serengeti II: Dynamics, Management, and Conservation of an Ecosystem.* University of Chicago Press, Chicago.

Sinclair, A.R.E. & Byrom, A.E. (2006) Understanding ecosystem dynamics for conservation of biota. *Journal of Animal Ecology* 75, 64–79.

Sinclair, A.R.E. & Norton-Griffiths, M. (eds.) (1979) *Serengeti: Dynamics of an Ecosystem.* University of Chicago Press, Chicago

Sinclair, A.R.E., Packer, C., Mduma, S.A.R. & Fryxell, J.M. (eds.) (2008) *Serengeti III: Human Impacts on Ecosystem Dynamics.* University of Chicago Press, Chicago.

Skarpe, C., Aarestad, P.A., Andreassen, H.P. et al. (2004) The return of the giants: ecological effects of an increasing elephant population. *Ambio* 33, 276–282.

Smith, R.J., Muir, R.D.J., Walpole, M.J. et al. (2003) Governance and the loss of biodiversity. *Nature* 426, 67–70.

Stam, B.R., Malechek, J.C., Bartos, D.L. et al. (2008) Effect of conifer encroachment into aspen stands on understory biomass. *Rangeland Ecology and Management* 61, 93–97.

Tausch, R.J. (1999) *Proceedings: Ecology and Management of Pinyon-Juniper Communities within the Interior West.* United States Department of Agriculture Forest Service, Rocky Mountain Research Station, Ogden, pp. 12–19, Proceedings RMRS-P-9.

Thirgood, S., Mosser, A., Tham, S. et al. (2004) Can parks protect migratory ungulates? The case of the Serengeti wildebeest. *Animal Conservation* 7, 113–120.

du Toit, J.T. (2004) Conserving tropical biodiversity: the arid end of the scale. *Trends in Ecology and Evolution* 19, 226.

du Toit, J.T. & Cumming, D.H.M. (1999) Functional significance of ungulate diversity in African savannas and the ecological implications of the spread of pastoralism. *Biodiversity and Conservation* 8, 1643–1661.

du Toit, J.T., Walker, B.H. & Campbell, B.M. (2004) Conserving tropical nature: current challenges for ecologists. *Trends in Ecology and Evolution* 19, 12–17.

Vandewalle, M. (2003) Historic and recent trends in the size and distribution of northern Botswana's elephant population. In: Vandewalle, M. (ed.) *Effects of Fire, Elephants, and Other Herbivores on the Chobe Riverfront Ecosystem.* Botswana Government Printer, Gaborone, pp. 7–16.

Wagner, F.H. (1988) *Predator Control in the Sheep Industry: The Role of Science in Policy Formation.* Iowa State University Press, Ames.

Wagner, F.H. (2006) *Yellowstone's Destabilized Ecosystem.* Oxford University Press, New York.

Walker, B.H. (1989) Diversity and stability in ecosystem conservation. In: Western, D. & Pearl, M.C. (eds.) *Conservation for the Twenty-First Century.* Oxford University Press, Oxford, pp. 121–130.

Wunder, S., Campbell, B., Frost, P.G.H. et al. (2008) When donors get cold feet: the community conservation concession in Setulang (Kalimantan, Indonesia) that never happened. *Ecology and Society* 13(1), 12 [online]. http://www.ecologyandsociety.org/vol13/iss1/art12/

$$\textbf{4}$$

Rangeland Conservation and Shrub Encroachment: New Perspectives on an Old Problem

Steven R. Archer

School of Natural Resources, University of Arizona, AZ, USA

Introduction

Grasslands, shrublands and savannas, collectively represented in the term *rangelands*, constitute ca. 50% of the Earth's land surface (Bailey 1996). Although typically characterized by low and highly variable annual rainfall, these geographically extensive arid and semi-arid landscapes represent 30–35% of terrestrial net primary productivity (NPP) (Field et al. 1998), contain more than 30% of the world's human population and support the majority of the world's livestock production (Safriel & Adeel 2005). As such, rangelands play an important role in global carbon, water and nitrogen cycles and human health (Campbell & Stafford Smith 2000). Their extensive airsheds and watersheds provide habitat for game and non-game wildlife and myriad ecosystem goods and services important to rapidly growing settlements and cities that may be geographically distant. Rangelands thus have considerable, multi-dimensional conservation value. Stewardship of vegetation composition, cover and production is the foundation of sustainable rangeland management. A key component of rangeland ecosystem management is maintaining vegetation within a desirable mix of herbaceous and woody plants (WPs).

Wild Rangelands: Conserving Wildlife While Maintaining Livestock in Semi-Arid Ecosystems,
1st edition. Edited by J.T. du Toit, R. Kock, and J.C. Deutsch.
© 2010 Blackwell Publishing

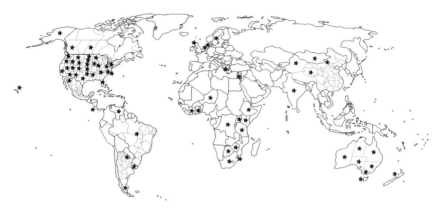

Figure 4.1 **Global distribution of reported woody plant encroachment. See Archer (2008) for a bibliography.**

One of the most striking land cover changes in rangelands worldwide over the past 150 years has been the proliferation of trees and shrubs at the expense of perennial grasses (Figure 4.1) (Archer 2008). In some cases, native WPs are increasing in stature and density within their historic geographic ranges; in other cases non-native WPs are becoming dominant (Hoffman & Moran 1988; Bossard & Rejmanek 1994; Bruce et al. 1997; Grice 2000; Katz & Shafroth 2003; Shafroth et al. 2005). In some areas, trees are expanding into grass- or shrub-dominated vegetation and in others they are in-filling savanna or mosaic woodland plant communities. Shifts in the abundance of woody and herbaceous vegetation represent fundamental alterations of habitat for animals (microbes, invertebrates and vertebrates) and hence marked alterations of ecosystem trophic structure. In arid and semi-arid regions, increases in the abundance of xerophytic shrubs at the expense of mesophytic grasses represent a type of desertification (e.g. Schlesinger et al. 1990; Havstad et al. 2006) often accompanied by accelerated rates of wind and water erosion (Wainwright et al. 2000; Gillette & Pitchford 2004; Breshears et al. 2009). In semi-arid and sub-humid areas, encroachment of shrubs and trees into grasslands and savannas may substantially promote primary production, nutrient cycling and accumulation of soil organic matter (Archer et al. 2001; Knapp et al. 2008a) but reduce stream flow, ground water recharge, livestock production and biological diversity. At the global scale, WP encroachment stands in

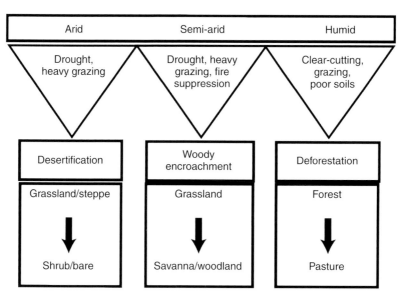

Figure 4.2   **Woody plant encroachment in arid lands represents a form of desertification, whereby xerophytic shrubs replace mesophytic grasses, often resulting in an increase in soil erosion. In higher rainfall zones, encroachment of trees and shrubs may enhance primary production and nutrient cycling but dramatically alter animal habitat and livestock-carrying capacity. The proliferation of WPs in the world's rangelands stands in stark contrast to the reductions in WP cover associated with deforestation (from Archer & Stokes 2000).**

stark contrast to deforestation (replacement of forest cover with pasture or cropland), which has received substantially more attention (Figure 4.2).

Although the conversion of grasslands and savannas to shrublands and woodlands has been formally documented and qualitatively observed in some areas, it should not be assumed that this transformation has been uniform or ubiquitous (e.g. Jacobs et al. 2008). A variety of techniques have been used to demonstrate shifts from grass to WP domination (Archer 1996). However, these represent local, site-specific assessments and should not be broadly extrapolated within or across bioclimatic zones or to topo-edaphically different areas. Indeed, repeat ground photography in western North America documents areas where shrublands have dominated landscapes since the

1800s (Humphrey 1987; Turner et al. 2003). Furthermore, many areas may have historically comprised of mixtures of woody and herbaceous vegetation (e.g. shrub-steppe or shrub- or tree-savannas) and efforts to eradicate WPs may be misguided (McKell 1977) and detrimental to native plants and wildlife (e.g. Knick et al. 2003).

Why has woody plant abundance increased on rangelands?

Where WP encroachment has been documented, the causes are a topic of active debate. Traditional explanations centre around the intensification of livestock grazing, changes in climate and fire regimes, the introduction of non-native woody species and declines (natural and human-induced) in the abundance of browsing animals (Box 4.1). Historical increases in atmospheric nitrogen deposition and atmospheric carbon dioxide (CO_2) concentration have also been suggested as potentially important drivers. It is likely that all of these have interacted to varying degrees in various locales. Hence, it is difficult to assign primacy to any one of them. The fact that the abundance of woody and herbaceous vegetation in dry lands is known to have shifted back and forth during the Holocene before the advent of significant human–environment interactions (e.g. Monger 2003) illustrates that large-scale, long-term factors like climate change and atmospheric CO_2 enrichment can drive shifts in grass–woody plant abundance. Furthermore, the fact that WP proliferation has been occurring in rangelands worldwide over the past 100 years hints that large-scale forcings may be at work. For example, it has been suggested that desert grasslands of the southwestern United States established and developed during the cooler, moister Little Ice Age period and were only marginally suited to the drier, warmer climatic conditions that have prevailed since that time (Neilson 1986). Thus, the grasslands present at the time of Anglo-European settlement may have been in a transient state, with livestock grazing merely accelerating a conversion to shrubland that was already underway and lagging (e.g. Cole 1985) behind changes in climate. However, in the few cases where site-specific, long-term climate and long-term vegetation records both occur, increases in WP abundance during the 1900s do not appear to be linked to climate change (e.g. Conley et al. 1992).

Box 4.1 **Causes of woody plant encroachment**

Causes for global increases in the abundance of trees and shrubs in dry lands are actively debated. There is likely no single-factor explanation for this widespread phenomenon. Most likely, it reflects drivers that vary locally or regionally, or from the interaction of more than one driver. Changes in a given driver may be necessary to tip the balance between woody and herbaceous vegetation, but may not be sufficient unless co-occurring with changes in other drivers. Potential causes for increases in WP abundance in rangelands include changes in the following:

Climate: Changes in the amount and seasonality of PPT can affect the balance between grasses and WPs. In semi-arid and sub-humid regions, increases in total PPT can enhance WP size and density; decreases in PPT can promote shifts from mesophytic grasses to xerophytic shrubs. Shifts from summer to winter PPT regimes can favour WPs. PPT effects at local scales are strongly mediated by soil texture and depth: WPs are favoured on relatively deep, well-drained soils and grasses on shallow, clayey soils.

Grazing: Utilization of grasses by folivores reduces their leaf and root biomass, making them more susceptible to other environmental stresses. Heavy grazing by livestock and utilization of grass seeds by granivores (rodents, ants) can cause shifts in herbaceous species composition to assemblages less effective at competitively excluding WP seedlings. Folivores and granivores may also be effective agents of WP seed dispersal in certain cases. Changes in soil properties and microclimate accompanying grazing may create conditions more favourable for WP establishment and less favourable for grass establishment.

Browsing: Preferential utilization of WPs by browsing herbivores may keep shrubs and trees from establishing or from reaching large sizes or high densities. WPs kept low in stature by browsers will be more susceptible to fire. Reductions in the abundance of browsers may remove a major constraint to WP dominance on sites that are otherwise climatically and edaphically suitable.

Fire regimes: Grasslands and savannas are typically characterized by high fine fuel loads and hence frequent fire that would either prevent WPs from establishing or prevent fire-tolerant WPs from gaining dominance. Grazing can reduce fine fuel abundance, thus reducing the frequency and intensity of fires that historically kept WPs suppressed. There is a strong synchrony between climate and fire, wherein fires occur during dry periods that follow wet periods that promote grass biomass (fine fuel) accumulation. Grazing can alter species composition and reduce fine fuel loads and continuity, thus disrupting this synchrony and providing opportunities for WP establishment and stand development.

Increases in atmospheric CO_2: There is some evidence that increases in atmospheric CO_2 concentrations since the industrial revolution may have favoured WPs that have the C_3 photosynthetic pathway over grasses that have the C_4 photosynthetic pathway (Morgan et al. 2007). However, WPs have numerous other adaptations that allow them to compensate for and overcome disadvantages that may be related to their photosynthetic pathway. Furthermore, the differential response of photosynthetic pathways to CO_2 fertilization cannot be invoked to explain increases in WP abundance in temperate bioclimatic zones where both grasses and shrubs possess the C_3 photosynthetic pathway.

Nitrogen deposition: A correlation between levels of N deposition and the extent of forest expansion into grasslands has been shown for the northern Great Plains of North America (Köchy & Wilson 2001).

For detailed reviews and discussions, see Archer (1994, 1995), Van Auken (2000) and Briggs et al. (2005).

The occurrence of fence-line contrasts (Figure 4.3) coupled with the fact that WP encroachment has occurred on some landscapes, but not others, within a biogeographic region (topo-edaphic features held equal), argues that local factors associated with land use are at play (Archer 1995, 1996). While livestock grazing and fire are local in their effects, intensification of livestock grazing has been globally widespread since the mid-1800s (Archer et al. 1995; Asner et al. 2004) and coincident with declines in fire frequency

Figure 4.3 *Acacia nilotica* invasion of Mitchell Grasslands in Queensland, Australia. Introduced from Africa to provide shade for livestock in these treeless plains, the plant has spread rapidly (see Kriticos et al. 2003). Fence line contrasts such as this suggest that local factors (grazing, seed dispersal and fire) are primary determinants of WP abundance. Factors operating at larger scales (e.g. climate, atmospheric CO_2 enrichment, nitrogen deposition) influence rates and dynamics of shrub encroachment and stand development, but may not be primary drivers.

owing to reductions in fine fuel mass and continuity (e.g. Savage & Swetnam 1990).

Traditional perspectives on woody plant encroachment

Traditional perspectives on WP encroachment have centred around impacts on livestock production, hydrology (ground water recharge and stream flow) and game management (wildlife valued for hunting) (Figure 4.4). While these areas of emphasis remain important, contemporary perspectives have been broadened to include impacts on biological diversity, ecosystem function (primary production and nutrient cycles) and land surface–atmosphere interactions. This section briefly reviews traditional perspectives. Contemporary perspectives are summarized in the sections Emerging perspectives on WP encroachment and Brush management in the twenty-first century.

Livestock and wildlife management

WP encroachment has long been of concern to rangeland managers (Leopold 1924). The basis for this concern centred around the adverse effects of WPs on forage production and livestock safety (e.g. WPs as cover for predators), health [e.g. as habitat for insect and arthropod pests and parasites such as ticks and tsetse flies (*Glossina* spp.) (Teel et al. 1997)] and handling (difficulty in gathering and moving animals increases with increasing WP

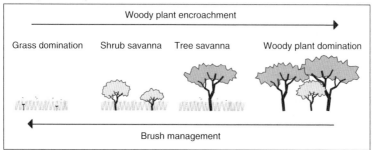

Traditional considerations

- livestock production
  - health
  - handling
  - forage production
- Hydrology
  - streamflow
  - groundwater recharge
- Wildlife
  - game management

Emerging considerations

- Primary production
- Carbon sequestration
- Biological diversity
- Land-surface atmospehere interactions
- Non-methane hydrocarbon emissions and trace gas emissions
- Wind and water erosion

Woody plant encroachment

Grass domination   Shrub savanna   Tree savanna   Woody plant domination

Brush management

Figure 4.4 **Traditional and contemporary perspectives on WP encroachment. Brush management practices have typically been implemented with consideration of traditional perspectives. Emerging considerations will reshape views on managing WP abundance to meet competing objectives.**

stature/cover/density). A preoccupation with forage production has also been the impetus to eliminate wildlife viewed as competitors with livestock. In some cases, these wildlife may have played an important role in keeping WPs in check; and their systematic elimination may have opened the door for WPs to increase in abundance (e.g. Weltzin et al. 1997).

The proliferation of WPs has often been associated with declines in forage availability, hence there is an extensive body of research on WP effects on grass composition and production (reviewed in Scholes & Archer 1997) and the development of management technologies and approaches for reducing WP cover/density (Vallentine 1971). Typically, these technologies centred on the use of prescribed fire, mechanical treatments (e.g. shredding, chaining, roller chopping, grubbing) and herbicides (Scifres 1980; Bovey 2001). Historically, these 'brush management' practices were applied indiscriminately across entire landscapes with the goal of eradicating WPs. Because the emphasis was narrowly focused on the utilitarian goal of maximizing livestock production, there was little consideration of impacts on wildlife, the ecological roles of trees and shrubs in ecosystems or the capacity of a given site to support alternate forms of vegetation.

Much rangeland research has aimed to understand how traditional brush management practices impact wildlife, as managers often strove for widespread, 'wall-to-wall' reductions in shrub cover to induce open grasslands deemed desirable for livestock production. Such practices were often to the detriment of wildlife valued for sport hunting. It is ironic that a change in one land use practice (intensification of livestock grazing) promoted a change in habitat (increased WP abundance) that was favourable to certain wildlife species valued for sport-hunting and that this, in turn, necessitated the instigation of another land use practice (brush management) to generate habitat more suitable for livestock and less so for certain classes of wildlife. In other cases, the desire to convert native shrubland and shrub-steppe to grassland to promote livestock production, even in areas where the former may have been the historically dominant community, can adversely affect native wildlife (e.g. Knick et al. 2003). Such cases illustrate trade-offs in resource management as well as the contrasting perspectives and priorities of stakeholders, user-groups and professional managers.

Stream flow and groundwater recharge

Arid and semi-arid rangelands are characterized by frequent drought, low and highly variable interannual rainfall and increasing demands on scarce water resources. A long-standing concern over WP proliferation in rangelands centres around the potential effects of trees and shrubs on stream flow and groundwater recharge. Although there is relatively little quantitative information regarding how increases in WP abundance may have changed the hydrological cycle, popular and grey literature articles have long promoted the notion that WPs are prolific users of water owing to their relatively large leaf area and deep roots. It is widely believed that aggressive shrub control can convert intermittent streams into perennial streams and increase rates of reservoir recharge. As a result, there has been substantial pressure to reduce WP cover on rangelands as a means of increasing water yield. Consequently, massive WP control and removal efforts have been, and continue to be, undertaken in arid and semi-arid environments.

The relationship between stream flow and forest cover is relatively well established [e.g. increased stream flow with timber harvest (Stedneck 1996);

decreased annual runoff and stream flow with afforestation (Trimble et al. 1987; Farley et al. 2005)]. In systems characterized by winter rainfall (e.g. chaparral), there is strong evidence that stream flow increases when woody cover is reduced (Hibbert 1983), but in other semi-arid environments the linkage is tenuous at best (Blackburn 1983). It is now clear that the effects of shrub removal on stream flow vary, depending on the traits of the WPs (their canopy and rooting architecture), climate (e.g. rainfall amount, seasonality, event sizes and intensities), soil type (deep vs. shallow; sandy vs. clayey) and geomorphology (Thurow & Hester 1997; Huxman et al. 2005; Wilcox et al. 2006). There may be little potential for increasing stream flow where annual precipitation (PPT) is less than 500 mm (Wilcox 2002; Wilcox et al. 2005). Thus, broad generalizations regarding shrub control effect on water yield should be viewed with suspicion.

In cases where rigorous measurements and evaluations have been undertaken, it appears that estimates of water savings realized by brush management may have been substantially overestimated (Dugas et al. 1998; Owens & Moore 2007). Studies indicating that brush management may not be achieving the desired outcomes with respect to water yield are accumulating (Belsky 1996; Wilcox & Thurow 2006). Estimates of the economic benefits of shrub control based solely on water salvage are therefore questionable. Even so, it may be desirable to manage shrub cover to promote wildlife habitat, biological diversity and soil health (Shafroth et al. 2005).

Emerging perspectives on woody plant encroachment

Biological diversity

The utilitarian world view of traditional rangeland managers (emphasizing livestock production, predator control and water yield) and traditional wildlife managers (emphasizing fur-bearing, charismatic and sport hunting species) dominated resource management until the emergence of conservation biology in the late 1970s. Conservation biologists work from the tenet that all species, not just those with known economic value or charismatic or aesthetic appeal, have intrinsic value via their contributions to ecosystem function and services. The growing appreciation of the value of ecosystem services and biodiversity has filled a void in traditional rangeland and wildlife management, this has led to new perspectives on WP encroachment in grasslands and savannas.

A conceptual model

Biodiversity, whether quantified as the genetic diversity of populations, species richness or the number of plant functional groups or animal guilds represented in an area, is strongly influenced by WP encroachment. Colonization of grasslands by WPs initially represents new species additions and hence promotes biodiversity directly, but their modification of soil properties, vertical vegetation structure and microclimate may also promote the ingress (via dispersal) and establishment of other plant and animal species (Figure 4.5).

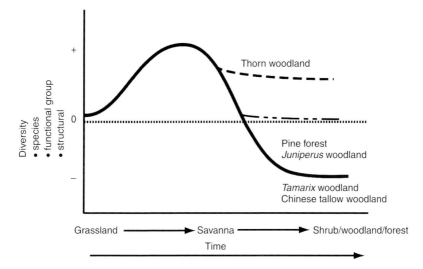

Figure 4.5   Conceptual model of changes in ecosystem biodiversity (species, growth form or structural) that potentially accompany WP proliferation in grasslands and savannas. Diversity is predicted to increase during early stages of WP encroachment, owing to the mixture of woody and herbaceous floral/faunal elements. Maximum diversity might be expected in savanna-like configurations where woody and herbaceous plants coexist or where the gains in new woody and herbaceous species outweigh the losses of the original herbaceous species. As woody plant abundance increases, loss of grassland components eventually occurs. In subtropical thorn woodland and dry forests with high WP species richness, a net increase in diversity may result. In other settings, there may be no net change in diversity, only a change in physiognomy. Where WPs form virtual monocultures with little or no understory, the loss of diversity may be profound. In all scenarios, regardless of the numerical changes in biodiversity, the existence of grassland and open savanna ecosystems and the plants and animals endemic to them is jeopardized.

The conceptual model in Figure 4.5 is based on numerical assessments of species, functional group and structural diversity. However, from the perspective of physiognomic diversity, WP encroachment is transformative. Grasslands become shrub or tree savannas; and shrub and tree savannas become shrublands or woodlands. Thus, even in cases where numerical diversity may be maintained or enriched by WP encroachment, there is a loss of grassland and savanna ecosystems and the plants and animals endemic to those systems. Thus, while 'brush management' has historically been advocated from the perspective of potential benefits for livestock production and hydrology, it should also be considered from the perspective of maintaining the existence of grassland and savanna ecosystems (Fulbright 1996) and mitigating impacts of non-native WPs on those ecosystems (e.g. Grice 2006). Exurban and agricultural development are widely recognized as threats to the existence of grassland and savanna ecosystem types worldwide. WP proliferation needs to be examined in this same light.

Above-ground biodiversity

Increasing plant diversity in rangelands has a multiplier effect on animal diversity by adding keystone structures and habitat heterogeneity (Tews et al. 2004) and providing nesting, perching and foraging sites and shelter against predators and extreme climatic conditions (Whitford 1997; Cooper & Whiting 2000; Valone & Sauter 2005; Blaum et al. 2007a). Indeed, numerous reptiles, birds and mammals appear to flourish in heterogeneous grass-dominated landscapes where scattered WPs provide up to 15% cover (Solbrig et al. 1996; Meik et al. 2002; Eccard et al. 2004; Bock et al. 2006; Thiele et al. 2008). For example, in the Kalahari region of southern Africa, scattered, large acacia trees provide vital resources and services for wildlife, including shade and shelter for antelope, canids and large ground foraging birds; places for leopards to hang prey out of reach of scavengers; perches for hunting, roosting, or nesting for owls, raptors and vultures; and cavities for birds, mammals and lizards (Dean et al. 1999; Eccard et al. 2004). In arid savanna rangelands, the diversity of small carnivores and their prey peaks at around 10–15% shrub cover (Blaum et al. 2007d). In the Chihuahuan Desert of North America, shrub-invaded sites harbour four times the number of ant foragers found at a relatively pristine desert grassland site, suggesting that ant diversity is enhanced by shrub encroachment and that several taxa benefit from it (Bestelmeyer 2005). The effects of WP encroachment vary among animal taxa and functional groups (e.g. Kazmaier et al. 2001), but as WP cover increases and habitat characteristics continue to shift, shrubland/woodland-adapted species would be expected to become favoured over grassland-adapted species (Box 4.2).

Box 4.2 **Avifauna and shrub encroachment**

- Grassland passerines are declining at a faster rate than any other bird group in North America (Peterjohn & Sauer 1999). WP encroachment associated with livestock grazing is among the contributing factors (Bakker 2003; Brennan and Kuvlesky, 2005).
- Passerine habitat becomes increasingly unsuitable when thresholds in WP cover and height are exceeded (Lloyd et al. 1998; Grant et al. 2004; Gottschalk et al. 2007) and as the extent and proximity to woody vegetation and woodland edge increases (Johnson 2001; Bakker et al. 2002; Fletcher and Koford 2002). Proximity to and abundance of WPs may influence food abundance and rates of predation and brood parasitism (e.g. Thiele et al. 2008).
- Species richness may be higher in shrublands than in the grasslands from which they developed (Whitford 1997), with 30% or more of the avian community differing between grassland and shrubland habitats (Pidgeon et al. 2001). However, declines in charismatic gallinaceous birds have been linked to the proliferation of trees in shrub-steppe (Crawford et al. 2004) and grasslands (Fuhlendorf et al. 2002).

- Sandhill cranes migrate between breeding grounds in the North American arctic and wintering grounds along the Gulf of Mexico coastline and in selected wetlands in the southwestern United States. The Platte River, which flows from the Rocky Mountains across the Great Plains and into the Missouri River, is an important stopover in this annual trek along the North American Central Flyway as it meets requirements for roosting, resting and energetic restoration. About 80% of the total sandhill crane population uses the Platte and North Platte Rivers during both fall and spring migration; and at peak times, 500,000 cranes may be concentrated in a 100-km stretch. This critical stopover habitat was historically a shallow, braided prairie river characterized by sandbars and marshes. Spring floods scoured the river basin and high summer flows prevented tree establishment. With the installation of dams and reservoirs, spring floods and summer flows were reduced and marshy and sandbar habitats are now largely dominated by forest (Johnson 1994). The effects of this dramatic and relatively recent habitat change on sandhill crane populations is not yet known.

Some grassland-obligate plants and animals may be affected immediately and negatively by WP encroachment. Even so, diversity may be maintained or enhanced if new species appear and co-occur with the more broadly adapted original species and if the displacement of grassland-obligate species is more than offset by the arrival of new species (e.g. Sauer et al. 1999; Blaum et al. 2007b,c). As WP cover and biomass continue to increase, the end result may be an overall gain in diversity, no net change in diversity, or a net loss in diversity (Figure 4.5). Qualitative observations suggest that tropical and subtropical grasslands may experience net gains in diversity with WP encroachment owing to large pools of tree and shrub species, large pools of herbaceous species capable of coexisting with WPs and large pools of invertebrates and passerine bird species. In other cases, the number of encroaching woody species may be very small and their traits such that most other plant species are eliminated subsequent to their establishment (e.g. Baez & Collins 2008; Knapp et al. 2008b). WP encroachment may then result in virtual monocultures of vegetation with concomitant impacts on faunal diversity. It should be kept in mind, however, that in each of these scenarios grassland ecosystems are being lost.

Below-ground biodiversity

Shifts from bacterial to fungal populations may accompany shifts from herbaceous to woody domination (e.g. Imberger & Chiu 2001; Purohit et al. 2002). Aanderud et al. (2008) found differences in gram-positive bacteria, Actinobacteria and fungi communities in soils below and between shrubs, ostensibly a response to shrub-induced changes in soil status.

Parasitic nematodes and nematodes feeding on bacteria and fungi in the immediate vicinity of plant roots are indicator taxa for changes in below-ground microbial communities. The maximum depth of occurrence of these organisms increased from 2.1 m in grasslands to 4.0 m in areas where WPs have replaced grasses and the composition of the nematode food web was reduced from five trophic groups to two (Jackson et al. 2002). Invaded woody sites also had lower bacterial and fungal species richness, primarily due to the loss of root-feeding species.

Biogeochemisty and ecosystem processes

Given the global extent of rangelands, shifts from grass to WP dominance have the potential to influence biogeochemical cycles and climate at local, regional

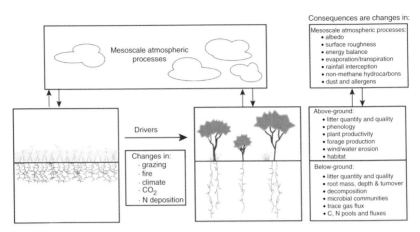

Figure 4.6 **Drivers of WP encroachment (see Box 4.1) and the potential consequences of ecosystem function and land surface–atmosphere interactions.**

and global scales through alterations of above- and below-ground productivity, quality and quantity of litter inputs, rooting depth and distribution, hydrology, microclimate and energy balance (Figure 4.6). Increased storage of soil organic carbon (SOC) is likely to occur if the NPP increases relative to the rate of organic matter decomposition. WPs produce lignin-rich structural tissues that decompose slowly and they are typically more deeply rooted than the grasses they displace (Jackson et al. 1996, 2000). Soils of sites dominated by WPs generally have higher carbon : nitrogen ratios and contain more complex organic compounds that are more resistant to decay (Liao et al. 2006). Under these conditions, organic matter will be more likely to accumulate. A significant proportion of this organic matter is distributed deep in the soil profile (Jobbágy & Jackson 2000) where decomposition and microbial activity are slow. Consequently, grassland → savanna → woodland cover changes have strong potential to increase ecosystem carbon storage and contribute to a global carbon sink (Wessman et al. 2005; Asner & Archer 2010).

In areas where WPs have displaced grasses, above-ground net primary production increases dramatically as a function of annual rainfall (Knapp et al. 2008a). Although these net accumulations in biomass represent carbon sequestration, they may be accompanied by increases in non-methane hydrocarbon and trace gas fluxes that potentially alter atmospheric chemistry and climate (in the section Land surface–atmosphere interactions). When the dominant woody species is a nitrogen-fixer, N accumulation can increase substantially (Geesing et al. 2000; Hughes et al. 2006), thus augmenting a key resource that, along with water, typically co-limits ecosystem productivity. In semi-arid, sub-humid and humid climate zones, tree plantations are attractive from both agroforestry and carbon sequestration perspectives (Nosetto et al. 2006). However, management intended to increase WP abundance is not without environmental costs, including (potentially) soil salinization, soil acidification and reductions in stream flow (Jackson et al. 2005).

A substantial majority of the carbon in grassland and savanna ecosystems is below ground (Schlesinger 1997), but it is not completely clear how grazing, climate and WP encroachment interact to affect gains and losses from these large pools. Despite consistent increases in above-ground carbon storage with woody vegetation encroachment (Knapp et al. 2008a) and dry land afforestation (e.g. Nosetto et al. 2006), the trends in SOC are highly variable, ranging from substantial losses (Jackson et al. 2002) to no net change (Wessman et al. 2005; Asner & Archer 2010) to large gains (Hibbard et al. 2003). Reasons for the variable response of SOC pools to shrub proliferation are not clear.

The ability to predict changes in landscapes characterized by mixtures of herbaceous and WPs began to emerge among the top priorities for global change research in the mid- to late 1990s (Houghton et al. 1999; Daly et al. 2000). In the United States, woody encroachment and 'thickening' of woody vegetation in dry land and montane ecosystems are believed to represent a potentially large, but highly uncertain, portion of the North American carbon sink (Schimel et al. 2000; Pacala et al. 2001; Houghton 2003a,b). WP proliferation in dry lands is also among the top priorities in Australia's greenhouse gas inventory effort (Noble 1997; Burrows et al. 1998; Fensham 1998), where the phenomenon has begun to influence policy deliberations as well (Gifford & Howden 2001; Henry et al. 2002). These facets add a new dimension to traditional discussions of impacts of WP encroachment that must be weighed against traditional concerns.

Land surface–atmosphere interactions

Atmospheric chemistry

Climate and atmospheric chemistry are directly and indirectly influenced by land cover via biophysical and biogeochemical aspects of land surface–atmosphere interactions. Shifts from grass to WP domination at a regional scale have the potential to influence biophysical aspects of land–atmosphere interactions by altering albedo, evapotranspiration, surface roughness, boundary layer conditions and dust loading. Such alterations potentially influence cloud formation and the amount, seasonality and intensity of PPT. Increases in carbon and nitrogen pools that occur when WPs proliferate in grasslands and savannas may be accompanied by increased emissions of trace gases (e.g. CO_2, nitrous oxide, methane) (McCulley et al. 2004; Sponseller 2007; McLain et al. 2008) and non-methane hydrocarbons (Monson et al. 1991; Guenther et al. 1995; Klinger et al. 1998; Geron et al. 2006). Emissions of such compounds influence atmospheric oxidizing capacity, heat retention capacity, greenhouse gas half-life, aerosol burdens and radiative properties. As a result, air quality (Monson et al. 1991) and energy balance can be affected.

Dust and pollen

Encroaching WPs can markedly influence the onset, duration, concentration and total production of pollen allergens affecting human health, both locally and at great distances (Levetin 1998). Conversion of grasslands to shrublands in arid lands (Figure 4.2) has also dramatically increased wind and water erosion and dust transportation (Wainwright et al. 2000; Gillette & Pitchford 2004; Breshears et al. 2009). Dust can potentially influence weather and climate by scattering and absorbing sunlight and affecting cloud properties, though the overall effect of mineral dust in the atmosphere is likely small compared to other human impacts (IPCC 2007). However, the mineral aerosols in dust originating from dry lands are thought to play a major role in ocean fertilization and CO_2 uptake (Blain et al. 2007), terrestrial soil formation and nutrient cycling (Chadwick et al. 1999; Neff et al. 2008) and public health (e.g. Mohamed & El Bassouni 2007). Dust deposition has been shown to decrease the albedo of alpine snowpack, thus accelerating melt and reducing snow-cover duration (Painter et al. 2007). In arid regions, erosion has been shown to increase sediment delivery and change flow conditions of rivers (Jepsen et al. 2003), impact water quality, riparian vegetation and water fauna (Cowley 2006), soil fertility and ecosystem processes (Valentin et al. 2005; Okin et al. 2006). Thus, the replacement of grasslands by shrublands in arid lands has potentially far-reaching ramifications.

Climate and weather

The physical linkage between the earth's surface and the atmosphere is such that changes in surface conditions can have a pronounced effect on weather and climate (Bryant et al. 1990; Pielke et al. 1998). Changes in ecosystem structure that occur when WPs replace grasses have the potential to influence local meteorology. Changes in vegetation height and patchiness would affect boundary layer conditions and aero-dynamic roughness; changes in leaf area and rooting depth would alter inputs of water vapour via transpiration; and changes in fractional ground cover, phenology and leaf habit (e.g. evergreen vs. deciduous) would affect albedo and soil temperature, thus influencing evaporation and latent and sensible heat exchange (e.g. Graetz 1991; Bonan 2002).

Effects of WP encroachment on mesoscale climate and local weather have not been assessed. However, evidence from tree-clearing studies suggest that decreases in WP cover can influence evapotranspiration, the incidence of convective storms and cloud formation (Jackson et al. 2007). Model simulations in tropical savannas indicate that clearing of woody vegetation could increase mean surface air temperatures and wind speeds, decrease PPT and humidity and increase the frequency the dry periods within the wet season (Hoffman & Jackson 2000). We might therefore expect increases in WP abundance to have the reverse effects on local weather and climate.

Brush management in the twenty-first century

A brief history

The narrow, agronomic goal of indiscriminately eradicating and controlling brush to promote livestock production dominated rangeland management from the 1940s through the 1970s (see the section Livestock production). There were, however, few success stories during this time. Herbicidal and mechanical treatments were relatively short-lived and by the 1980s the price of fossil fuels and labour had escalated to the point where the economic viability of traditional brush management for livestock production was difficult to demonstrate. Broad generalizations regarding shrub control effects on water budgets also came into question (see the section Stream flow and groundwater recharge).

Perceived negative effects of WP encroachment on livestock production (see the section Livestock production), stream flow and groundwater recharge (see the section Stream flow and groundwater recharge) were typically the basis for management practices aimed at reducing the cover and density of WPs on rangelands. During the post-World War II era, heavy equipment and chemicals were readily available in the United States and Australia and were used on a large scale for what commonly became known as 'brush' or 'woody weeds' management in the United States and Australia; and as 'bush clearing' in Africa. Our understanding of ecosystem processes and ecosystem goods and services was in its infancy during this period and few environmental regulations were in place. Applied research in range science focused on the development and application of herbicides and mechanical techniques such as roller chopping, chaining and grubbing (Scifres 1980; Bovey 2001), often with the goal of eradicating shrubs. Techniques for chemical and mechanical brush control developed in the United States and Australia were widely exported. Brush management during this period was often applied indiscriminately, even in areas where shrublands/woodlands may have been the natural vegetation based on soils and prevailing climate and disturbance regimes. These large-scale applications were often detrimental to wildlife.

Efforts aimed at widespread eradication in the 1940s and 50s gave way to efforts aimed at selective control and containment in the 1960s and 70s. By this time, it was clear there were no 'silver bullets' for brush management. Herbicide and mechanical treatments were relatively short-lived and shrub cover often returned to pre-treatment levels (or higher) within 5–15 years depending on climate, the shrub seed bank and the capacity of shrubs to vegetatively regenerate (sprout from dormant buds that survived the treatment). The necessity of re-treating landscapes at relatively high frequencies made brush management non-sustainable and difficult to justify when the cost of brush management exceeded revenues generated from subsequent livestock production or did not improve stream flow, etc. During this same period, ecological research was beginning to shed light on the notion that the structure of ecosystems on a given site could assume various 'states' depending on climate-disturbance interactions and that land use practices could cause ecosystems to transition from one state to another (Chapter 2). In the case of WP encroachment, ecosystems were transitioning from grassland states to shrubland or woodland states, wherein WPs fundamentally alter soils, seed banks and ecosystem processes in ways that re-enforce their persistence, thus making it difficult to use traditional brush management approaches to push the system back to a persistent or stable grassland state (Archer 1989).

Basic and applied research from the 1940s to the 1970s led to the realization that brush management practices

- were treating symptoms (the shrubs) rather than root causes of land cover change (disruption of historic grazing and fire regimes);
- must be conducted in concert with progressive livestock grazing management (Chapter 9);
- when applied in an indiscriminate 'wall-to-wall' fashion to promote livestock production
  - were detrimental to many wildlife populations;
  - adversely impacted non-target plants and animals;
  - led to an undesirable 'homogenization' and loss of biological diversity;
  - put landscapes at risk for catastrophic soil erosion; and
  - were not cost-effective.
- generally short-lived, with shrubs re-establishing dominance in 5–10 years.

Collectively, these realizations led to the development of integrated brush management systems (IBMS) in the 1980s (Scifres et al. 1985; Brock 1986; Hamilton et al. 2004; Noble & Walker 2006).

Integrated brush management systems (IBMS)

IBMS are based on long-term planning processes that move away from a purely livestock production perspective and towards the management of rangelands for multiple uses and values. The IBMS planning process begins by identifying the multiple management goals and objectives for a site. These might include increasing forage production, maintaining or promoting wildlife habitat, augmenting stream flow or ground water recharge, controlling pests, pathogens or invasive species, maintaining scenic value, reducing wildfire risk or preserving grassland and savanna ecosystem types. Specific objectives are refined upon a comprehensive inventory of ecosystem components (plants, animals, soils), projecting the responses of those components to brush treatment alternatives and considering the effects of treatment alternatives on management goals on other sites (Hanselka et al. 1996). Brush management techniques (herbicidal, mechanical, prescribed burning) differ with respect to environmental impacts, implementation costs, efficacies and treatment longevities. Thus, the IBMS approach advocates consideration of the type and timing of a given brush management technology and makes explicit allowances for consideration of the type and timing of follow-up treatments. This, in turn, requires knowledge of how woody and herbaceous plants will respond and how soils, topography and livestock and wildlife management might mediate plant responses. The IBMS approach is therefore inherently interdisciplinary and dependent on the active collaboration of a diverse group of management professionals.

In the IBMS approach, brush management techniques can be targeted for certain portions of a landscape and distributed across landscapes in both time and space such that mosaics of vegetation structures, patch sizes, shapes and age-states are created. This, in turn, would promote the co-occurrence of suites of insect, reptile, mammalian and avian species with diverse habitat requirements (Jones et al. 2000). The logistics of planning and applying spatially heterogeneous brush management practices at appropriate scales is facilitated by advances in geomatics [e.g. global positioning system (GPS), geographic information system (GIS), remote sensing imagery] and landscape ecology that allow habitat and population data to be readily linked over large areas. Thus, a low diversity shrubland or woodland developing on a grassland site can be transformed into a diverse patchwork of grassland–savanna–shrubland communities using a spatial placement of landscape treatments that promotes biological diversity (Scifres et al. 1988; Fulbright 1996). Examples of the IBMS approach are accumulating (Teague et al. 1997; Grant et al. 1999; Paynter & Flanagan 2004; Ansley & Castellano 2006; Ansley & Wiedemann 2008) and expert system tools are available to assist land managers in selecting appropriate brush management practices and techniques (e.g. Hanselka et al. 1996). Special adaptations of the IBMS approach for small landholdings also exist (McGinty & Ueckert, 2001).

Currently, robust guidelines for predicting the effect of brush management on biological diversity do not exist (Box 4.3). To date, there have been only widely scattered studies on this topic. Most of these have been short-term,

Box 4.3 **Avifauna and brush management**

In areas where grasslands and savannas have been transformed into shrublands or woodlands, restoration of grassland habitats could potentially improve grassland bird populations. However, there are numerous caveats:

- Grassland bird populations have been observed to increase after roller chopping of shrub-invaded prairies (Fitzgerald & Tanner 1992), but such responses may require that shrubs be removed in sufficient amounts and in appropriate configurations (Fulbright & Guthery 1996).
- Herbicides applied at rates intended to kill shrubs may decrease habitat quality, whereas light applications that decrease canopy cover and increase grassland forb production may promote nesting and foraging opportunities (e.g. Crawford et al. 2004).
- In regions that provide both breeding and wintering habitat, nesting and foraging needs must be accommodated when designing brush management programs (Wilkins et al. 2002).
- Restoration of ecosystem processes (e.g. fire and hydrological regimes) and mitigation of non-native species impacts will be required to sustain bird populations post-treatment.

Shrub management dilemmas can result when grassland bird populations decline simultaneously with increases in shrub-associated species (Sauer et al. 1999). In these cases the functional composition is quite different, but overall avian diversity may change little. Thus, from a numerical biological diversity perspective, WP encroachment may be of little consequence. However, from the perspective of conserving and preserving grassland ecosystems and the plants and animals endemic to them, such shifts in composition are of great importance.

In the case of sandhill cranes (Box 4.2), the development of hardwood forest provides habitat corridors that allow some passerines historically restricted to Eastern Deciduous Forest to extend westward into the Great Plains. Management of trees in the river basin to maintain habitat for sandhill cranes would reduce habitat for these birds and, in doing so, cause consternation among those who find forest habitats and the birds associated with them aesthetically pleasing.

The spread of the non-native shrub salt cedar (*Tamarix* spp.) in North America has created habitat favourable for the endangered southwestern willow fly-catcher (*Empidonax traillii extimus*). Thus, while some land management agencies and conservation groups have targeted this shrub for eradication, others are seeking to protect it as critical habitat. Resolution of these competing perspectives is complicated by a limited understanding of the effects of this invasive shrub on overall ecosystem goods and services (Cohn 2005).

Shrub management trade-offs result when species co-occurring within an area have contrasting habitat requirements. For example, in the Edwards Plateau of Texas, sympatric populations of golden-cheeked warblers (*Dendroica chrysoparia*) and black-capped vireos (*Vireo atricapillus*) are threatened. Golden-cheeked warblers require dense, old-growth *Juniperus* stands for nesting (Kroll 1980), whereas black-capped vireos prefer scrub oak (Grzybowski et al. 1986). It is likely that neither of these community types would have been very abundant pre-settlement, when fire maintained the vegetation as grassland and open oak savanna. Management aimed at maintaining or promoting present-day oak scrub and old-growth juniper habitats conflict with each other, as the oak scrub community requires periodic disturbance, whereas the old-growth juniper requires minimal disturbance. Hence, selective brush management practices (including no brush management) must be spatially distributed across landscapes if habitats suitable for both species are to be maintained.

small-scale and narrowly focused on a target organism or group of organisms. Given below is a brief survey of some observations.

- In a 2-year study of herbicide effects on biodiversity of mesquite shrublands in the southern Great Plains of North America, vegetation species richness and evenness, Shannon's index, beta diversity and the proportion of rare plant species did not differ between controls and treated sites. Rodent and avian relative frequency, richness and diversity were comparable as well (Nolte & Fulbright 1997; Peterson 1997).
- Jones et al. (2000) examined how native herpetofauna in tallgrass prairie forest ecotone communities were influenced by vegetation types derived

from combinations of herbicide applications and prescribed burning. Relative total abundance and species richness of herpetofauna were similar on all treatments. However, differences were apparent by taxonomic group. In general, amphibians were most abundant in untreated and herbicide-only sites; lizards were most abundant on the untreated sites and snakes were most abundant on sites receiving herbicide and fire.

- Brush management can be advantageous to wild ungulates that are common in grassland habitats [e.g. wildebeest (*Connochaetes taurinus*), zebra (*Equus quagga*)], while not being detrimental to resident dwellers of dense woodland [e.g. impala (*Aepyceros melampus*), kudu (*Tragelaphus stresiceros*)] (Ben-Shaher 1992). However, the response of ungulates to bush clearing may be weak, and therefore clearance might not be justified if intended to propagate grazers.

- When contemplating brush management, consideration should be given to the role that WPs may play in stabilizing protein intake of domestic or wild herbivores (see the section on Traditional perspectives on woody plant encroachment) and providing other vital resources and services for wildlife, including shade, shelter and perching, roosting and nesting structures (see the section on Biological diversity).

- In systems where shrubs regenerate vegetatively, brush management may increase stem densities and favour less desirable over more desirable browse species (Fulbright & Beasom 1987). However, creative use of low-intensity fire and herbicides can help promote a savanna physiognomy (Ansley et al. 1997, 2003).

- Use of brush management to restore biodiversity or ecosystem goods and services begs the question 'What is the desired level of WP abundance?' The answer is highly dependent on management goals and objectives. In a survey of ranchers' willingness to participate in a brush control cost-sharing program in Texas, respondents estimated that current brush cover on their land averaged 41%, which contrasted with their preference that brush cover average 27% (Thurow et al. 2000). This expression of preferred brush cover was similar to an independent estimate by a panel of experts that ranch livestock and deer-hunting lease value would be maximized at 30% brush cover.

- Brush management has the potential to create conditions favourable for the establishment and growth of weeds and invasive species (Young et al. 1985; Belsky 1996; Bates et al. 2007) that, among other unintended consequences, can have adverse affects on biodiversity (Figure 4.7). In many cases, exotic grasses are planted in conjunction with brush management. When seeding

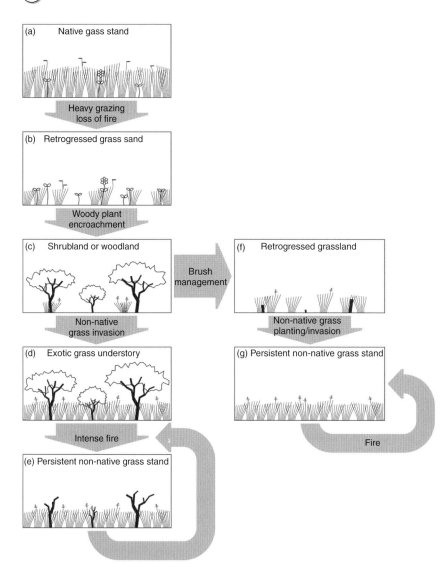

Figure 4.7   **Steps in the conversion of diverse native perennial grassland to shrubland or woodland, then to low-diversity non-native grassland. Only recently have the potential downsides of steps D–G been recognized. Graphic art by Rob Wu.**

of non-native grasses is successful, the result is often a persistent, long-lived monoculture of exotic vegetation. While this may be valued for livestock production, these plants may represent threats to the biodiversity of native plants and animals (McClaran & Anable 1992; Williams & Baruch 2000; Schussman et al. 2006). When seeding is not successful, lands may remain barren and exposed to wind and water erosion.

Woody plants in future environments

Climate change and CO_2 enrichment

Climate, atmospheric CO_2 enrichment and nitrogen deposition likely inter-acted with changes in the grazing and fire regimes over the past 100+ years to influence the rates and dynamics of changes in WP abundance, while not driving it *per se* (e.g. Bahre 1991; Archer et al. 1995) (Box 4.1). It remains to be seen how projected changes in these factors will affect the future distribution of woody and herbaceous vegetation. Given that WP encroachment has been ongoing in many parts of the world since the early 1900s, it stands to reason that some landscapes may have already reached their carrying capacity for WP cover (Browning et al. 2008).

In areas still characterized by grasslands and open savannas, continued increases in atmospheric CO_2 concentrations may favour the growth of WPs (Polley 1997; Morgan et al. 2007), thus allowing them to escape more readily from mortality factors associated with periodic fire (Bond et al. 2003) and drought (Polley et al. 2002). Predictions of higher temperatures, increased rainfall variability and more frequent and severe droughts in arid lands may induce WP mortality in areas where their abundance has increased in recent decades (Bowers 2005; Breshears et al. 2005; Fensham et al. 2009) or it may further promote the replacement of mesophytic grasses by xerophytic shrubs better adapted to such conditions. In high latitude areas, warming has already caused substantial increases in WP abundance (Sturm et al. 2005). In regions where grassland and forest biomes intersect, trees have expanded into grasslands in recent decades. However, with climate warming, this trend may reverse, with trees giving way to grasses in response to increases in the frequency, intensity or duration of drought and the abundance of tree pathogens and diseases. In semi-arid and sub-humid areas projected to get more rainfall, WPs would be expected to be favoured over grasses (Scanlan & Archer 1991), particularly if accompanied by a shift toward more winter PPT

(Brown et al. 1997; Briggs et al. 2005). However, invasion by exotic grasses could change the entire picture (Figure 4.7).

Invasive grasses

Non-native grasses often act as 'transformer species' (Richardson et al. 2000; Grice 2006) in that they change the character, condition, form or nature of a natural ecosystem over substantial areas. Land use and climate markedly influence the probability, rate and pattern of alien species invasion, and future change for each of these drivers will interact to strongly impact scenarios of plant invasion and ecosystem transformation (Sala et al. 2000; Walther et al. 2002; Hastings et al. 2005). Changes in ecosystem susceptibility to invasion by non-native plants may be expected with changes in climate (Ibarra-Flores et al. 1995; Mau-Crimmins et al. 2006), CO_2 (Smith et al. 2000, Nagel et al. 2004) and nitrogen deposition (Fenn et al. 2003).

Once established, non-native annual and perennial grasses can generate massive, high-continuity fine fuel loads that predispose arid lands to fires, in some cases more frequent and intense than in the ecosystems in which those species evolved (Williams & Baruch 2000). The result is the potential for desert scrub, shrub-steppe and desert grassland/savanna biotic communities to be quickly and radically transformed into monocultures of invasive grasses over large areas (Figures 4.7 and 4.8), and further promoting desertification (Ravi et al. 2009). This is already well underway in the cold desert region of North America (Knapp 1998) and is in its early stages in hot deserts (Arriaga et al. 2004; Salo 2005; Mau-Crimmins et al. 2006; Franklin et al. 2006). By virtue of their profound impact on the fire regime and hydrology, invasive plants in arid lands will very likely trump direct climate impacts on native vegetation where they gain dominance.

The good, the bad and the ugly

Regions prone to WP encroachment present a novel series of dilemmas and challenges to mitigation. The proliferation of WPs in ecosystems grazed by livestock typically invokes land management actions to minimize its cover using herbicidal, mechanical, or prescribed burning techniques (see the section on Traditional perspectives on woody plant encroachment). The

Figure 4.8   **Mojave desert scrub near Las Vegas, NV (foreground); and area invaded by the non-native annual grass red brome (*Bromus rubra*, background) following a fire that carried from desert floor upslope into pinyon–juniper woodlands. This exotic grass has instigated a positive disturbance (fire) feedback that now threatens the biodiversity and function of this desert scrub ecosystem (Figure 4.7) and constitutes a new ignition source for fire in upper elevation forests. Photo: T.E. Huxman.**

recent realization that WP proliferation may substantially promote ecosystem carbon stocks in the geographically extensive dry lands (see the section on Biogeochemisty and ecosystem processes) may trigger new land use drivers as industries seek opportunities to acquire carbon credits to offset CO_2 emissions. WP proliferation in grasslands and savannas may therefore shift from being an economic liability in the context of extensive livestock production systems to a source of income in a carbon sequestration context. If so, unexpected incentives may result as land management shifts to promote, rather than deter, WP encroachment as programs for emissions trading and carbon credits (e.g. the Chicago Climate Exchange and the European Union Greenhouse Gas Emission Trading Scheme) gain momentum. A recent survey in Texas gives hope that unexpected or unintended incentives may not materialize. Olenick et al. (2005) found that landowners favourably viewed programs that would reduce WP cover in an effort to increase water yields or to improve wildlife habitat, but they disapproved of programs that would encourage the proliferation of WPs to promote carbon sequestration.

The potential benefits of carbon sequestration should be carefully weighed against costs in the form of reductions in livestock production, the possibility of reductions in stream flow and groundwater recharge, extirpation of plants and animals characteristic of grasslands/savannas (loss of biological diversity) to the extent that such ecosystems could become locally or regionally extinct, and increases in trace gas production, non-methane hydrocarbon emissions and wind/water erosion. Quantification of these trade-offs is a current challenge that must be addressed to appropriately evaluate the impacts of various management and policy scenarios.

Conclusions

Rangelands support the majority of the world's livestock production and provide important wildlife habitat. Their airsheds and watersheds influence the health and well-being of the 30% of the world's human population that lives in them. WP proliferation has emerged as a major issue in managing many of the world's rangelands. This phenomenon is a stark contrast to deforestation, which has received considerably more attention. Traditional concerns about this change in land cover were narrowly focused on impact related to livestock production and management of wildlife valued for sport hunting. We have only recently begun to explore the implications of this widespread change in land cover on ecosystem goods and services that arise from the changes in biogeochemical cycles and land surface–atmosphere interactions that accompany the conversion of grasslands to shrublands or woodlands.

In arid and semi-arid regions, WP encroachment represents desertification and is accompanied by accelerated rates of water and wind erosion. The latter has known impacts on ocean productivity and snow pack persistence. Pollen and dust production triggered by encroaching trees and shrubs can contribute to human health problems across large areas far downwind. Thus, the replacement of grasslands by shrublands in arid lands has potentially far-reaching ramifications.

In semi-arid and sub-humid regions, primary production, nutrient cycling and soil organic matter accumulation may be promoted by WP proliferation but possibly (though not necessarily) at the expense of stream flow, ground water recharge, livestock production and biological diversity. Shifts from grass to WP domination have major implications for the global carbon cycle and carbon sequestration. Available evidence generally indicates increases in the above-ground NPP and carbon pools; but these are at risk for rapid depletion by fire or drought. Effects on the much larger below-ground carbon pools are equivocal, with available evidence suggesting that soil carbon pools may increase, decrease or remain unchanged. Thus, we cannot yet reliably predict consequences of WP encroachment for ecosystem carbon storage. Tree and shrub proliferation also has the potential to stimulate trace gas and volatile organic carbon emissions from plants and soils and thus influence greenhouse gas concentration, lifespan and tropospheric ozone production.

Increases in WP abundance has fundamentally altered land surface–atmosphere interactions to potentially influence weather and climate. However, robust generalizations of the regional and global consequences of these phenomena are not yet possible. From a conservation biology perspective, WP proliferation represents a major threat to the preservation of grassland and savanna ecosystems. In some cases, WPs form virtual monocultures; in other cases they may have no net impact or they may promote numerical diversity. However, in such cases, we are losing grassland and savanna ecosystem types and the plants, animals and microbes endemic to them.

It is now clear that policy and management issues related to grazing land conservation extend well beyond the traditional concerns of livestock production and game management (wildlife valued for sport hunting) to include potential effects on hydrology, carbon sequestration, biological diversity, atmospheric chemistry and the climate system. The research community is challenged with quantifying and monitoring these varied impacts and the management community with devising approaches for creating or maintaining woody–herbaceous mixtures in spatial arrangements that satisfy competing conservation objectives. The latter will require spatially explicit integrated brush, weed and invasive (non-native) plant management approaches that articulate the type and timing and spatial location of follow-up treatments.

It is important to keep in mind that our understanding of ecosystems and the causes and consequences of WP encroachment in grasslands and savannas is largely based upon modern observations. Near-term shifts in climate, coupled with non-native species introductions, increases in atmospheric CO_2 concentrations and nitrogen deposition are likely to trigger a reshuffling of organisms. Novel communities with a composition unlike any found today are a very likely possibility. These novel communities will present novel challenges and will require new perspectives for conservation and management (Seastedt et al. 2008).

References

Aanderud, Z.T., Shuldman, M.I., Drenovsky, R.E. & Richards, J.H. (2008) Shrub-interspace dynamics alter relationships between microbial community composition and belowground ecosystem characteristics. *Soil Biology and Biochemistry* 40, 2206–2226.

Ansley, R.J. & Castellano, M.J. (2006) Strategies for savanna restoration in the southern Great Plains: effects of fire and herbicides. *Restoration Ecology* 14, 420–428.

Ansley, R.J., Kramp, B.A. & Jones, D.L. (2003) Converting mesquite thickets to savanna through foliage modification with clopyralid. *Journal of Range Management* 56, 72–80.

Ansley, R.J., Kramp, B.A. & Moore, T.R. (1997) Development and management of mesquite savanna using low intensity prescribed fires. *Proceedings, Fire Effects on Rare and Endangered Species and Habitats Conference.* International Association of Wildland Fire, Coeur d'Alene, pp. 155–161.

Ansley, R.J. & Wiedemann, H.T. (2008) Reversing the woodland steady state: vegetation responses during restoration of *Juniperus*-dominated grasslands with chaining and fire. In: Van Auken, O.W. (ed.) *Western North American Juniperus Community: A Dynamic Vegetation Type.* Springer, New York, pp. 272–292.

Archer, S. (1989) Have southern Texas savannas been converted to woodlands in recent history? *American Naturalist* 134, 545–561.

Archer, S. (1994) Woody plant encroachment into southwestern grasslands and savannas: rates, patterns and proximate causes. In: Vavra, M., Laycock, W. & Pieper, R. (eds.) *Ecological Implications of Livestock Herbivory in the West.* Society for Range Management, Denver, pp. 13–68.

Archer, S. (1995) Herbivore mediation of grass-woody plant interactions. *Tropical Grasslands* 29, 218–235.

Archer, S. (1996) Assessing and interpreting grass-woody plant dynamics. In: Hodgson, J. & Illius, A.W. (eds.) *The Ecology and Management of Grazing Systems.* CAB International, Wallingford, Oxon, United Kingdom, pp. 101–143.

Archer, S. (2008) *Proliferation of Woody Plants in Grasslands and Savannas: A Bibliography.* University of Arizona, http://ag.arizona.edu/research/archer/.

Archer, S., Boutton, T.W. & Hibbard, K.A. (2001) Trees in grasslands: biogeochemical consequences of woody plant expansion. In: Schulze, E.-D., Heimann, M., Harrison, S. et al. (eds.) *Global Biogeochemical Cycles in the Climate System.* Academic Press, San Diego, pp. 39–46.

Archer, S., Schimel, D.S. & Holland, E.A. (1995) Mechanisms of shrubland expansion: land use, climate or CO_2? *Climatic Change* 29, 91–99.

Archer, S. & Stokes, C.J. (2000) Stress, disturbance and change in rangeland ecosystems. In: Arnalds, O. & Archer, S. (eds.) *Rangeland Desertification*. Kluwer Academic Publishers, Dordrecht, The Netherlands, pp. 19–38.

Arriaga, L., Castellanos, A.E., Moreno, E. & Alaron, J. (2004) Potential ecological distribution of alien invasive species and risk assessment: a case study of buffelgrass in arid regions of Mexico. *Conservation Biology* 18, 1504–1514.

Asner, G. & Archer, S. (2010) Livestock and the global carbon cycle. In: Steinfeld, H., Mooney, H., Schneider, F. & Neville, L.E. (eds.) *Livestock in a Changing Landscape, Volume 1, Drivers, Consequences and Responses*. Island Press, Washington, D.C., pp. 69–82.

Asner, G.P., Elmore, A.J., Olander, L.P., Martin, R.E. & Harris, A.T. (2004) Grazing systems, ecosystem responses and global change. *Annual Review of Environment and Resources* 29, 261–299.

Baez, S. & Collins, S.L. (2008) Shrub invasion decreases diversity and alters community stability in northern Chihuahuan desert plant communities. *PLoS ONE* 3, e2332. DOI:10.1371/journal.pone.0002332.

Bahre, C.J. (1991) *A Legacy of Change: Historic Human Impact on Vegetation of the Arizona Borderlands*. University of Arizona Press, Tucson.

Bailey, R.G. (1996) *Ecosystem Geography*. Springer, New York.

Bakker, K.K. (2003) *The Effect of Woody Vegetation on Grassland Nesting Birds: An Annotated Bibliography*. United States Fish and Wildlife Service Habitat and Population Evaluation Team, Fergus Falls.

Bakker, K.K., Naugle, D.E. & Higgins, K.E. (2002) Incorporating landscape attributes into models for migratory grassland bird conservation. *Conservation Biology* 16, 1–10.

Bates, J.D., Miller, R.E. & Svejcar, T. (2007) Long-term vegetation dynamics in a cut western juniper woodland. *Western North American Naturalist* 67, 549–561.

Belsky, A.J. (1996) Western juniper expansion: is it a threat to arid northwestern ecosystems? *Journal of Range Management* 49, 53–59.

Ben-Shaher, R. (1992) The effects of bush clearance on African ungulates in a semi-arid nature reserve. *Ecological Applications* 2, 95–101.

Bestelmeyer, B.T. (2005) Does desertification diminish biodiversity? Enhancement of ant diversity by shrub invasion in south-western USA. *Diversity and Distributions* 11, 45–55.

Blackburn, W.H. (1983) Influence of brush control on hydrologic characteristics of range watersheds. In: McDaniel, K.W. (ed.) *Proceedings, Brush Management Symposium*. Texas Tech Press, Lubbock, pp. 73–88.

Blain, S., Queguiner, B. & Armand, L. (2007) Effect of natural iron fertilization on carbon sequestration in the Southern Ocean. *Nature* 446, 414–417.

Blaum, N., Rossmanith, E., Fleissner, G. & Jeltsch, F. (2007a) The conflicting importance of shrubby landscape structures for the reproductive success of the yellow mongoose (*Cynictis penicillata*). *Journal of Mammalogy* 88, 194–200.

Blaum, N., Rossmanith, E. & Jeltsch, F. (2007b) Land use affects rodent communities in Kalahari savannah rangelands. *African Journal of Ecology* 45, 189–195.

Blaum, N., Rossmanith, E., Popp, A. & Jeltsch, F. (2007c) Shrub encroachment affects mammalian carnivore abundance and species richness in semiarid rangelands. *Acta Oecologica: International Journal of Ecology* 31, 86–92.

Blaum, N., Rossmanith, E., Schwager, M. & Jeltsch, F. (2007d) Responses of mammalian carnivores to land use in arid savanna rangelands. *Basic and Applied Ecology* 8, 552–564.

Bock, C.E., Jones, Z.F. & Bock, J.H. (2006) Abundance of cottontails (*Sylvilagus*) in an exurbanizing southwestern savanna. *Southwestern Naturalist* 51, 352–357.

Bonan, G.B. (2002) *Ecological Climatology: Concepts and Applications.* Cambridge University Press, Cambridge, UK.

Bond, W.J., Midgley, G.F. & Woodward, F.T. (2003) The importance of low atmospheric CO_2 and fire in promoting the spread of grasslands and savannas. *Global Change Biology* 9, 973–982.

Bossard, C.C. & Rejmanek, M. (1994) Herbivory, growth, seed production and resprouting of an exotic invasive shrub. *Biological Conservation* 67, 193–200.

Bovey, R.W. (2001) *Woody Plants and Woody Plant Management: Ecology, Safety and Environmental Impact.* Marcel Dekker, New York.

Bowers, J.E. (2005) Effects of drought on shrub survival and longevity in the northern Sonoran Desert. *Journal of the Torrey Botanical Society* 132, 421–431.

Brennan, L.A. & Kuvlesky, W.P. (2005) North American grassland birds: an unfolding conservation crisis?. *Journal of Wildlife Management* 69, 1–13.

Breshears, D.D., Cobb, N.S., Rich, P.M. et al. (2005) Regional vegetation die-off in response to global-change-type drought. *Proceedings of the National Academy of Sciences of the United States of America* 102, 15144–15148.

Breshears, D.D., Whicker, J.J., Zou, C.B., Field, J.P. & Allen, C.D. (2009) Aeolian sediment transport in undisturbed and disturbed dryland ecosystems along the grassland-forest continuum: a conceptual framework spanning gradients of woody plants. *Geomorphology* 105, 28–38.

Briggs, J.M., Knapp, A.K., Blair, J.M. et al. (2005) An ecosystem in transition: causes and consequences of the conversion of mesic grassland to shrubland. *BioScience* 55, 243–254.

Brock, J.H. (1986) The growing need for integrated brush management. *Rangelands* 7, 212–214.

Brown, J.H., Valone, T.J. & Curtin, C.G. (1997) Reorganization of an arid ecosystem in response to recent climate change. *Proceedings of the National Academy of Sciences of the United States of America* 94, 9729–9733.

Browning, D., Archer, S., Asner, G., McClaran, M. & Wessman, C. (2008) Woody plants in grasslands: post-encroachment stand dynamics. *Ecological Applications* 18, 928–944.

Bruce, K.A., Cameron, G.N., Harcombe, P.A. & Jubinsky, G. (1997) Introduction, impact on native habitats and management of a woody invader, the Chinese tallow tree, *Sapium sebiferum* (L) *Roxb. Natural Areas Journal* 17, 255–260.

Bryant, N.A., Johnson, L.F., Brazel, A.J., Balling, R.C., Hutchinson, C.F. & Beck, L.R. (1990) Measuring the effect of overgrazing in the Sonoran Desert. *Climatic Change* 17, 243–264.

Burrows, W.H., Compton, J.F. & Hoffmann, M.B. (1998) Vegetation thickening and carbon sinks in the grazed woodlands of north-east Australia. In: Dyason, R., Dyason, L. & Gadsen, R. (eds.) *Proceedings, Australian Forest Growers Conference*, Australian Forest Growers, Lismore, pp. 305–316.

Campbell, B.D. & Stafford Smith, D.M. (2000) A synthesis of recent global change research on pasture and rangeland production: reduced uncertainties and their management implications. *Agriculture, Ecosystems & Environment* 82, 39–55.

Chadwick, O.A., Derry, L.A., Vitousek, P.M., Huebert, B.J. & Hedin, L.O. (1999) Changing sources of nutrients during four million years of ecosystem development. *Nature* 397, 491–497.

Cohn, J.P. (2005) Tiff over tamarisk: can a nuisance be nice, too?. *BioScience* 55, 648–655.

Cole, K. (1985) Past rates of change, species richness and a model of vegetation inertia in the Grand Canyon, Arizona. *American Naturalist* 125, 289–303.

Conley, W., Conley, M.R. & Kart, T.R. (1992) A computational study of episodic events and historical context in long-term ecological processes: climate and grazing in the northern Chihuahuan Desert. *Coenoses* 7, 55–60.

Cooper, W.E. & Whiting, M.J. (2000) Islands in a sea of sand: use of Acacia trees by tree skinks in the Kalahari Desert. *Journal of Arid Environments* 44, 373–381.

Cowley, D.E. (2006) Strategies for ecological restoration of the Middle Rio Grande in New Mexico and recovery of the endangered Rio Grande silvery minnow. *Reviews in Fisheries Science* 14, 169–186.

Crawford, J.A., Olson, R.A., West, N.E. et al. (2004) Synthesis paper: ecology and management of sage-grouse and sage-grouse habitat. *Journal of Range Management* 57, 2–19.

Daly, C., Bachelet, D., Lenihan, J.M., Neilson, R.P., Parton, W. & Ojima, D. (2000) Dynamic simulation of tree-grass interactions for global change studies. *Ecological Applications* 10, 449–469.

Dean, W.R.J., Milton, S.J. & Jeltsch, F. (1999) Large trees, fertile islands and birds in arid savanna. *Journal of Arid Environments* 41, 61–78.

Dugas, W.A., Hicks, R.A. & Wright, P. (1998) Effect of removal of *Juniperus ashei* on evapotranspiration and runoff in the Seco Creek watershed. *Water Resources Research* 34, 1499–1506.

Eccard, J.A., Meyer, J. & Sundell, J. (2004) Space use, circadian activity pattern and mating system of the nocturnal tree rat *Thallomys nigricauda*. *Journal of Mammalogy* 85, 440–445.

Farley, K.A., Jobbágy, E.G. & Jackson, R.B. (2005) Effects of afforestation on water yield: a global synthesis with implications for policy. *Global Change Biology* 11, 1565–1576.

Fenn, M.E., Baron, J.S., Allen, E.B. et al. (2003) Ecological effects of nitrogen deposition in the western United States. *BioScience* 53, 404–420.

Fensham, R.J. (1998) Resolving biomass fluxes in Queensland woodlands. *Climate Change Newsletter* 10, 13–15.

Fensham, R.J., Fairfax, R.J. & Ward, D.P. (2009) Drought-induced tree death in savanna. *Global Change Biology* 15, 380–387.

Field, C.B., Behrenfeld, M.J., Randerson, J.T. & Falkowski, P. (1998) Primary production of the biosphere: integrating terrestrial and oceanic components. *Science* 281, 237–240.

Fitzgerald, S.M. & Tanner, G.W. (1992) Avian community response to fire and mechanical shrub control in south Florida. *Journal of Range Management* 45, 396–400.

Fletcher, R.J. & Koford, R.R., Jr. (2002) Habitat and landscape associations of breeding birds in native and restored grasslands. *Journal of Wildlife Management* 66, 1011–1022.

Franklin, K.A., Lyons, K., Nagler, P.L. et al. (2006) Buffelgrass (*Pennisetum ciliare*) land conversion and productivity in the plains of Sonora, Mexico. *Biological Conservation* 127, 62–71.

Fuhlendorf, S.D., Woodward, A.J.W., Leslie, D.M., Jr. & Shackford, J.S. (2002) Multi-scale effects of habitat loss and fragmentation on lesser prairie-chicken populations of the US southern Great Plains. *Landscape Ecology* 17, 617–628.

Fulbright, T.E. (1996) Viewpoint: a theoretical basis for planning woody plant control to maintain species diversity. *Journal of Range Management* 49, 554–559.

Fulbright, T.E. & Beasom, S.L. (1987) Long-term effects of mechanical treatment on white-tailed deer browse. *Wildlife Society Bulletin* 15, 560–564.

Fulbright, T.E. & Guthery, F.S. (1996) Mechanical manipulation of plants. In: Krausman, P.R. (ed.) *Rangeland Wildlife*. Society for Range Management, Denver.

Geesing, D., Felker, P. & Bingham, R.L. (2000) Influence of mesquite (*Prosopis glandulosa*) on soil nitrogen and carbon development: implications for global carbon sequestration. *Journal of Arid Environments* 46, 157–180.

Geron, C., Guenther, A., Greenberg, J., Karl, T. & Rasmussen, R. (2006) Biogenic volatile organic compound emissions from desert vegetation of the southwestern US. *Atmospheric Environment* 40, 1645–1660.

Gifford, R.M. & Howden, S.M. (2001) Vegetation thickening in an ecological perspective: significance to national greenhouse gas inventories and mitigation policies. *Environmental Science and Policy* 4, 59–72.

Gillette, D.A. & Pitchford, A.M. (2004) Sand flux in the northern Chihuahuan Desert, New Mexico, USA and the influence of mesquite-dominated landscapes. *Journal of Geophysical Research-Earth Surface* 109, F04003.

Gottschalk, T.K., Ekschmitt, K. & Bairlein, F. (2007) Relationships between vegetation and bird community composition in grasslands of the Serengeti. *African Journal of Ecology* 45, 557–565.

Graetz, R.D. (1991) The nature and significance of the feedback of changes in terrestrial vegetation on global atmospheric and climatic change. *Climatic Change* 18, 147–173.

Grant, T.A., Madden, E. & Berkey, G.B. (2004) Tree and shrub invasion in northern mixed-grass prairie: implications for breeding grassland birds. *Wildlife Society Bulletin* 32, 807–818.

Grant, W.E., Hamilton, W.T. & Quintanilla, E. (1999) Sustainability of agroecosystems in semi-arid grasslands: simulated management of woody vegetation in the Rio Grande Plains of southern Texas and northeastern Mexico. *Ecological Modelling* 124, 29–42.

Grice, A. (2000) Weed management in Australian rangelands. In: Sindel, B.M. (ed.) *Australian Weed Management Systems*. R.G. and F.J. Richardson, Meredith, Victoria, pp. 429–458.

Grice, A.C. (2006) The impacts of invasive plant species on the biodiversity of Australian rangelands. *The Rangeland Journal* 28, 1–27.

Grzybowski, J., Clapp, R. & Marshall, J.J. (1986) History and current population status of the black-capped vireo in Oklahoma. *American Birds* 40, 1151–1161.

Guenther, A., Hewitt, C., Erickson, D. et al. (1995) A global model of natural volatile organic compound emissions. *Journal of Geophysical Research* 100, 8873–8892.

Hamilton, W.T., McGinty, A., Ueckert, D.N., Hanselka, C.W. & Lee, M.R. (2004) *Brush Management: Past, Present, Future*. Texas A&M University Press, College Station.

Hanselka, C.W., Hamilton, W.T., Rector, B.S. (1996) *Integrated Brush Management Systems for Texas*. Texas Agricultural Extension Service, College Station.

Hastings, A., Cuddington, K., Davies, K. et al. (2005) The spatial spread of invasions: new developments in theory and evidence. *Ecology Letters* 8, 91–101.

Havstad, K.M., Huenneke, L.F. & Schlesinger, W.H. (eds.) (2006) *Structure and Function of a Chihuahuan Desert Ecosystem: The Jornada Basin Long-Term Ecological Research Site*. Oxford University Press, New York.

Henry, B.K., Danaher, T., McKeon, G.M. & Burrows, W.H. (2002) A review of the potential role of greenhouse gas abatement in native vegetation management in Queensland's rangelands. *Rangeland Journal* 24, 112–132.

Hibbard, K., Schimel, D., Archer, S., Ojima, D. & Parton, W. (2003) Grassland to woodland transitions: integrating changes in landscape structure and biogeochemistry. *Ecological Applications* 13, 911–926.

Hibbert, A.R. (1983) Water yield improvement potential by vegetation management on western rangelands. *Water Resources Bulletin* 19, 375–381.

Hoffman, J. & Moran, V. (1988) The invasive weed *Sesbania punicea* in South Africa and prospects for its biological control. *South African Journal of Science* 84, 740–742.

Hoffman, W.A. & Jackson, R.B. (2000) Vegetation-climate feedbacks in the conversion of tropical savanna to grassland. *Journal of Climate* 13, 1593–1602.

Houghton, R.A. (2003a) Why are estimates of the terrestrial carbon balance so different? *Global Change Biology* 9, 500–509.

Houghton, R.A. (2003b) Revised estimates of the annual net flux of carbon to the atmosphere from changes in land use and land management 1850–2000. *Tellus* 55B, 378–390.

Houghton, R.A., Hackler, J.L. & Lawrence, K.T. (1999) The US carbon budget: contributions from land-use change. *Science* 285, 574–578.

Hughes, R.F., Archer, S., Asner, G.P., Wessman, C.A., McMurtry, C. & Nelson, J. (2006) Changes in aboveground primary production and carbon and nitrogen pools accompanying woody plant encroachment in a temperate savanna. *Global Change Biology* 12, 1733–1747.

Humphrey, R.R. (1987) *90 years and 535 miles: Vegetation Changes Along the Mexican Border*. University of New Mexico Press, Albuquerque.

Huxman, T.E., Wilcox, B.P., Breshears, D.D. et al. (2005) Ecohydrological implications of woody plant encroachment. *Ecology* 86, 308–319.

Ibarra-Flores, A., Cox, J.R., Martin, M.H., Crowl, T.A. & Call, C.A. (1995) Predicting buffelgrass survival across a geographical and environmental gradient. *Journal of Range Management* 48, 53–59.

Imberger, K.T. & Chiu, C. (2001) Spatial changes of soil fungal and bacterial biomass from a sub-alpine coniferous forest to grassland in a humid, sub-tropical region. *Biology and Fertility of Soils* 33, 105–110.

Intergovernmental Panel on Climate Change (IPCC) (2007) In: Solomon, S., Qin, D., Manning, M. et al. (eds.) *Climate Change 2007: The Physical Science Basis. Contribution of Working Group I to the Fourth Assessment Report of the Intergovernmental Panel on Climate Change* Cambridge University Press, Cambridge, New York.

Jackson, R.B., Banner, J.L., Jobbágy, E.G., Pockman, W.T. & Wall, D.H. (2002) Ecosystem carbon loss with woody plant invasion of grasslands. *Nature* 418, 623–626.

Jackson, R.B., Canadell, J., Ehleringer, J.R., Mooney, H.A., Sala, O.E. & Schulze, E.D. (1996) A global analysis of root distributions for terrestrial biomes. *Oecologia* 108, 389–411.

Jackson, R.B., Farley, K.A., Hoffmann, W.A., Jobbágy, E.G. & McCulley, R.L. (2007) Carbon and water tradeoffs in conversions to forests and shrublands. In: Canadell, J.G., Pataki, D.E. & Pitelka, L.F. (eds.) *Terrestrial Ecosystems in a Changing World*. Springer-Verlag, Berlin, pp. 236–246.

Jackson, R.B., Jobbágy, E.G., Avissar, R. et al. (2005) Trading water for carbon with biological carbon sequestration. *Science* 310, 1944–1947.

Jackson, R.B., Schenk, H.J., Jobbágy, E.G. et al. (2000) Belowground consequences of vegetation change and their treatment in models. *Ecological Applications* 10, 470.

Jacobs, B.F., Romme, W.H. & Allen, C.D. (2008) Mapping "old" vs. "young" pinyon-juniper stands with a predictive topo-climatic model. *Ecological Applications* 18, 1627–1641.

Jepsen, R., Langford, R., Roberts, J. & Gailani, J. (2003) Effects of arroyo sediment influxes on the Rio Grande River channel near El Paso, Texas. *Environmental and Engineering Geoscience* 9, 305–312.

Jobbágy, E.G. & Jackson, R.B. (2000) The vertical distribution of soil organic carbon and its relation to climate and vegetation. *Ecological Applications* 10, 423–436.

Johnson, D.H. (2001) Habitat fragmentation effects on birds in grasslands and wetlands: a critique of our knowledge. *Great Plains Research* 11, 211–231.

Johnson, W.C. (1994) Woodland expansion in the Platte River, Nebraska: patterns and causes. *Ecological Monographs* 64, 45–84.

Jones, B., Fox, S.F., Leslie, D.M., Engle, D.M. & Lochmiller, R.L. (2000) Herpetofaunal responses to brush management with herbicide and fire. *Journal of Range Management* 53, 154–158.

Katz, G.L. & Shafroth, P.B. (2003) Biology, ecology and management of *Elaeagnus angustifolia* L. (Russian olive) in western North America. *Wetlands* 23, 763–777.

Kazmaier, R., Hellgren, E. & Ruthven, D. (2001) Habitat selection by the Texas tortoise in a managed thornscrub ecosystem. *Journal of Wildlife Management* 65, 653–660.

Klinger, L., Greenberg, J., Guenther, A. et al. (1998) Patterns in volatile organic compound emissions along a savanna-rainforest gradient in Central Africa. *Journal of Geophysical Research* 102, 1443–1454.

Knapp, A.K, Briggs, J.M., Collins, S.L. et al. (2008a) Shrub encroachment in North American grasslands: shifts in growth form dominance rapidly alters control of ecosystem carbon inputs. *Global Change Biology* 14, 615–623.

Knapp, A.K., McCarron, J.K., Silletti, G.A. et al. (2008b) Ecological consequences of the replacement of native grassland by *Juniperus virginiana* and other woody plants. In: Van Auken, O.W. (ed.) *Western North American Juniperus Communities: A Dynamic Vegetation Type*. Springer, New York, pp. 156–159.

Knapp, P.A. (1998) Spatio-temporal patterns of large grassland fires in the Intermountain West, USA. *Global Ecology and Biogeography Letters* 7, 259–272.

Knick, S.T., Dobkin, D.S., Rotenberry, J.T., Schroeder, M.A., Haegen, W.M.V. & van Riper III, C. (2003) Teetering on the edge or too late? Conservation and research issues for avifauna of sagebrush habitats. *The Condor* 105, 611–634.

Köchy, M. & Wilson, S.D. (2001) Nitrogen deposition and forest expansion in the northern Great Plains. *Journal of Ecology* 89, 807–817.

Kriticos, D.J., Sutherst, R.W., Brown, J.R., Adkins, S.W. & Maywald, G.F. (2003) Climate change and the potential distribution of an invasive alien plant: *Acacia nilotica* ssp. *indica* in Australia. *Journal of Applied Ecology* 40, 111–124.

Kroll, J.C. (1980) Habitat requirements of the golden-cheeked warbler: management implications. *Journal of Range Management* 33, 60–65.

Leopold, A. (1924) Grass, brush, timber and fire in southern Arizona. *Journal of Forestry* 22, 1–10.

Levetin, E. (1998) A long-term study of winter and early spring tree pollen in the Tulsa, Oklahoma atmosphere. *Aerobiologia* 14, 21–28.

Liao, J.D., Boutton, T.W. & Jastrow, J.D. (2006) Storage and dynamics of carbon and nitrogen in soil physical fractions following woody plant invasion of grassland. *Soil Biology & Biochemistry* 38, 3184–3196.

Lloyd, J., Mannan, R.W., Destefano, S. & Kirkpatrick, C. (1998) The effects of mesquite invasion on a southeastern Arizona grassland bird community. *Wilson Bulletin* 110, 403–408.

Mau-Crimmins, T., Schussman, H.R. & Geiger, E.L. (2006) Can the invaded range of a species be predicted sufficiently using only native-range data? Lehmann lovegrass (*Eragrostis lehmanniana*) in the southwestern United States. *Ecological Modelling* 193, 736–746.

McClaran, M.P. & Anable, M.E. (1992) Spread of introduced Lehmann lovegrass along a grazing intensity gradient. *Journal of Applied Ecology* 29, 92–98.

McCulley, R.L., Archer, S.R., Boutton, T.W., Hons, F.M. & Zuberer, D.A. (2004) Soil respiration and nutrient cycling in wooded communities developing in grassland. *Ecology* 85, 2804–2817.

McGinty, A. & Ueckert, D.N. (2001) The Brush Busters success story. *Rangelands* 23, 3–8.

McKell, C.M. (1977) Arid land shrubs: a neglected resource. *Agricultural Mechanization in Asia*, Winter, 28–36.

McLain, J.E.T., Martins, D.A. & McClaran, M.P. (2008) Soil cycling of trace gases in response to mesquite management in a semiarid grassland. *Journal of Arid Environments* 72, 1654–1665.

Meik, J.M., Jeo, R.M., Mendelson III, J.R. & Jenks, K.E. (2002) Effects of bush encroachment on an assemblage of diurnal lizard species in central Namibia. *Biological Conservation* 106, 29–36.

Mohamed, A.-M. & El Bassouni, K. (2007) Externalities of fugitive dust. *Environmental Monitoring and Assessment* 130, 83–98.

Monger, H.C. (2003) Millennial-scale climate variability and ecosystem response at the Jornada LTER site. In: Greenland, D., Goodin, D.G. & Smith, R.C. (eds.) *Climate Variability and Ecosystem Response at the Long-Term Ecological Research Sites*. Oxford University Press, New York, pp. 341–369.

Monson, R.K., Guenther, A.B. & Fall, R. (1991) Physiological reality in relation to ecosystem- and global-level estimates of isoprene emission. In: Sharkey, T.D., Holland, E.A. & Mooney, H.A. (eds.) *Trace Gas Emissions by Plants*. Academic Press, New York, pp. 185–207.

Morgan, J.A., Milchunas, D.G., LeCain, D.R., West, M. & Mosier, A.R. (2007) Carbon dioxide enrichment alters plant community structure and accelerates shrub growth in the shortgrass steppe. *Proceedings of the National Academy of Sciences of the United States of America* 104, 14724–14729.

Nagel, J.M., Huxman, T.E., Griffin, K.L. & Smith, S.D. (2004) CO_2 enrichment reduces the energetic cost of biomass construction in an invasive desert grass. *Ecology* 85, 100–106.

Neff, J., Ballantyne, A., Farmer, G. et al. (2008) Increasing eolian dust deposition in the western United States linked to human activity. *Nature Geoscience* 1, 189–195.

Neilson, R.P. (1986) High resolution climatic analysis and southwest biogeography. *Science* 232, 27–34.

Noble, I. (1997) *The Contribution of "Vegetation Thickening" to Australia's Greenhouse Gas Inventory*. Australian Department of the Environment, Sport & Territories, Canberra.

Noble, J.C. & Walker, P. (2006) Integrated shrub management in semi-arid woodlands of eastern Australia: a systems-based decision support model. *Agricultural Ecosystems* 88, 332–359.

Nolte, K.R. & Fulbright, T.E. (1997) Plant, small mammal and avian diversity following control of honey mesquite. *Journal of Range Management* 50, 205–212.

Nosetto, M., Jobbágy, E. & Paruelo, J. (2006) Carbon sequestration in semi-arid rangelands: comparison of *Pinus ponderosa* plantations and grazing exclusion in NW Patagonia. *Journal of Arid Environments* 66, 142–156.

Okin, G.S., Gillette, D.A. & Herrick, J.E. (2006) Multi-scale controls on and consequences of aeolian processes in landscape change in arid and semi-arid environments. *Journal of Arid Environments* 65, 253–275.

Olenick, K.L., Kreuter, U.P. & Conner, J.R. (2005) Texas landowner perceptions regarding ecosystem services and cost-sharing land management programs. *Ecological Economics* 53, 247–260.

Owens, M.K. & Moore, G.W. (2007) Saltcedar water use: realistic and unrealistic expectations. *Rangeland Ecology and Management* 60, 553–557.

Pacala, S.W., Hurtt, G.C., Baker, D. et al. (2001) Consistent land- and atmosphere-based US carbon sink estimates. *Science* 292, 2316–2320.

Painter, T.H., Barrett, A.P., Landry, C. et al. (2007) Impact of disturbed desert soils on duration of mountain snow cover. *Geophysical Research Letters* 34, L12502.

Paynter, Q. & Flanagan, G.J. (2004) Integrating herbicide and mechanical control treatments with fire and biological control to manage an invasive wetland shrub, *Mimosa pigra*. *Journal of Applied Ecology* 41, 615–629.

Peterjohn, B.G. & Sauer, J.R. (1999) Population status of North American species of grassland birds from the North American breeding bird survey, 1966–1996. *Studies in Avian Biology* 19, 27–44.

Peterson, R. (1997) Comment: plant, small mammal and avian diversity following control of honey mesquite. *Journal of Range Management* 50, 443.

Pidgeon, A.M., Mathews, N.E., Benoit, R. & Nordheim, E.V. (2001) Response of avian communities to historic habitat change in the northern Chihuahuan Desert. *Conservation Biology* 15, 1772–1788.

Pielke, R.A., Avissar, R., Raupach, M., Dolman, A.J., Zeng, X.B. & Denning, A.S. (1998) Interactions between the atmosphere and terrestrial ecosystems: influence on weather and climate. *Global Change Biology* 4, 461–475.

Polley, H.W. (1997) Implications of rising atmospheric carbon dioxide concentration for rangelands. *Journal of Range Management* 50, 562–577.

Polley, H.W., Tischler, C.R., Johnson, H.B. & Derner, J.D. (2002) Growth rate and survivorship of drought: CO_2 effects on the presumed trade off in seedlings of five woody legumes. *Tree Physiology* 22, 383–391.

Purohit, U., Mehar, S.K. & Sundaramoorthy, S. (2002) Role of *Prosopis cineraria* on the ecology of soil fungi in Indian desert. *Journal of Arid Environments* 52, 17–27.

Ravi, S., D'Odorico, P., Collins, S.L. & Huxman, T.E. (2009) Can biological invasions induce desertification? *New Phytologist* 181, 512–515.

Richardson, D.M., Allsopp, N., D'Antonio, C.M., Milton, S.J. & Rejmánek, M. (2000) Plant invasions – the role of mutualisms. *Biological Reviews* 75, 65–93.

Safriel, U. & Adeel, Z. (2005) Dryland systems. In: Hassan, R., Scholes, R. & Ash, N. (eds.) *Ecosystems and Human Well-Being: Current State and Trends. Volume 1, The Millennium Ecosystem Assessment Series*. Island Press, Washington, D.C., pp. 623–662.

Sala, O.E., Chapin, F.S., Armesto, J.J. et al. (2000) Global biodiversity scenarios for the year 2100. *Science* 287, 1770–1774.

Salo, L.F. (2005) Red brome (*Bromus rubens* subsp. *madritensis*) in North America: possible modes for early introductions, subsequent spread. *Biological Invasions* 7, 165–180.

Sauer, J.R., Hines, J.E., Thomas, I., Fallon, J. & Gough, G. (1999) *The North American Breeding Bird Survey: Results and Analysis 1966–1998*, Version 98.1. United States Geological Survery, Patuxent Wildlife Research Center, Laurel.

Savage, M. & Swetnam, T.W. (1990) Early 19th-century fire decline following sheep pasturing in a Navajo ponderosa pine forest. *Ecology* 71, 2374–2378.

Scanlan, J.C. & Archer, S. (1991) Simulated dynamics of succession in a North American subtropical *Prosopis* savanna. *Journal of Vegetation Science* 2, 625–634.

Schimel, D., Melillo, J., Tian, H.Q. et al. (2000) Contribution of increasing CO_2 and climate to carbon storage by ecosystems in the United States. *Science* 287, 2004–2006.

Schlesinger, W.H. (1997) *Biogeochemistry: An Analysis of Global Change*. Academic Press, New York.

Schlesinger, W.H., Reynolds, J.F., Cunningham, G.L. et al. (1990) Biological feedbacks in global desertification. *Science* 247, 1043–1048.

Scholes, R.J. & Archer, S.R. (1997) Tree-grass interactions in savannas. *Annual Review of Ecology and Systematics* 28, 517–544.

Schussman, H., Geiger, E., Mau-Crimmins, T. & Ward, J. (2006) Spread and current potential distribution of an alien grass, *Eragrostis lehmanniana* Nees, in the southwestern USA: comparing historical data and ecological niche models. *Diversity & Distributions* 12, 582–592.

Scifres, C.J. (1980) *Brush Management: Principles and Practices for Texas and the Southwest*. Texas A&M University Press, College Station.

Scifres, C.J., Hamilton, W.T., Conner, J.R. et al. (1985) *Integrated Brush Management Systems for South Texas: Development and Implementation*. Texas Agricultural Experiment Station, College Station.

Scifres, C.J., Hamilton, W.T., Koerth, B.H., Flinn, R.C. & Crane, R.A. (1988) Bionomics of patterned herbicide application for wildlife habitat enhancement. *Journal of Range Management* 41, 317–321.

Seastedt, T.R., Hobbs, N.T. & Suding, K.N. (2008) Management of novel ecosystems: are novel approaches required? *Frontiers in Ecology and Environment* 6, 547–553.

Shafroth, P.B., Cleverly, J.R., Dudley, T.L. et al. (2005) Control of tamarix in the western United States: implications for water salvage, wildlife use and riparian restoration. *Environmental Management* 35, 231–246.

Smith, S.D., Huxman, T.E., Zitzer, S.F. et al. (2000) Elevated CO_2 increases productivity and invasive species success in an arid ecosystem. *Nature* 408, 79–82.

Solbrig, O.T., Medina, E. & Silva, J.F. (eds.) (1996) *Biodiversity and Savanna Ecosystem Processes: A Global Perspective*. Springer-Verlag, New York.

Sponseller, R.A. (2007) Precipitation pulses and soil CO_2 flux in a Sonoran Desert ecosystem. *Global Change Biology* 13, 426–436.

Stedneck, J. (1996) Monitoring the effects of timber harvest on annual water yield. *Journal of Hydrology* 176, 79–95.

Sturm, M., Schimel, J., Michaelson, G. et al. (2005) Winter biological processes could help convert Arctic tundra to shrubland. *BioScience* 55, 17–26.

Teague, R., Borchardt, R., Ansley, J. et al. (1997) Sustainable management strategies for mesquite rangeland: the Waggoner Kite project. *Rangelands* 19, 4–8.

Teel, P.D., Marin, S., Grant, W.E. & Stuth, J.W. (1997) Simulation of host-parasite-landscape interactions: influence of season and habitat on cattle fever tick (*Boophilus* sp.) population dynamics in rotational grazing systems. *Ecological Modelling* 97, 87–97.

Tews, J., Brose, U., Grimm, V. et al. (2004) Animal species diversity driven by habitat heterogeneity/diversity: the importance of keystone structures. *Journal of Biogeography* 31, 79–92.

Thiele, T., Jeltsch, F. & Blaum, N. (2008) Importance of woody vegetation for foraging site selection in the southern pied babbler (*Turdoides bicolor*) under two different land use regimes. *Journal of Arid Environments* 72, 471–482.

Thurow, T.L. & Hester, J.W. (1997) How an increase or reduction in juniper cover alters rangeland hydrology. Juniper Symposium Proceedings. Texas Agricultural Experiment Station Technical Report, College Station, pp. 4–21.

Thurow, T.L., Thurow, A.P., Garriga, M.D. (2000) Policy prospects for brush control to increase off-site water yields. *Journal of Range Management* 53, 23–31.

Trimble, S.W., Weirich, F.H. & Hoag, B.L. (1987) Reforestation and the reduction of water yield on the Southern Piedmont since circa 1940. *Water Resources Research* 23, 425–437.

Turner, R.M., Webb, R.H., Bowers, J.E. & Hastings, J.R. (2003) *The Changing Mile Revisited: An Ecological Study of Vegetation Change with Time in the Lower Mile of an Arid and Semiarid Region.* University of Arizona Press, Tucson.

Valentin, C., Poesen, J. & Li, Y. (2005) Gully erosion: impacts, factors and control. *Catena* 63, 132–153.

Vallentine, J.F. (1971) *Range Development and Improvements.* Brigham Young University Press, Provo.

Valone, T.J. & Sauter, P. (2005) Effects of long-term cattle exclosure on vegetation and rodents at a desertified arid grassland site. *Journal of Arid Environments* 61, 161–170.

Van Auken, O.W. (2000) Shrub invasions of North American semiarid grasslands. *Annual Review of Ecology and Systematics* 31, 197–215.

Wainwright, J., Parsons, A.J. & Abrahams, A.D. (2000) Plot-scale studies of vegetation, overland flow and erosion interactions: case studies from Arizona and New Mexico. *Hydrological Processes* 14, 2921–2943.

Walther, G.R., Post, E., Convey, P. et al. (2002) Ecological responses to recent climate change. *Nature* 416, 389–395.

Weltzin, J.F., Archer, S. & Heitschmidt, R.K. (1997) Small-mammal regulation of vegetation structure in a temperate savanna. *Ecology* 78, 751–763.

Wessman, C., Archer, S., Johnson, L. & Asner, G. (2005) Woodland expansion in US grasslands: assessing land-cover change and biogeochemical impacts.

In: Gutman, G., Janetos, A.C., Justice, C.O. et al. (eds.) *Land Change Science: Observing, Monitoring and Understanding Trajectories of Change on the Earth's Surface*. Springer, Heidelberg, pp. 185–208.

Whitford, W.G. (1997) Desertification and animal biodiversity in the desert grasslands of North America. *Journal of Arid Environments* 37, 709–720.

Wilcox, B.P. (2002) Shrub control and stream flow on rangelands: a process based viewpoint. *Journal of Range Management* 55, 318–326.

Wilcox, B.P., Owens, M.K., Dugas, W.A., Ueckert, D.N. & Hart, C.R. (2006) Shrubs, stream flow and the paradox of scale. *Hydrological Processes* 20, 3245–3259.

Wilcox, B.P., Owens, M.K., Knight, R.W. & Lyons, R.K. (2005) Do woody plants affect stream flow on semiarid karst rangelands?. *Ecological Applications* 151, 127–136.

Wilcox, B.P. & Thurow, T.L. (2006) Emerging issues in rangeland ecohydrology: vegetation change and the water cycle. *Rangeland Ecology and Management* 59, 220–224.

Wilkins, R.N., Hejl, S.J., Magness, D.R. & Bedford, T.L. (2002) Wildlife response to brush management. In: Arringto, D.A., Conner, R., Dugas, W. et al. (eds.) *Ecosystem and Wildlife Implications of Brush Management Systems Designed to Improve Water Yield*. Technical Report 201, Texas Water Research Institute, Texas Agricultural Experiment Station, College Station, pp. 103–164.

Williams, D.G. & Baruch, Z. (2000) African grass invasion in the Americas: ecosystem consequences and the role of ecophysiology. *Biological Invasions* 2, 123–140.

Young, J., Evans, R. & Rimby, C. (1985) Weed control and revegetation following western juniper control. *Weed Science* 33, 513–517.

5

Health and Disease in Wild Rangelands

Richard Kock[1], *Mike Kock*[2], *Sarah Cleaveland*[3]
and Gavin Thomson[4]

[1]Conservation Programme Zoological Society of London,
London, UK
[2]Wildlife Conservation Society of New York, NY, USA
[3]University of Glasgow, Glasgow, Scotland
[4]TadScientific CC, Pretoria, South Africa

Introduction

Population health and disease are important components of ecosystem function (Cook et al. 2004) and therefore relevant to the conservation of wild rangelands. Although infection and disease are natural biological events, early ecologists did not consider disease a significant factor in population dynamics. More recently, however, especially where anthropogenic impacts have disrupted ecosystems and biodiversity loss has altered human and animal disease ecology (Chivian & Bernstein 2004), populations have become vulnerable to disease impacts. Novel diseases or disease events have emerged with sometimes devastating consequences on human and animal health, biodiversity and ecosystems, ultimately affecting global economies given the high cost of controlling disease. The cost of a bird flu pandemic was estimated at $167 billion (Asian Developmental Bank 2005). These effects may be direct, such as the affliction of a species or population that leads to its decline or extinction (Cleaveland et al. 2002; Woodroffe et al. 2004; Dixon 2005),

Wild Rangelands: Conserving Wildlife While Maintaining Livestock in Semi-Arid Ecosystems,
1st edition. Edited by J.T. du Toit, R. Kock, and J.C. Deutsch.
© 2010 Blackwell Publishing

or indirect, such as impacts on humans or domestic livestock resulting in restrictions on land use and animal movement (e.g. fencing), which in turn disrupt ecological processes and cause species decline (Bengis et al. 2002). Díaz et al. (2006) posit that adverse effects of diseases on biodiversity remove an ecosystem's buffers against environmental change (e.g. temperature, precipitation and pathogens). And as climate change moves from science fiction to fact, emerging disease is even suggested as a key indicator of this man-made crisis (Harvell et al. 2002).

Wild rangelands have absorbed introduced diseases and suffered human- and livestock-induced ecological disruption for millennia; as such they provide an ideal platform on which to examine the disease's role in ecosystem function and resilience and responses of pathogens to changes in biodiversity, species composition and abundance. Changes to rangelands, particularly over the last century, have included human settlement, land degradation and increased populations of humans, livestock and even certain (alien/feral) wildlife. These alterations raise the potential for disease creating new transmission pathways, increasing contact rates and changing parasite flow (Polley 2005). The situation is predicted to worsen as climate change and drought reduce water availability and concentrate populations in many regions.

This chapter focuses on Africa, where most of the remaining truly wild rangelands occur, but references other continents and ecosystems. The chapter first describes key diseases and some impacts and issues that arise from them relevant to conservation. Second, the importance of disease in rangelands' landuse planning is considered using the contemporary example of the developing transboundary conservation areas in southern Africa. Third, the chapter reviews and proposes solutions to problems arising from economic and trade effects of epidemic livestock disease and their likely impact on wildlife conservation in rangelands, using sub-Saharan Africa as the case study. The final section reflects on trends for disease research, veterinary intervention on rangelands, potential impacts on biodiversity conservation and landuse decisions.

Diseases at the rangelands' wildlife–livestock–human interface

In the so-called developed world, landuse practice and disease control have resolved or reduced some of the major multi-host infectious diseases, such as rinderpest, foot-and-mouth disease (FMD), tuberculosis (TB) [bovine tuberculosis and brucellosis (BTB)], brucellosis and rabies, to low levels. This has been achieved by veterinary controls among livestock and wildlife species and improvements in public health practice, modern medicines and immuno-prophylaxis. Reducing free-ranging wildlife hosts and disease vectors and restricting contact between species have also helped control such infections as nagana, plague (*Pasteurella pestis*), theileriosis and malaria. Global changes, however, call into question the sustainability of this status quo, even in developed countries. TB, transmissible spongiform encephalopathies (TSEs), swine fevers and avian influenza (AI) are an increasing global threat, whilst other less familiar diseases emerge with new production systems and changes in land use or management. AI is widespread in wild bird populations but is of little consequence, while highly pathogenic strains present only in domestic poultry, have evolved and now threaten global health. In recent years, a number of epizootics have been recorded as causing disease and mortality in human, animal and wild bird populations. Other emerging diseases include Rift Valley Fever and Ebola virus infection (Africa), Nipah/Hendra viruses (Asia) and haemorrhagic diseases (Australia).

Transboundary animal diseases (TADs) and zoonotic infections in less developed countries remain widespread and highly intractable, particularly in the tropics. Some similar infections resulted from co-evolution between microorganisms and the many ungulates present (Allison 1982), whereas others were introduced from other continents. These regions have weaker public and veterinary health systems than developed countries and are home to more contact between humans and animals and among animal species. Land degradation likely alters disease dynamics as well, with dire consequences for man and beast. These diseases threaten wild rangelands, as human and economic imperatives drive radical disease control at the expense of wildlife and ecosystems.

Among the most ancient and productive of wild ecosystems, African rangelands are a microcosm of multi-species diseases and their issues. One reason is that Africa's environment has remained relatively stable over hundreds of thousands of years, without sterilizing events such as ice ages that occurred elsewhere. This has made African rangelands particularly conducive to co-evolution. African buffalo (*Syncerus caffer*) act as maintenance host for South African types (SATs) of FMD viruses (Hedger 1972); the resultant division of land between wildlife and livestock has changed African rangelands (Thomson et al. 2003). FMD has also influenced veterinary policy and rangeland commerce due to the perceived threat to livestock industries of developed countries. FMD directly causes few negative effects on wildlife populations, with few clinical syndromes reported, though in Asia, FMD possibly contributes to the saiga antelope's decline because of its apparent susceptibility and exposure to livestock (Morgan et al. 2006). FMD is also thought to have caused mortalities in mountain gazelles (*Gazella gazella*) in Israel in 1985 (Shimshony et al. 1986) and 2006, as well as in captive gazelle and oryx in the United Arab Emirates (J. Mwanzia, personal communication, 2007) and Saudi Arabia (O. Mohamed, personal communication, 2008).

Rinderpest has a long history on African rangelands and potentially devastating impacts on wildlife, cattle and ecosystems (Rossiter 2004). Being one of the first diseases to be associated with humans' alteration of pathogen transmission pathways, the disease's introduction via livestock to herds of African wildlife caused the great cattle plague of 1889–1905. Cattle, buffalo, *Tragelaphine* antelope, wild suids and wildebeest *Connochaetus taurinus* declined precipitously, while humans starved after losing cattle and wildlife resources. The disappearance of antelope and wild suids apparently caused the extinction of tsetse flies (*Glossina* spp.) in certain areas for lack of hosts and may have contributed to such wildlife distribution anomalies as the formation of isolated metapopulations of species such as sable antelope (*Hippotragus niger*) and greater kudu (*Tragelaphus strepsiceros*) (Rossiter 2000).

The changing ecology of the Serengeti following rinderpest's eradication illustrates the importance of disease in populations and ecosystems. In the first half of the twentieth century, cyclical rinderpest outbreaks depressed wildebeest numbers to approximately 250,000 animals. Once the disease was controlled, however, wildebeest populations rebounded to over 2.5 million in a few decades; this recovery spelt benefits for grassland, predator and economic systems. Rinderpest strains usually associated with mortality in cattle have been easy to eradicate, but at least one milder cattle strain has persisted in East Africa, causing continued wildlife losses (Kock 2006). It is only in recent times that the epidemiology of the disease in buffalo and other wildlife species has been elucidated, confirming the belief that under current wildlife population levels wild species are not competent maintenance hosts but possible vectors for cattle rinderpest (Kock, 2008). This supports the rationale for eradication of rinderpest and current evidence supports the probability that the infection is now absent from Africa as a result of cattle vaccination over decades (Lubroth, J, personal communication, 2009). Canine distemper, another morbillivirus infection, has afflicted and killed many endangered rangeland carnivores such as African wild dogs (*Lycaon pictus*). There is evidence that canine distemper's virulence has increased, causing an epidemic in the lion (*Panthera leo*) population of the Serengeti

National Park (Roelke-Parker et al. 1996). Evidence also suggests that increased contact between wild animals and domestic dogs caused canine distemper outbreaks in the Serengeti–Mara (Cleaveland et al. 2000).

In eastern and southern Africa, wild pigs and the eyeless tampan (*Ornithodoros* spp.) maintain the African Swine Fever (ASF) virus (Thomson 1985). There is no proven strategy to eradicate this virus from wildlife. This fact has precluded the use of wild rangelands for domestic swine, a potentially positive implication of this disease for conservation, given the free-ranging or feral pigs' destructive nature. Another example is classical swine fever (CSF), which poses problems in Europe, with persistence in wild suids and conflict between wild and domestic animals.

Anthrax is endemic in many parts of the world and periodic 'natural' outbreaks have been documented in wildlife populations, for example, Etosha National Park, Namibia (Lindeque & Turnbull 1994) and Kruger National Park, South Africa (Hugh-Jones & de Vos 2002). This multi-species disease is fatal in most hosts and is considered an ecologically important driver in controlling populations, though the disease has become uncommon in the northern hemisphere. Large-scale outbreaks have become more frequent, however, as a result of declining veterinary services and poor cattle vaccination practice. The emergence of anthrax at the wildlife–livestock interface has occurred over the last decade in and around Luangwa Valley National Park, Zambia; Serengeti National Park, Tanzania; Mago National Park, Ethiopia; Malilangwe Wildlife Reserve, Zimbabwe; Queen Elizabeth National Park, Uganda and Samburu National Reserve, Kenya. Species affected includes hippopotamus (*Hippopotamus amphibious*), greater and lesser kudu (*Tragelaphus imberbis*), African buffalo (*Syncerus caffer*) and impala (*Aepyceros melampus*) among other ungulates; most recently, the disease was confirmed among the critically endangered Grevy's zebra (*Equus grevyii*). With species populations increasingly fragmented and isolated, anthrax is one of the more important wildlife diseases on African rangeland that poses an extinction threat.

BTB is prominent on rangelands, but the percentage of human TB cases caused by BTB on rangelands is not established. In 1997, the prevalence of TB infection in humans was estimated at 32% of the global population (Dye et al. 1999). This was a shock and the situation has only deteriorated since (Corbett et al. 2003). Now found worldwide, BTB was probably introduced to Africa during the colonial era and is a key disease at the wildlife–livestock interface (Michel 2005). The disease has become endemic in buffalo populations in South Africa (Bengis 1999), Tanzania (Cleaveland et al. 2005) and Uganda, as well as in Kafue lechwe (*Kobus leche kafuensis*) populations in Zambia. In some areas, buffalo and lechwe have become true maintenance hosts, and sporadic spillover of infection has been documented in other herbivores, cattle, primates, carnivores and other species. BTB is one of the most serious wildlife health problems to confront conservationists and veterinary regulatory officials in South Africa (Keet et al. 1996; Bengis 1999; Michel et al. 2006). The potential problems associated with persistent infection in wildlife reservoirs is highlighted by ongoing difficulties in controlling the disease elsewhere, such as in England, where badgers are a source of persistent infection to cattle (Krebs et al. 1997); in Europe where wild boar host BTB (Hermoso de Mendoza et al. 2006) and in New Zealand, where possums maintain the infection (O'Neil & Pharo 1995). In African rangelands, the long-term effects of this chronic progressive disease on wildlife host populations at sustained high prevalence rates are unknown. Preliminary evidence suggests that it may negatively affect population structure and dynamics in buffalo and lion, although other species like badgers and possum continue to thrive in its presence (Rodwell et al. 2001). Recent work on the role of genetics in moderating the infection rates and disease prevalence of TB and other infections in populations is intriguing, with evidence that higher levels are associated with homozygosity in populations (Acevedo-Whitehouse et al. 2003, 2005).

Rabies remains a problem in many developing countries. The situation with sylvatic (wildlife) rabies is complex with, for example, many different variants associated with a range of carnivore species in southern Africa (Nel et al. 1993; von Teichman et al. 1995). The domestic dog is the major reservoir for canid viruses, identified as the main threat to many endangered species, such as the Ethiopian wolf (Randall et al. 2004) and African wild dog (Gascoyne et al. 1993). Transmission from domestic dogs poses the major threat to public health – source of the vast majority of human rabies cases and exposures in Africa. Wildlife reservoirs or epidemics do not appear to pose a major threat to humans and are rarely implicated. For canid strains of rabies, phylogenetic analysis and historical accounts support a common lineage in both past and present reservoirs for rabies in Europe (Smith et al. 1993; von Teichman et al. 1995). Some of these, such as European rabies – the so-called canine 'street' virus – were introduced to rangelands. The extinction of a carnivore will adversely impact the ecology of some rangeland ecosystems, as has occurred with the loss of the Ethiopian wolf in Afro-Alpine habitats of Ethiopia and of wolves (*Canis lupus*) in Europe, India and North America. The explosion of deer populations in Europe and North America and the nilgai antelope in India are also contemporary examples of the impact of predator loss. The parallel increase in tick populations from the increase in herbivore hosts has resulted in the emergence of Lyme disease in humans. Throughout much of rural Africa, poverty and underdevelopment are significant factors in rabies persistence, where domestic dog populations are poorly managed and have a high turnover, providing ideal conditions for epidemic diseases. Although dependent on owners, government subsidies or donors, dog vaccination programmes have been shown to be effective in reducing the incidence of canine rabies and human exposure, but these measures are costly and may be difficult to sustain in the long term. These donor-dependent dog vaccination programmes are designed on developed country models and have unfortunately proven ineffective in eradicating disease in many less-developed countries to date. Under these conditions, targeted wildlife vaccination may be more effective in protecting endangered species and ecosystems on rangelands, particularly as low-coverage vaccination strategies are likely to be effective in reducing population extinction risks (Haydon et al. 2006). Oral bait technology proved effective in controlling fox rabies in Europe; oral vaccines and bait-delivery systems have been tested successfully in African wild dog populations in southern Africa (Knobel et al. 2003) and are under trial in Ethiopian wolves. However, constraints to their application in rangeland settings still remain, both for targeting endangered wildlife populations and for supplementing domestic dog vaccination strategies. For example, in areas with mixed scavenger and carnivore populations, the oral-bait approach may be less effective because of uptake of bait by non-target species.

Trypanosomosis is considered the single most important factor in limiting agriculture and livestock development in many African rangelands, which results in large unexploited areas of bush. As such, the disease shapes landuse patterns and prevents encroachment into wildlife-rich areas. From a conservation perspective, this can be viewed as positive, but it has also generated a negative attitude to protected areas. The public health and economic importance of the disease requires development of disease control strategies that, at the very least, allow for integrated landuse in buffer zones surrounding wildlife-protected areas. Efforts to control the disease through eradication of wildlife hosts have largely failed. Newer techniques of sterile fly release, insecticidal targets and spraying (with low environmental persistence) can result in eradication of the tsetse fly vector in specific localized and isolated situations (e.g. Okavango Delta, Botswana and Zanzibar Island, Tanzania) (Vreysen 2000; Sharma et al. 2001). However, these top-down control schemes are extremely expensive and unlikely to be sustainable for many of the poorer countries of sub-Saharan Africa (Maudlin 2006). Recent research suggests that simple solutions that can be carried out by farmers [e.g. restricted-application treatment of cattle (Torr et al. 2005); treatment at livestock markets to prevent disease spread (Fèvre et al. 2005)] provide real opportunities for effective disease management and control.

Echinococcosis or hydatid disease is a parasitic disease perpetuated by predator–prey relationships of definitive mammalian hosts (carnivores) and intermediate hosts (including ungulates, rodents and humans). *Echinococcus granulosus*, which causes cystic echinococcosis in humans, and *Echinococcus multilocularis*, the cause of alveolar hydatid disease, are persistent problems in many parts of the world, particularly in pastoral communities of Africa and Asia, as well as South America and Australia/New Zealand. The disease burden in pastoral and nomadic communities can be extremely high. Ultrasound abdominal screening surveys detect cysts in 5–19% of pastoralists among the Turkana of north-west Kenya and in Tibetan nomadic populations (MacPherson et al. 1989; Li et al. 2005). Studies indicate that a wide range of potential predator–prey cycles of *E. granulosus* may exist in Africa (Jenkins & MacPherson 2003). However, the public health significance of these findings still needs investigation.

Human population growth and expanding routes for disease transmission through animal or animal product trade and human movements are associated with several new emerging diseases.

Human immunodeficiency virus 1 (HIV-1) and human immunodeficiency virus 2 (HIV-2) evolved from a chimpanzee (*Pan troglodytes*) strain and a sooty mangabey (*Cercocebus torquatus*) strain of Simian immunodeficiency viruses respectively (Hahn et al. 2000). The emergence of human immunodeficiency virus-acquired immunodeficiency syndrome (HIV-AIDS) in the twentieth century appears to be the result of a complex set of largely ecological and social changes in Africa, including the following: human populations expanding, deforestation and the bushmeat trade, rural displacement, urbanization and poverty, altered sexual behaviour and increased local and international trade (Bengis et al. 2004b). HIV infection is probably becoming one of the biggest zoonotic pandemics in recent human history.

Highly pathogenic AI (subtype H5N1), in recent years, has resulted in hundreds of millions of poultry dying or being slaughtered and high mortalities amongst wild birds of certain species in Eurasia on a few occasions. There has been spillover infection with human deaths and a number of other mammalian species affected, but in low numbers. Since 1918, Spanish Flu epidemic (subtype H1N1) was believed to have originated from birds (Alexandra 2000); the potential threat posed by H5N1 is widely agreed to be significant. The ability of this virus to become endemic in different ecosystems and amongst rangeland bird species is unknown but it appears to circulate, at times, undetected and mainly in domestic water fowl (Kilpatrick et al. 2006). The epidemiologic role of wild birds in this pandemic remains largely undetermined (United Nations Environment Programme 2008), but where it has appeared, it has been in temperate and wet ecosystems, so rangelands are probably reasonably safe.

In 2002–3, a novel viral respiratory disease, severe acute respiratory syndrome (SARS) emerged in humans in Southeast Asia. This disease was found to be caused by a novel coronavirus, which circulated in farmed palm civets (*Paguma larvata*) sold for food. The disease in humans was frequently life-threatening and directly contagious and became rapidly disseminated by international travel.

Another example of the bird–mammal interface (involving vector mosquitoes) is with West Nile Virus that spread across the North American continent far more rapidly than expected, with significant mortality in wild birds, horses and people. Hantavirus emerged from its rodent reservoir recently, infecting humans and other species and it is postulated that this is due to a changing environment and contact rates (Zeier et al. 2005). Another example of changing transmission pathways for disease is shown with the transmissible spongiform encaphalopathies caused by prion agents (with zoonotic potential). These diseases emerged most probably with the manufacture and feeding of livestock and farmed deer with compound pellets using animal-derived proteins. Chronic wasting disease (CWD), the most novel of these, is currently spreading amongst cervids in the North American rangelands; transmission appears to be horizontal through contact. The ecological and economic consequences of CWD and its spread remain to be determined; moreover, public health implications remain a question of intense interest (Williams & Miller 2002). Lyme disease has become a significant and spreading public health threat in the Americas and Eurasia, which has been associated with changing landscapes and climate. The infection cycle is dependent on the deer tick, *Ixodes scapularis*, in the northeastern and North Central United States; *Ixodes pacificus* in the western United States; the sheep tick, *I. ricinus*, in Europe and the taiga tick, *Ixodes persulcatus*, in Asia, illustrating the complex dynamics between species and vectors on rangeland.

Other infections at the interface on rangelands remain poorly understood, such as Rift Valley Fever, which is highly sporadic but a devastating haemorrhagic disease amongst wildlife, livestock and people. The role of the mosquito vector and maintenance host is established but alternate reservoirs, perhaps in rodents and other wildlife species, remain to be proven. The impact of climate change on this disease is likely to be significant as the opportunities for vector expansion increase. The epidemiology of heartwater, theileriosis and malignant catarrhal fever (MCF), as well as BTB in a mixed species system is well documented, but these diseases in livestock and humans respectively are often under-reported, especially in poor countries, but they are significant issues at the interface.

Perceptions and tolerance of disease in protected areas and their buffer zones

How important are diseases to politicians, policy makers and communities who live within these large rangelands, and is health considered important in the overall rural livelihoods and conservation picture? With the former, it will be dependent on whether the particular disease has an indirect or direct impact on health, in its broadest sense, which includes livestock, and in terms of socio-economic impacts. It is much easier for people to understand disease impacts if it affects their livelihoods, their ability to market livestock and products at a local and regional level and, on a larger scale, internationally. The latter is of great importance to policy makers, politicians and disease control authorities and can have major impacts on wildlife (Kock et al. 2002; Kock 2005). It is much more difficult to deal with and quantify the impacts of disease on human and ecosystem health, as well as peoples' perceptions of this, especially when dealing with diseases at the interface (especially zoonotic diseases) or with diseases such as HIV/AIDS and bird flu where there are political machinations, social stigmas and economic perturbations. A more holistic approach is needed when considering disease issues across large rangeland landscapes, beginning with a broad brush and then focusing in on specific issues but always considering the 'one health' aspect (See Figure 5.1).

   In southern Africa, with the recent rapid growth in the tourism industry, the transboundary management of natural resources, particularly of water and wildlife, and the associated development of transfrontier conservation areas (TFCAs) have become a major focus of attention. Twenty potential and existing TFCAs have been identified in the Southern African Development Community (SADC) region[1], involving 12 continental African member

[1] SADC comprises Angola, Botswana, the Democratic Republic of Congo, Lesotho, Madagascar, Malawi, Mauritius, Mozambique, Namibia, South Africa, Swaziland, United Republic of Tanzania, Zambia and Zimbabwe.

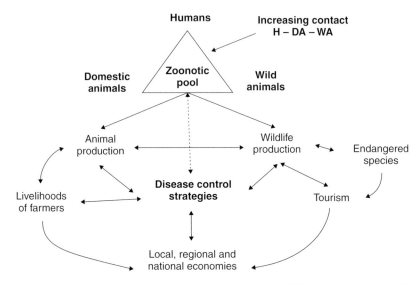

Figure 5.1 **Conceptual diagram of the linkages among wildlife (WA), livestock (DA) and human (H) diseases (one health approach) and the potential implications of disease control strategies for livelihoods and conservation (D. Cumming, personal communication, 2006).**

states. The TFCAs include many national parks (NPs), neighbouring game reserves, hunting areas and conservancies, mostly occurring within a matrix of land under traditional communal tenure; altogether the proposed TFCAs cover about 120 million hectares. The development of these TFCAs has proceeded at the political level regardless of the reality of the situation on the ground. Certainly, in terms of the established Greater Limpopo TFCA and planned Kavango mid-Zambezi TFCA in southern Africa, the management of wildlife and livestock diseases, including zoonotic diseases, within the envisaged larger transboundary landscapes remains unresolved and an emerging issue of major concern to livestock production, export markets and other sectors in the region. The dynamics of increasing interactions at the interface between animal health, ecosystem goods and services on the one hand and human health and well-being on the other are also poorly understood, with the result that policy development is compromised by a lack of appropriate information and understanding of the complex systems and issues involved.

Wildlife and livestock disease issues include those that can directly affect human health and will likely have a significant impact on the future development of sustainable land uses, transboundary natural resource management, biodiversity conservation and human livelihoods in the marginal lands of southern Africa. Around 65% of southern Africa is semi-arid to arid, where extensive livestock and wildlife production systems are the most suitable and potentially sustainable forms of land use. The need to arrest desertification and enhance the capacity of these marginal areas to generate wealth and sustain improved human livelihoods is of paramount importance to the region. There does not appear to be an existing formal policy on animal health and disease control for the TFCAs being developed and this needs to be addressed sooner rather than later in the interest of both regional biodiversity conservation and development.

To illustrate the complexity of these issues, it would be appropriate to examine two TFCAs in southern Africa, namely, the Great Limpopo Transfrontier Conservation Area (GLTFCA) and the Kavango–Zambezi Transfrontier (KAZA) TFCA.

Greater Limpopo TFCA

The international treaty to establish the Great Limpopo Transfrontier Park (GLTP) was signed by the presidents of Mozambique, South Africa and Zimbabwe in December 2002. Agreement was reached on creating a TFCA that encompassed the GLTP and the intervening matrix of conservancies and wildlife ranches on freehold land, together with the communal farming areas (Figure 5.2). The precise boundaries of this vast TFCA remain undefined, but the primary land use in the matrix is expected to be wildlife-based tourism with reasonably unimpeded movement of wildlife and tourists. The GLTFCA covers about 10 million hectares or 100,000 km^2. The area includes several land use/land tenure regimes including NPs, state and private safari and hunting areas, conservancies and game ranches on freehold land, small-scale agro-pastoral farming areas under communal tenure, large-scale commercial irrigation schemes and smaller irrigation schemes within the communal areas. About 35% of the area comprises state-protected areas and a further approximately 10% is freehold land under wildlife. Most of the remaining land, the matrix between the designated NPs, is under communal tenure with varying forms of small-scale agro-pastoralism.

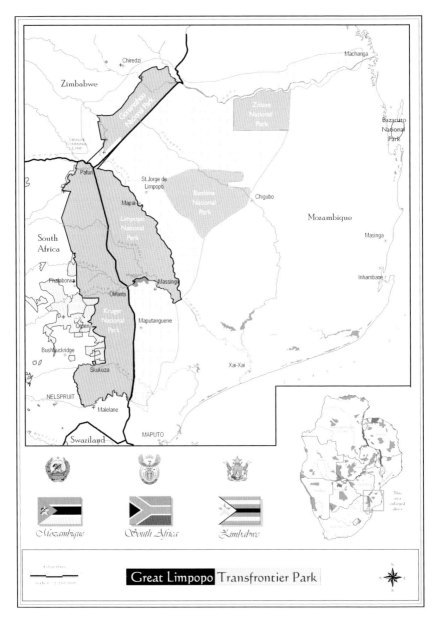

Figure 5.2 **Map showing the GLTFCA (Bengis, 2005).**

The control and containment of livestock diseases has relied heavily on fences and the monitoring and control of domestic and wild animal movements and translocations (Scott Wilson Group 2000; Bengis et al. 2004a). Removing barriers to wildlife and livestock therefore has major implications for animal health and disease control within the GLTFCA. It could also have wider implications for disease control in the three countries concerned (Bengis 2005). The GLTFCA covers land of diverse tenure and use in all three countries and because of the large 'edge effect' within each country, the animal health and land management strategies within the GLTFCA will have major implications for livestock disease control, production and export markets in each country.

The extraordinary conservation and economic opportunities represented by this transboundary concept are matched in magnitude by the management challenges such a landuse complex poses – not the least of which relates to the management of biologically and economically important diseases that may be transmitted between wildlife and livestock and for some pathogens, to people.

Some of the animal health issues presently of greatest concern in the GLTFCA are the following:

- The breakdown of controls for FMD in Zimbabwe and its spread (including all three SAT serotypes represented by two to three topotypes of FMD), many of the outbreaks occur within the southeastern sector of the country.
- The possible re-invasion of tsetse fly and trypanosomosis – there is also evidence of a return of tsetse fly to the Save–Rundi junction area of the Gonarezhou National Park in Zimbabwe.
- The northward spread of BTB in South Africa's Kruger National Park, with buffalo and other species involved, for which there is already evidence documented via South African National Parks research. The possible entry of this disease into Zimbabwe and its status in Mozambique are of course of great concern.

It is critical to note that many of the diseases historically or currently impacting on the Great Limpopo ecosystem have been or are essentially alien invasive species, which have previously impacted on biodiversity or have the potential to do so. Historically, the major external shocks in the region to both livestock and wildlife in the mid- to late 1800s were in fact from introduced diseases mostly discussed above, including rinderpest, contagious bovine pleuropneumonia (CBPP), edible commodities for (ECF) rabies and BTB. A novel FMD topotype, or strain, from northwestern Zimbabwe was also introduced when buffalo were translocated from Hwange and Matusadona National Parks to Gonarezhou National Park in the late 1990s. HIV/AIDS is a significant constraint to the development of TFCAs, through the increased susceptibility of communities to diseases such as TB and malaria. This is particularly true in the case of the GLTFCA where the interaction between wildlife and rural communities may pose significant risks regarding health. Today, BTB is pervasive in much of South Africa's Kruger National Park but, as yet, has not been thoroughly studied in the wildlife of Zimbabwe's Gonarezhou National Park nor Mozambique's Limpopo National Park. It has not been found in a preliminary survey of cattle in the intervening Sengwe Communal Land to the north of the Limpopo River (C. Foggin, personal communication, 2005). Although hard data on the incidence of zoonotic diseases in GLTFCA communities is largely unavailable, the high incidence of HIV/AIDS in the region makes zoonoses (diseases transmissible between animals and man) like BTB that much more of a threat. With the recent development of cases of human extreme drug resistant tuberculosis (XDRTB) in parts of South Africa, often related to HIV/AIDS, the potential for further developments with TB, including the possible increase in extra-pulmonary TB associated with BTB, must be a concern.

BTB in wildlife reduces the potential to resolve the disease in livestock and is probably one of the most serious wildlife health issues currently confronting conservationists and veterinary regulatory officials in several African countries (Bengis 2005).

Kavango mid-Zambezi TFCA

Historically, major disease outbreaks have occurred in Africa from cross-border movement of livestock either for trade or in a search for grazing, particularly with pastoralist communities. The KAZA region has not been immune to this with diseases such as CBPP having a major socio-economic impact on rural livelihoods and negative impacts on wildlife through fence construction in Botswana (Amanfu 2000). Within the proposed KAZA TFCA there are major sustainable livelihood issues with 68% of the people, within the region, who live below the poverty line. Many of the diseases mentioned above were identified as having the potential to impact on human, livestock and wildlife health within the KAZA area (Osofsky et al. 2005a) and therefore, impact the development of the TFCA. Owing to strong disease control policies, FMD, CBPP and Trypanosomosis were prominent as constraints on development of the TFCA. There are significant potential animal health implications and challenges that should be expected when extending the geographic range of certain animal pathogens and disease vectors. At least two of the countries have access to lucrative export markets for their meat and therefore have to give priority attention to various measures to prevent the introduction of diseases that could threaten their agricultural economies. Without international boundary barrier fences and with biological bridges being formed through contiguous wildlife populations, any contagious/infectious disease or vector present in any one of the participating countries or areas will predictably spread throughout the entire TFCA and beyond (Bengis 2005). These concerns must be addressed within a broader environmental context that considers not only biodiversity conservation but also the long-term provision of key environmental goods and services.

In the development of TFCAs in southern Africa, little attention has been paid to significant constraints on the ground related to diseases. These need to be addressed sooner rather than later. In this context, health is an important part of the conservation and rural development toolkit and should be utilized to promote wise use of natural resources and address disease issues. Indeed, the link between diseases and protected areas management was recognized as an important emerging issue at the World Parks Congress held in Durban, South Africa, in 2003 (Osofsky et al. 2005b). Health is a key area where holistic, integrated multidisciplinary approaches across landscapes need to be adopted to ensure long-term conservation success. Figure 5.3 shows graphically how land use becomes complicated as competing economic sectors determine the boundaries established, often through fencing for reasons of disease control. A significant factor here is elephants and their movements. Although they are not directly important in terms of disease, they can be in terms of disease control and human health with destruction of fences and human–elephant conflict impacting on health.

Economic effects of livestock diseases: likely conservation impacts

The number and diversity of wild ungulates in sub-Saharan Africa have co-evolved with an array of micro-parasites, some of which affect sympatric

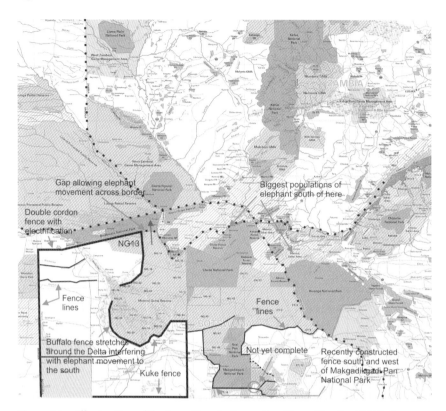

Figure 5.3   **Illustration of the fence lines in the region (from Martin, 2005). Courtesy of John Hanks, adapted by M. Kock.**

domestic livestock adversely (Bengis et al. 2004b). Diseases caused by infectious agents such as the South African types of FMD- MCF-, ASF- and African horse-sickness (AHS) viruses are classic examples that may have large-scale effects on domestic animal species that are phylogenetically closely related to their natural hosts. The effects of these infections are especially severe for exotic breeds of livestock introduced, sometimes misguidedly, to improve the productivity of indigenous breeds. Conversely, infectious agents have also been imported into Africa with exotic domestic livestock and have spilled over into wildlife populations with unfortunate consequences. Rinderpest virus is an example of the latter category of agents. So, transmission of infectious agents between wildlife and domestic livestock occurs in both directions.

As a consequence of these events, sub-Saharan Africa has come to be recognized as the repository of an array of dangerous infections of ungulates that the rest of the world is anxious to avoid importing because of the severe economic consequences they potentially have. The result has been a set of trade embargoes (in effect non-tariff barriers) on livestock and livestock products derived from sub-Saharan Africa. This has largely excluded African livestock producers, pastoralists in particular, from the most lucrative markets and contributed to rural poverty in the subcontinent.

The need for trade in livestock and livestock products in pastoral areas

Poverty among pastoral communities is an escalating problem caused, to a large extent, by growing numbers of people occupying fixed areas of arid land with limited carrying capacity for ungulates and, therefore, the people dependent upon them (Schwartz 1993). Contributing to this has been the exclusion of pastoral people from natural resources through establishment of wildlife protected areas in rangelands (DFID 2002). Perhaps 6 million people – 25% of the population occupying such arid regions of sub-Saharan Africa – are entirely dependent upon livestock (Thornton et al. 2003). Historically, these people and their livestock have shared this rangeland with large numbers of wildlife, but only in East Africa is this still the situation and even there in increasingly limited areas.

Improving the livelihoods of pastoralists is a complex problem, but an obvious contributor to better living standards could be increasing trade in the only valuable commodity many of these people possess, namely, livestock and products derived from livestock. To be successful, this would require that both the volume produced (to ensure surpluses over and above domestic requirement) and price achieved for the traded commodities would need to increase. This, in turn, implies intensification of production and access to higher value markets than is currently the case for most pastoralists. There are other vital impediments to trade in livestock and their products, such as lack of infrastructure, difficulties with transport and multiple taxation along trade routes, that would also need to be addressed (Aklilu 2002). Lack of any significant plant for processing livestock products is a particular problem when it comes to the prospect of beneficiation.

The ecological impact of increasing trade in livestock commodities derived from pastoral areas and whether it can be achieved without detrimental effects on the environment is a vital question in relation to significantly increasing livestock production in arid areas of sub-Saharan Africa, which inevitably involves at least an element of intensification. Part of this question relates to the interaction between increasing livestock numbers in such areas and wildlife. With the increasing pressure on the available land, it is inevitable that competition among livestock and wildlife for grazing and water is becoming concomitantly greater. This is quite apart from the two-way transmission of serious animal diseases that are the cause of many of the major non-tariff barriers that inhibit exports of livestock commodities from sub-Saharan Africa (Bengis et al. 2004a).

Ideally, livestock and wildlife could be managed such that their combined production is exploited for the benefit of the local inhabitants (e.g. production of livestock products integrated with eco-tourism). This has long been an ideal of integrated development, but has only very rarely been realized.

What would be needed to increase export of livestock commodities to high-value markets from localities where domestic livestock and wildlife coexist? Export of livestock and livestock products from regions of the world endemically infected with trade-influencing TADs has traditionally required eradication of trade-influencing TADs as a precondition. The problem with this approach is that the epidemiologies of most of these diseases preclude this requirement, considering the technologies and resources available currently. There are exceptions and rinderpest seems to be a disease that has now reached the stage of final eradication from Africa (and the world) although unambiguous proof is lacking. For the other diseases mentioned above, as well as a number of others, there is no such prospect within the foreseeable future (Thomson et al. 2004).

To some extent, this poor position, insofar as trade is concerned, has been successfully addressed by 'zonation'. The idea is that it is often possible to confine specific diseases to a geographic zone within a country (i.e. a region of that country), while the rest of the country can be proven to be free of the infection in question. So, for example, a mechanism exists for official recognition of zonal freedom from FMD (either with or without vaccination) to be accorded by the Office international des epizooties (OIE) (International Office for Epizootic Disease Control) – the world animal health organization. This avenue has been exploited successfully by a number of countries in southern Africa (e.g. Botswana and Namibia) which, as a consequence, have been able to export beef to the European Union. The reason these countries have been able to do this is that the 'infected zones' have effectively been isolated from the 'free' zone of the country using elaborate fencing systems. To a large extent, this has been achieved because pastoralism is not a significant lifestyle in southern Africa. Furthermore, wildlife areas have been clearly separated from areas of domestic livestock production by fences and other animal movement control measures.

In the pastoral areas of East Africa by contrast, where fencing is both philosophically and practically unacceptable, quite apart from issues around land tenure, zonation in accordance with international standards has not been possible. Nevertheless, various countries in East Africa have, or are in the process of, unilaterally establishing disease-free zones to facilitate regional and international trade in livestock commodities. It remains to be seen whether these attempts will be successful but the probability of success is low because the approaches espoused do not generally fulfil internationally accepted criteria. The basic problem is the difficulty of controlling livestock and wildlife movement into the 'export zone'. This means that efforts by the World Trade Organization (WTO), through the Doha round of talks, to significantly lower tariff barriers and reduce agricultural subsidies in the developed world may not benefit countries with significant pastoral communities. This is because the constraints posed by non-tariff barriers, particularly in relation to animal diseases, will remain, irrespective of whether tariff barriers are softened or not.

It has become increasingly evident that, apart from zoning, there are other ways of rendering exported livestock commodities acceptably safe in respect of animal infections and their trade effects. One of these developments is the concept of 'compartmentalization' whereby the establishments involved in the chain of production, as well as conduits between them, can be effectively isolated from the surrounding environment where animals infected with TADs may be present. A second and potentially more useful development is provided by the fact that some commodities are inherently incapable of transmitting particular infectious agents. For example, cow's milk does not contain infectious quantities of the prion that causes bovine spongiform encephalopathy (BSE or mad cow disease). Therefore, to prove that the area of production is free from BSE is pointless where export of milk and milk products are concerned. Processing of commodities derived from livestock (e.g. cooking, to use a simple example) can also be highly effective in ensuring that products derived from livestock are rendered safe for a whole range of infectious agents, including those important to human food safety. This approach has been referred to as *commodity-based trade*, whereby the specific commodity largely defines the risks that may or may not be associated with it (Thomson et al. 2004).

The increasing acceptance of concepts such as compartmentalization and commodity-based trade means that the position of pastoralists in relation to international trade in livestock commodities is not as hopeless as it seemed once. However, precisely how these systems can be effectively applied in practice requires further consideration.

A further problem associated with producing ECF export is rapidly escalating standards required in respect of food safety. This is particularly so for livestock producers whose livestock is continually moving in search of new grazing or water. This means that the delivery of raw products such as cattle or milk to abattoirs and dairies is logistically complicated.

These international animal health and food safety requirements collectively fall under the WTO's Agreement on the Application of Sanitary and Phytosanitary Measures (SPS Agreement). In the case of animal health, the standards are determined by the Office international des épizooties (OIE), the world organization for animal health. International standards for food safety are set by the Codex Alimentarius Commission which is jointly administered by the World Health and the Food and Agriculture Organizations of the United Nations. Increasingly, however, large trading blocks (e.g. the European Union) and even multinational companies and supermarkets are influencing minimum standard determination, especially in respect of food safety (Thomson et al. 2004).

Likely future developments

Lowering of tariff barriers and reducing agricultural subsidies in favour of developing countries seem to be issues that need to be addressed, judging by ongoing reports in the media. Consequently, positive developments in this direction seem increasingly probable, despite the fact that the extent and speed of change are far less predictable. These developments, combined with more flexible approaches to combat the spread of dangerous animal infections derived from livestock, are likely to provide export opportunities that were unthinkable even a few years ago. The desired effect will hopefully be achieved, thereby addressing poverty effectively.

However, a consequence of such developments may be that wildlife in pastoral areas will be seen as detrimental to trade in livestock commodities because they greatly complicate the control of animal diseases and are therefore perceived as increasingly incompatible with the attainment of OIE standards and thus economic development. This trend will also likely be exacerbated by the complications that arise from zoonoses[2] because human food safety is rapidly becoming the most important issue in the trade of commodities derived from livestock. Examples from the developed world such as AI (H5 & H7 types) provide precedents for this.

The only way to ensure that this does not happen is to make certain that the returns derived from the presence of wildlife in pastoralist areas at least match those likely to increasingly accrue from trade in commodities derived from domestic species.

Future direction for interventions on health and disease at the interface on rangelands

Identifying and managing disease threats in rangeland communities clearly poses enormous scientific, practical and political challenges. This is primarily because most of the important diseases of concern to public health, livestock economies and wildlife conservation involve multiple host species (Cleaveland et al. 2001), often with complex infection dynamics. Ecological factors that affect landuse, together with changes in human and domestic animal demographics and movement patterns, have been identified as the major factors associated with human disease emergence (Woolhouse & Gowtage-Sequeria 2005). Similar ecological and anthropogenic factors also affect the emergence of livestock and wildlife diseases. With rapidly changing patterns of contact among wildlife, livestock and people, rangeland interface areas are set to become high-risk sites for pathogen transmission and disease emergence.

Unravelling the complex dynamics of multi-host pathogens and quantifying disease threats can be enormously problematic, as has been typified by studies into the epidemiology of BTB in badgers and cattle in the United Kingdom. These problems are particularly acute in sub-Saharan Africa, where detection of pathogens is notoriously difficult, especially in wildlife. Generating reliable case-surveillance data is hampered by the practical and logistic problems of finding, collecting and storing appropriate samples under field conditions, as well as a lack of species-specific diagnostic tests. Poor surveillance and misdiagnosis of zoonoses in the human population also represents a serious

---

[2] Animal infections transmissible from animals to humans.

problem. Although much attention has focused on high-profile emerging zoonoses, such as HPAI and SARS, most endemic zoonoses remain severely neglected. Not only does the lack of data limit our understanding of the epidemiology and impacts of zoonotic diseases but it also leads to the perception that zoonoses are insignificant, which results in further neglect.

Despite these difficulties, recent technical advances, together with the growing recognition of the need for integrated approaches to human and animal health (the 'one health' 'one medicine' and 'ecosystem health' approaches) (Zinsstag et al. 2005; Osofsky et al. 2005b) provide optimism that progress can be made by tackling many of the interface disease problems facing rangelands.

For example, new methods for detecting trypanosome infection include polymerase chain reaction (PCR)-based diagnostic methods that distinguish human and non-human infective forms of *Trypanosoma brucei* (the cause of human sleeping sickness and *nagana* in cattle) (Picozzi et al. 2005). This tool has led to the identification of cattle as the reservoir of infection in Uganda, and disease control measures have now been designed accordingly (Welburn et al. 2001). In contrast to the Ugandan situation, human-infective forms of *T. brucei* in the Serengeti are most frequently associated with wildlife (warthogs) (Kaare et al. 2007), suggesting that different risk factors and maintenance patterns may be operating in these wildlife areas that are more species-rich and, therefore, different approaches to control will be needed.

Identifying transmission pathways is important in understanding disease risks as well as in designing appropriately targeted disease control measures, but remains highly problematic for multi-host pathogens in interface systems. However, advances in analytical techniques provide powerful new tools that can generate important insights. For example, analysis of high-resolution genetic sequence data has the potential to provide valuable detailed information on transmission pathways and genetic diversity. Recent phylogenetic analyses of foot-and-mouth virus topotypes highlighted the genetic complexity of virus types in sub-Saharan Africa, the importance of molecular characterization to trace the origin of FMD outbreaks, as well as the need for immunologically relevant vaccines for disease control (Bastos et al. 2003). New statistical analyses of sequence data from Serengeti rabies isolates have been used to construct transmission networks that demonstrate the predominance of dog-to-wildlife patterns of transmission, consistent with the view that dogs are the reservoir of rabies in the area (Kaare et al. 2009). New analytical approaches have also been developed to analyse contact-tracing data and reconstruct the history of transmission events – for example, the 2001

FMD epidemic in the United Kingdom (Haydon et al. 2003) and the 2003–4 rabies outbreak in the Serengeti (K. Hampson, personal communication, 2007) – allowing control measures to be targeted most appropriately.

Where wildlife hosts have been identified as important sources and/or reservoirs of livestock and human diseases, the options for disease control in wildlife are limited. In addition to our lack of knowledge of infection dynamics in wild animal populations, there remain considerable technical, ethical, political and financial difficulties associated with disease control in wildlife. Where wildlife is the source of zoonoses, traditional control methods have often been used to eliminate the source of infection (e.g. through culling). The culling of palm civets by the Chinese authorities, following the identification of civets as potential sources of SARS infection, demonstrated how ingrained this response has become. However, we now have a much better understanding of the ecological consequences of culling wild populations and recognize that, far from containing disease risks, this can often exacerbate disease spread. 'Perturbation effects', such as increased dispersal, heightened aggression and heightened susceptibility, have been documented in several instances, including in jackal populations in Zimbabwe where culling has been used to control rabies (McKenzie 1993) and in badgers in the United Kingdom that have been culled to control bovine TB (Donnelly et al. 2006; Woodroffe et al. 2006). In addition to being counter-productive for disease control, such approaches often meet strong public resistance and run counter to the need to maintain both local and global diversity. Greater understanding of the role of 'liaison' hosts, such as the livestock hosts that facilitate transmission of Nipah virus from bat reservoirs to humans (Breed et al. 2006), is clearly important to develop effective and ecologically robust approaches towards the control of wildlife zoonoses.

A further consequence of wildlife involvement in human diseases is the potential threat to the wildlife tourism industry. The economic damage, caused by a decline in visitors to countries suffering from SARS and Ebola virus, clearly highlights this problem. Equally clear from the SARS outbreak is the critical importance of early detection and reporting, with open exchange and dissemination of epidemiological data essential to allow rapid response measures that will reduce the scale of epidemics. Balancing these needs presents a dilemma for wildlife managers, particularly in countries dependent on wildlife tourism for economic development.

Although disease interventions have usually been adopted with the primary goal of reducing disease risks, interventions can additionally generate powerful information about disease reservoirs, transmission dynamics and cost-effectiveness of control strategies. The mass vaccination of domestic dogs surrounding the Serengeti National Park has thus been used not only to reduce disease threats for the human and wild carnivore populations but also to identify disease reservoirs and inform the design of national rabies control strategies (Kaare et al. 2009). For example, the reduction in demand for human post-exposure vaccination in the wake of mass dog vaccination has the potential to generate economic benefits to the public health sector that could help sustain control strategies.

Multidisciplinary approaches are clearly needed to tackle the problems of multi-host pathogens in rangelands, and international coordination is essential where disease risk is associated with unrestricted animal movements. The 'one health', 'one medicine' and 'ecosystem health' approaches have gone a long way towards enhancing collaborations between sectors. However, there is still a need for concurrent studies in human, domestic animal and wildlife populations – to improve surveillance and diagnosis, to understand spatial and temporal relationships between animal and human diseases and to identify risk factors for disease emergence. The rangelands interface provides important opportunities to pioneer these approaches.

References

Acevedo-Whitehouse, K., Gulland, F., Greig, D. & Amos, W. (2003) Inbreeding: disease susceptibility in California sea lions. *Nature* 422, 35.

Acevedo-Whitehouse, K., Vicente, J., Gortazar, C., Hofle, U., Fernandez-de-Mera, I.G. & Amos, W. (2005) Genetic resistance to bovine tuberculosis in the Iberian wild boar. *Molecular Ecology* 14, 3209–3217.

Aklilu, Y. (2002) *An Audit of the Livestock Marketing Status in Kenya, Ethiopia and Sudan*. Organization of African Unity/Interafrican Bureau for Animal Resources, Nairobi.

Alexandra, D.J. (2000) A review of avian influenza in different bird species. *Veterinary Microbiology* 73, 3–13.

Allison, A.C. (1982) Co–evolution between hosts and infectious disease agents and its effects on virulence in population biology of infectious diseases. In: Anderson, R.M. & May, R.M. (eds.) *Population Biology of Infectious Diseases: Dahlem Workshop Reports Life Sciences Research Report 25*. Springer-Verlag, New York, pp. 245–267.

Amanfu, W., Food and Agriculture Organization of the United Nations (FAO)/World Organisation for Animal Health (OIE)/Organisation of African Unity (OAU)/ International Atomic Energy Agency (IAEA) Consultative Group on Contagious

Bovine Pleuropneumonia. (CBPP) (2000) Contagious bovine pleuropneumonia surveillance: the experience of Botswana. *Report of the Second Meeting of the FAO/OIE/OAU/IAEA Consultative Group on Contagious Bovine Pleuropneumonia (CBPP); Reviving Progressive Control of CBPP in Africa*. Session 2, 24–26 Oct 2000. Food and Agriculture Organization of the United Nations, Rome, pp. 75–78.

Asian Development Bank (2005) *Avian Influenza and the Risk of an Influenza Pandemic*. Asian Development Bank, Manila.

Bastos, A.D.S., Haydon, D.R., Sangare, O., Boshoff, C.I., Edrich, J.L. & Thomson, G.R. (2003) The implications of virus diversity within the SAT 2 serotype for control of foot-and-mouth disease in sub-Saharan Africa. *Journal of General Virology* 84, 1595–1606.

Bengis, R.G. (1999) Tuberculosis in free-ranging mammals. In: Fowler, M.E. & Miller, R.E. (eds.) *Zoo and Wildlife Medicine: Current Therapy 4*. W.B. Saunders Company, Philadelphia, pp. 101–114.

Bengis, R.G. (2005). Transfrontier conservation area initiatives in sub-Saharan Africa: some animal health challenges. In: Osofsky, S.A. (ed.) Cleaveland, S., Karesh, W.B., Kock, M.D., Nyhus, P.J., Starr, L. & Yang, A. (assoc. eds.) *Conservation and Development Interventions at the Wildlife/Livestock Interface: Implications for Wildlife, Livestock and Human Health*. Occasional Paper of the IUCN Species Survival Commission No. 30. International Union for Conservation of Nature and Natural Resources (IUCN), Gland, pp. 15–19.

Bengis, R.G., Kock, R.A. & Fischer, J. (2002) Infectious animal diseases: the wildlife/livestock interface. *Revue Scientifique et Technique International des Epizooties* 21, 53–66.

Bengis, R.G., Kock, R.A., Thomson, G.R. & Bigalke, R.D. (2004a) Infectious animal diseases in sub-Saharan Africa: the wildlife/livestock interface. In: Coetzer, J.A.W. & Tustin, R.C. (eds.) *Infectious Diseases of Livestock*, 2nd edn. Oxford University Press Southern Africa, Cape Town, pp. 225–238.

Bengis, R.G., Leighton, F.A., Fischer, J.R., Artois, M., Mörner, T. & Tate, C.M. (2004b) The role of wildlife in emerging and re-emerging zoonoses. *Revue Scientifique et Technique International des Epizooties* 23, 497–511.

Breed, A,C., Field, H.E., Epstein, J.H. & Daszak, P. (2006) Emerging henipaviruses and flying foxes–conservation and management perspectives. *Biological Conservation* 131, 211–220.

Chivian, E. & Bernstein, A.S. (2004) Embedded in nature: human health and biodiversity. *International Journal of Infectious Disease* 8, 5–12.

Cleaveland, S., Appel, M.G.J., Chalmers, W.S.K., Chillingworth, C., Kaare, M. & Dye, C. (2000) Serological and demographic evidence for domestic dogs as a source of canine distemper virus infection for Serengeti wildlife. *Veterinary Microbiology* 72, 217–227.

Cleaveland, S., Hess, G., Dobson, A.P. et al. (2002) The role of pathogens in biological conservation. In: Hudson, P.J., Rizzoli, A., Grenfell, B.T., Heesterbeck, H. & Dobson, A.P. (eds.) *Ecology of Wildlife Diseases.* Oxford University Press, Oxford, pp. 139–150.

Cleaveland, S., Laurenson, M.K. & Taylor, L.H. (2001) Diseases of humans and their domestic mammals: pathogen characteristics, host range and the risk of emergence. *Philosophical Transactions of the Royal Society of London Series B, Biological Sciences* 356, 991–999.

Cleaveland, S., Mlengeya, T., Kazwala, R.R. et al. (2005) Tuberculosis in Tanzanian wildlife. *Journal of Wildlife Diseases* 41, 446–453.

Cook, A., Jardine, A. & Weinstein, P. (2004) Using human disease outbreaks as a guide to ecosystem level interventions. *Environmental Health Perspectives* 112, 1143–1146

Corbett, E.L., Watt, C.J., Walker, N. et al. (2003). The growing burden of tuberculosis: global trends and interactions with the HIV epidemic. *Archives of Internal Medicine* 163, 1009–1021.

Department For International Development (DFID-UK) (2002) *Wildlife and Poverty Study.* Livestock and Wildlife Advisory Group, Rural Livelihoods Department, London.

Díaz, S., Fargione, J., Chapin, F.S. & Tilman, D. (2006) Biodiversity loss threatens human well-being. *PLoS Biology* 4(8), e277, http://biology.plosjournals.org/perlserv/?request=get–document&doi=10.1371%2Fjournal.pbio.0040277.

Dixon, B. (2005) Leaping to extinction. *Lancet Infectious Diseases* 5, 738.

Donnelly, C.A., Woodroffe, R., Cox, D.R. et al. (2006) Positive and negative effects of widespread badger culling on tuberculosis in cattle. *Nature* 439, 843–846.

Dye, C., Scheele, S., Dolin, P., Pathania, V. & Raviglione, M.C. (1999) Global burden of tuberculosis: estimated incidence, prevalence and mortality by country. WHO Global Surveillance and Monitoring Project. *JAMA: The Journal of the American Medical Association* 282, 677–686.

Fèvre, E.M., Picozzi, K., Fyfe, J. et al. (2005) A burgeoning epidemic of sleeping sickness in Uganda. *Lancet* 366, 745–747.

Gascoyne, S.C., King, A.A., Laurenson, M.K., Borner, M., Schildger, B. & Barrat, J. (1993) Aspects of rabies infection and control in the conservation of the African wild dog (*Lycaon pictus*) in the Serengeti Region, Tanzania. *Onderstepoort Journal of Veterinary Research* 60, 415–420.

Hahn, B.H., Shaw, G.M., De Cock, K.M. & Sharp, P.M. (2000) AIDS as a zoonosis: scientific and public health implications. *Science* 287, 607–614.

Harvell, C.D., Mitchell, C.E., Ward, J.R. et al. (2002) Climate warming and disease risks for terrestrial and marine biota. *Science* 296, 2158–2162.

Haydon, D.T., Chase-Topping, M., Shaw, D.J. et al. (2003) The construction and analysis of epidemic trees with reference to the 2001 U.K. foot-and-mouth outbreak. *Proceedings of the Royal Society of London Series B, Biological Sciences* 270, 121–127.

Haydon, D.T., Randall, D.A., Matthews, L. et al. (2006) Low-coverage vaccination strategies for the conservation of endangered species. *Nature* 443, 692–695.

Hedger, R.S. (1972) Foot and mouth disease and the African buffalo (*Syncerus caffer*). *Journal of Comparative Pathology* 82, 19–28.

Hermoso de Mendoza, J., Parra, A., Tato, A. et al. (2006) Bovine tuberculosis in wild boar *Sus scrofa*, red deer *Cervus elaphus* and cattle *Bos Taurus*, in a Mediterranean ecosystem 1992–2004. *Preventive Veterinary Medicine* 74, 239–247.

Hugh-Jones, M.E. & de Vos, V. (2002) Anthrax and wildlife. *Revue Scientifique et Technique International des Epizooties* 21, 359–383.

Jenkins, D.J. & MacPherson, C.N.L. (2003) Transmission ecology of Echinococcus in wildlife in Australia and Africa. *Parasitology* 127, S63–S72.

Kaare, M., Lembo, T., Hampson, K. et al. (2009) Rabies control in rural Africa: evaluating strategies for effective domestic dog vaccination. *Vaccine* 27, 152–160.

Kaare, M., Picozzi, K., Mlengeya, T. et al. (2007) Sleeping sickness–a re-emerging disease in the Serengeti? *Travel Medicine and Infectious Disease* 5, 117–124.

Keet, D.F., Kriek, N.P.J., Penrith, M-L., Michel, A. & Huchzermeyer, H.F.A.K. (1996) Tuberculosis in buffaloes (*Syncerus caffer*) in the Kruger National Park: spread of the disease to other species. *Onderstepoort Journal of Veterinary Research* 63, 239–244.

Kilpatrick, A.M., Chmura, A.A., Gibbons, D.W., Fleischer, R.C., Marra, P.P. & Daszak, P. (2006) Predicting the global spread of H5N1 avian influenza. *Proceedings of the National Academy of Sciences of the United States of America* 103, 19368–19373.

Knobel, D.L., Liebenberg, A. & du Toit, J.T. (2003) Seroconversion in captive African wild dogs (*Lycaon pictus*) following administration of a chicken head bait/SAG-2 oral rabies vaccine combination. *Onderstepoort Journal of Veterinary Research* 70, 73–77.

Kock, R.A. (2005) What is this infamous "Wildlife/Livestock Disease Interface?" A review of current knowledge for the African continent. In: Osofsky, S.A. (ed.) Cleaveland, S., Karesh, W.B., Kock, M.D., Nyhus, P.J., Starr, L. & Yang, A. (assoc. eds.) *Conservation and Development Interventions at the Wildlife/Livestock Interface: Implications for Wildlife, Livestock and Human Health*. Occasional Paper of the IUCN Species Survival Commission No. 30. International Union for Conservation of Nature and Natural Resources (IUCN), Gland, pp. 1–13.

Kock, R.A. (2006) Rinderpest and wildlife. In: Barrett, T., Pastoret, P.-P. & Taylor, W. (eds.) *Rinderpest and Peste des Petits Ruminants: Virus Plagues of Large and Small Ruminants (Biology of Animal Infections)*. Academic Press, London, pp. 143–162.

Kock, R.A. (2008) *The Role of Wildlife in the Epidemiology of Rinderpest in East and Central Africa 1994–2004: A Study Based on Serological Surveillance and Disease Investigation*. Thesis for the Degree of Doctor of Veterinary Medicine, University of Cambridge, England.

Kock, M.D., Mullins, G.R. & Perkins, J.S. (2002) Wildlife health, ecosystems and rural livelihoods in Botswana. In: Aguirre, A.A., Ostfeld, R.S., Tabor, G.M., House, C.

& Pearl, M.C. (eds.) *Conservation Medicine: Ecological Health in Practice*. Oxford University Press, New York, pp. 265–275.

Krebs, J., Anderson, R.M., Clutton-Brock, T., Donnelly, C.A., Morrison, W.I. & Young, D. (1997) *Bovine Tuberculosis in Badgers and Cattle*. MAFF Publications, London.

Li, T.Y., Qiu, J.M., Yang, W. et al. (2005) Echinococcosis in Tibetan populations, western Sichuan Province, China. *Emerging Infectious Diseases* 11, 1866–1873.

Lindeque, P.M. & Turnbull, P.C.B. (1994) Ecology and epidemiology of anthrax in the Etosha National Park, Namibia. *Onderstepoort Journal of Veterinary Research* 61, 71–83.

MacPherson, C.N.L., Spoerry, A., Zeyhle, E., Romig, T. & Gorfe, M. (1989) Pastoralists and hydatid-disease–an ultrasound scanning prevalence survey in east-Africa. *Transactions of the Royal Society of Tropical Medicine and Hygiene* 83, 243–247.

Martin, R.B. (2005) The Influence of veterinary control fences on certain wild large mammal species in the Caprivi, Namibia. In: Osofsky, S.A., Cleaveland, S., Karesh, W.B., Kock, M.D., Nyhus, P.J., Starr, L. and Yang, A. (eds.) *Conservation and Development Interventions at the Wildlife/Livestock Interface: Implications for Wildlife, Livestock and Human Health*. IUCN, Gland, Switzerland and Cambridge, United Kingdom. pp. 27–39.

Maudlin, I. (2006) African trypanosomiasis. *Annals of Tropical Medicine and Parasitology* 100, 679–701.

McKenzie, A.A. (1993) Biology of the black-backed jackal (*Canis mesomelas*) with reference to rabies. *Onderstepoort Journal of Veterinary Research* 60, 367–371.

Michel, A.L. (2005) Tuberculosis–what makes it a significant player at the wildlife/livestock/human interface? In: Osofsky, S.A. (ed.) Cleaveland, S., Karesh, W.B., Kock, M.D., Nyhus, P.J., Starr, L. & Yang, A. (assoc. eds.) *Conservation and Development Interventions at the Wildlife/Livestock Interface: Implications for Wildlife, Livestock and Human Health*. Occasional Paper of the IUCN Species Survival Commission No. 30. International Union for Conservation of Nature and Natural Resources (IUCN), Gland, pp. 47–50.

Michel, A.L., Bengis, R.G., Keet, D.F. et al. (2006) Wildlife tuberculosis in wildlife conservation areas: implications and challenges. *Veterinary Microbiology* 112, 91–100.

Morgan, E.R., Lundevold, M., Medley, G.F., Shaikenov, B.S., Torgerson, P.R. & Milner-Gulland, E.J. (2006) Assessing risks of disease transmission between wildlife and livestock: the saiga antelope as a case study. *Biological Conservation* 131, 244–254.

Nel, L.H., Thomson, G.R. & von Teichman, B.F. (1993) Molecular epidemiology of rabies virus in South Africa. *Onderstepoort Journal of Veterinary Research* 60, 301–306.

O'Neil, B.D. & Pharo, H.J. (1995) The control of bovine tuberculosis in New Zealand. *New Zealand Veterinary Journal* 43, 249–255.

Osofsky, S.A. (ed.) Cleaveland, S., Karesh, W.B., Kock, M.D., Nyhus, P.J., Starr, L. & Yang, A. (assoc. eds.) (2005a) *Conservation and Development Interventions at the*

Wildlife/Livestock Interface: Implications for Wildlife, Livestock and Human Health. Occasional Paper of the IUCN Species Survival Commission No. 30. International Union for Conservation of Nature and Natural Resources (IUCN), Gland.

Osofsky, S.A., Kock, R.A., Kock, M.D. et al. (2005b) Building support for protected areas using a 'one health' perspective. In: McNeely, J.A. (ed.) *Friends for Life: New Partners in Support of Protected Areas.* International Union for Conservation of Nature and Natural Resources, Gland, pp. 65–79.

Picozzi, K., Fèvre, E.M., Odiit, M. et al. (2005) Sleeping sickness in Uganda: a thin line between two fatal diseases. *British Medical Journal* 331, 1238–1241.

Polley, L. (2005) Navigating parasite webs and parasite flow: emerging and re-emerging parasitic zoonoses of wildlife origin. *International Journal of Parasitology* 35, 1279–1294.

Randall, D.A., Williams, S.D., Kuzmin, I.V. et al. (2004) Rabies in endangered Ethiopian wolves. *Emerging Infectious Diseases* 10, 2214–2217.

Rodwell, T.C., Whyte, I.J. & Boyce, W.M. (2001) Evaluation of population effects of bovine tuberculosis in free-ranging African buffalo (*Syncerus caffer*). *Journal of Mammology* 82, 231–238.

Roelke-Parker, M.E., Munson, L., Packer, C. et al. (1996) A canine distemper virus epidemic in Serengeti lions (*Panthera leo*). *Nature* 381, 172.

Rossiter, P.B. (2000) Morbilliviral diseases: rinderpest. In: Williams, E.S. & Barker, I.K. (eds.) *Infectious Diseases of Wild Mammals.* Iowa State University Press, Ames, pp. 37–45.

Rossiter, P.B. (2004) Rinderpest. In: Coetzer, J.A.W. & Tustin, R.C. (eds.) *Infectious Diseases of Livestock*, 2nd edn. Oxford University Press Southern Africa, Cape Town, pp. 629–659.

Schwartz, H.J. (1993) Pastoral production systems in the dry lowlands of eastern Africa. In: Baumann, M.P.O., Janzen, J. & Schwartz, H.J. *Pastoral Production in Central Somalia.* Deutsche Gesellschaft für Technische Zusammenarbeit (GTZ) GmbH, Eschborn, pp. 1–15.

Scott Wilson Group. (2000) *Environmental Assessment of Veterinary Fences in Ngamiland: Summary Report.* Scott Wilson Group plc, Environment and Development Group, London.

Sharma, S.P., Losho, T.C., Malau, M. et al. (2001) The resurgence of trypanosomosis in Botswana. *Journal of the South African Veterinary Association* 72, 232–234.

Shimshony, A., Orgad, U., Baharav, D. et al. (1986). Malignant foot-and-mouth disease in mountain gazelles. *Veterinary Record* 119, 175–176.

Smith, J.S., Yager, P.A. & Orciari, L.A. (1993) Rabies in wild and domestic carnivores of Africa: epidemiological and historical associations determined by limited sequence analysis. *Onderstepoort Journal of Veterinary Research* 60, 307–314.

von Teichman, B.F., Thomson, G.R., Meredith, C.D. & Nel, L.H. (1995) Molecular epidemiology of rabies virus in South Africa–evidence for 2 distinct virus groups. *Journal of General Virology* 76, 73–82.

Thomson, G.R. (1985) The epidemiology of African swine fever: the role of free-living hosts in Africa. *Onderstepoort Journal of Veterinary Research* 52, 201–209.

Thomson, G.R., Tambi, E.N., Hargreaves, S.K. et al. (2004). International trade in livestock and livestock products: the need for a commodity based approach. *Veterinary Record* 155, 429–433.

Thomson, G.R., Vosloo, W., Bastos, A.D.S. (2003) Foot and mouth disease in wildlife. *Virus Research* 91, 145–161.

Thornton, P.K., Kruska, R.L., Henninger, N. et al. (2003) *Mapping Livestock and Poverty in the Developing World.* International Livestock Research Institute, Nairobi.

Torr, S.J., Hargrove, J.W. & Vale, G.A. (2005) Towards a rational policy for dealing with tsetse. *Trends in Parasitology* 21, 537–541.

United Nations Environment Programme (UNEP) Convention on Migratory Species (CMS) (2008) UNEP/CMS/Resolution 9.8: responding to the challenge of emerging and re-emerging diseases in migratory species, including highly pathogenic avian influenza H5N1. 9th Meeting of the Conference of the Parties to the Convention on Migratory Species, 1–5 December 2008, Rome.

Vreysen, M.J.B. (2000) Glossina austeni (Diptera: Glossinidae) eradicated on the island of Unguja, Zanzibar, using the sterile insect technique. *Journal of Economic Entomology* 93, 123–135.

Welburn, S.C., Picozzi, K., Fèvre, E.M. et al. (2001) Identification of human-infective trypanosomes in animal reservoir of sleeping sickness in Uganda by means of serum-resistance-associated (SRA) gene. *Lancet* 358, 2017–2019.

Williams, E.S. & Miller, M.W. (2002) Chronic wasting disease in deer and elk in North America. *Revue Scientifique et Technique International des Epizooties* 21, 305–316.

Woodroffe, R., Cleaveland, S., Courtenay, O., Laurenson, M.K. & Artois, M. (2004) Infectious disease in the management and conservation of wild canids. In: Macdonald, D.W. & Sillero-Zubiri, C. (eds.) *The Biology and Conservation of Wild Canids.* Oxford University Press, Oxford, pp. 123–142.

Woodroffe, R., Donnelly, C.A., Jenkins, H.E. et al. (2006) Culling and cattle controls influence tuberculosis risk for badgers. *Proceedings of the National Academy of Sciences of the United States of America* 103, 14713–14717.

Woolhouse, M.E.J. & Gowtage-Sequeria, S. (2005) Host range and emerging and reemerging pathogens. *Emerging Infectious Diseases* 11, 1842–1847.

Zeier, M., Handermann, M., Bahr, U. et al. (2005) New ecological aspects of hantavirus infection: a change of a paradigm and a challenge of prevention–a review. *Virus Genes* 30, 157–180.

Zinsstag, J., Schelling, E., Wyss, K. & Mahamat, M.B. (2005) Potential of cooperation between human and animal health to strengthen health systems. *Lancet* 366, 2142–2145.

6

Contemporary Views of Human–Carnivore Conflicts on Wild Rangelands

Alexandra Zimmermann[1,2], *Nick Baker*[3], *Chloe Inskip*[1],
John D.C. Linnell[4], *Silvio Marchini*[2], *John Odden*[4],
Gregory Rasmussen[2] *and Adrian Treves*[5]

[1]North of England Zoological Society, Chester Zoo, UK
[2]Wildlife Conservation Research Unit, University of Oxford, UK
[3]University of Queensland, Brisbane, Australia
[4]Norwegian Institute for Nature Research, Norway
[5]Nelson Institute for Environmental Studies,
University of Wisconsin-Madison, WI, USA

Introduction

Conflicts between wildlife and people pose a challenge of increasing concern to conservation scientists (Woodroffe et al. 2005). Human–wildlife conflict arises when the behaviour of wild animal species poses a direct and recurring threat to the livelihood or safety of a community and, in response, persecution of the species ensues. Persecution – the persistent killing, chasing or other harassment of a species – in the context of human–wildlife conflict differs from hunting in that the hunter seeks a product (meat, trophy, sport), while in conflict the aim is to menace or eradicate the animal or species. Retaliation

Wild Rangelands: Conserving Wildlife While Maintaining Livestock in Semi-Arid Ecosystems,
1st edition. Edited by J.T. du Toit, R. Kock, and J.C. Deutsch.
© 2010 Blackwell Publishing

against wildlife occurs in numerous scenarios, including when commercial ranchers and subsistence pastoralists lose livestock to predation, farmers lose crops to wildlife, hunters and carnivores compete over game species or ordinary villagers live in fear of animal attacks, as can happen for predatory, territorial, or defensive reasons (Conover 2002). Populations of many species, especially large carnivores, have declined significantly as a result of such conflict.

Terrestrial, large-bodied carnivore families are particularly susceptible to conflict with people because these animals require the spaces and resources compromised by increasing human dominance on landscapes. Conflict often occurs where natural prey is depleted and livestock provide a good alternative food source, or where habitat is disturbed, such as on the edges of protected areas where the likelihood of animal–human encounters increases, creating population sinks (Woodroffe & Ginsberg 1998). Livestock management practices (e.g. level of protection, general condition of stock) also affect conflict (Hoogesteijn 2003; Ogada et al. 2003). An individual predator's condition, including health, age and territory, has also been correlated with depredation, although not proven to predispose carnivores to prey on livestock (cf. Rabinowitz 1986; Linnell et al. 1999; Miquelle et al. 1999; Wydeven et al. 2004).

The order Carnivora contains 11 families and around 260 species (Macdonald 2001). Human–wildlife conflicts tend to involve cats (Felidae), dogs (Canidae), bears (Ursidae) and hyaenas (Hyaenidae). It is not widely known or documented that conflicts typically involve smaller-bodied species such as raccoons (Procyonidae), weasels (Mustelidae), skunks (Mephitidae), civets (Viverridae) or mongooses (Herpestidae); the exception is the wolverine, a predator of sheep and semi-domestic reindeer in northern Eurasia. Even within the cat, dog, bear and hyaena families, certain species are more prone than others to conflicts with people. Likelihood of conflict appears to be a function of carnivores' body size and proximity to human-dominated landscapes.

Understanding the dynamics of a human–wildlife conflict is challenging in itself; resolving conflict is even more difficult. For local communities, the economic impacts of coexistence, as well as perceptions and attitudes, influence tolerance of carnivores (Woodroffe & Ginsberg 1998; Sillero-Zubiri & Laurenson 2001; Hussain 2003; Treves & Naughton-Treves 2005; Zimmermann et al. 2005). This plays a critical role in conservation, which requires positive attitudes and participation on the part of involved communities. Conflict resolution requires a balance of practical solutions, outreach and the best available information on both the ecology of the species concerned and the social psychology of the people affected.

Human-carnivore conflicts occur in many different habitats, but rangelands have always been particularly prone to conflicts, as this is often where carnivores and their prey, and people and their livestock occur together in the same space and are very likely to encounter one another. In some areas, natural prey species are reduced in numbers, in others domestic livestock are semi-free-ranging and poorly protected and thereby present a much easier catch for the carnivore than the quicker and more alert wild ungulate. In most cases it appears to be a combination of these factors that leads carnivores into depredation and resentful ranchers or pastoralists into retaliatory or preventative persecution. In some rangelands the predators are persecuted to the point of near-eradication, which presents ethical as well as ecological issues. Carnivores are fundamentally important top-down regulators of herbivore communities and so their removal from an ecosystem to which they are adapted has wide implications for ecosystem structure and function (Treves & Karanth 2003).

For rangelands already threatened by agricultural conversion, overgrazing, invasive species and more, human–wildlife conflicts add yet another dimension of complexity to conservation (in Foreword by Sinclair and Schaller). In this chapter, we aim to illustrate some of the diverse pressures, responses and dynamics shaping this issue. The six case studies we highlight from around the globe demonstrate the range of regions and socio-economic conditions in which conflicts occur.

Hunter versus predator: wolves in North America

Thirty-two years ago, the last of the grey wolves (*Canis lupus*) in the lower 48 states of the United States were declared endangered by the Federal Government. They had been eliminated from all but a tiny area around Lake Superior because they competed with humans for livestock and game. Today, the western Great Lakes region of the United States holds around 3500 wolves across human-transformed areas of Minnesota (MN), Wisconsin (WI) and Michigan (MI) (USFWS 2007). In 2007, the US government proposed removing federal endangered/threatened species protections for this grey wolf population. The scenario above is not entirely rosy. Returning the management of wolves to the range states raises concerns about renewed extirpation campaigns. For example, proposals to hunt wolves in Wisconsin are alarming wolf preservationists. Transfer of authority is more than a shift in institutional responsibility for wolf management; it signals a change in emphasis from restoration to coexistence. Yet wolves continue to attack domestic animals and the definition of coexistence by the many interest groups differ.

In the western Great Lakes region, wolf predation on domestic animals – mainly beef cattle and hunting dogs – kept pace with population growth through 2000 (Fritts et al. 1992; Treves et al. 2002). But such predation incidents may have begun to outpace wolf population increase in the past 7 years. Domestic animal depredation and other real and perceived conflicts generate opposition to wolf conservation, especially in rural areas with high proportions of livestock producers and hunters (Naughton-Treves et al. 2003; Treves et al. 2007). Opponents of wolf recovery may retaliate illicitly. More than half of Wisconsin's adult wolf mortality is caused by people, much of it, apparently, intentional (Wydeven et al. 2001). More commonly, opponents call for stricter control of suspected problem packs and for regulated public hunting of wolves.

Neither selective lethal control nor public hunting has a strong record of effectiveness in preventing future conflicts or eliminating culprits selectively (Treves & Naughton-Treves 2005). Wolf predation on livestock continues despite approximately 20 years of legal, lethal control in MN and of several years in WI and MI, as well as continuous illicit killing (Wydeven et al. 2001). Yet perceptions have changed about problem carnivores.

The traditional view that any carnivore will kill livestock if given the opportunity has given way in the face of scientific evidence (Treves et al. 2002; Wydeven et al. 2004). An alternative explanation is that problem individuals arise spontaneously, so large carnivore populations will always have problem individuals. Opinions differ about the genesis of such problem carnivores. Some maintain that exposure to livestock carcasses, garbage or other human food sources leads to problem behaviours (Andelt & Gipson 1979; Jorgensen 1979). Another conjecture is that non-culprit carnivores are being killed at rates too high to exert efficient selection against problem individuals, so we see no diminution of conflicts. A recent review found high error rates in lethal control operations on wolves, bears and coyotes (Treves & Naughton-Treves 2005). In sum, for various reasons, we should not expect a decline in livestock predation by wolves or any other large carnivore in the western Great

Lakes region until managers and stakeholders act in a systematic manner to understand and prevent factors that may underlie conflicts.

Even if livestock predation by wolves could be reduced to a minimum, hunters might still object, in force, to the presence of this competitor (Hammill 2005). Hunters may find less game in their habitual shooting grounds. Preventing a large carnivore from pursuing its wild prey is more complex than preventing predation on livestock (Treves & Naughton-Treves 2005). Hence interventions to reduce conflicts between hunters and wolves will take ingenuity and changes in human behaviour. Some groups address the issue through efforts at hunter education, while others advocate a change in status of wolves from protected species to game species. Numerous challenges lie ahead before wolves are subject to regulated public hunting, but in the Lake Superior Region with its strong, popular tradition of hunting, this intervention may gather momentum now that states have the authority.

Finally, when one discusses the future of wolf recovery, managers often lower their voices and allude to drastic changes if a wild wolf attacks a human. Other regions' experiences warn us to prepare for such an attack (Rajpurohit 1998; Linnell & Bjerke 2002). It could precipitate a dramatic change in wolf management policy. Swift justice would be demanded and the response might not be limited to the suspected culprit or its pack. Pre-empting disproportional retaliation should be an explicit management objective. In addition to a better understanding of the genesis and behaviour of problem carnivores, we need greater effort and support for social scientific research aimed at understanding tolerance, anthropogenic mortality and the events that precipitate political backlashes against large carnivores.

Developing a risk assessment protocol: wild dogs in Africa

Colonialism in Africa ensured that the traditionally negative European perception of the wolf was transferred to the African wild dog (*Lycaon pictus*) wherever agriculture was established. Wild dogs were declared vermin and a bounty put on their heads in both ranchlands and national parks (Bere 1956; Childes 1988). Justification for this was self-defence, preservation of game populations and protection of domestic stock, though rarely was the extent of damage verified or quantified. In Zimbabwe, ranchers and government agencies systematically persecuted wild dogs from 1904 to 1988, by which time populations of wild dogs were at risk of being extirpated. The perception persisted, however, of wild dogs as voracious cattle- and game-killers partly based on rare occurrences of predation on livestock. In the early 1990s, a spate of incidents in the cattle-ranching region of Nyamandlovu, Zimbabwe revived the conflict between humans and wild dogs, but presented the opportunity to develop a better understanding of the real rather than the perceived threat and a new approach for alleviating human–wild dog conflict. The primary objective in developing a conflict protocol was ensuring a viable number of dogs for maintaining genetic diversity through natural dispersal (Fuller et al. 1992) and active meta-population management (Mills et al. 1998).

A critical first step in reducing negative perceptions of wild dogs was discerning and publicizing the actual extent to which they posed a threat to livestock. To accomplish this, the researchers aimed to determine the accuracy of claims of wild dogs killing cattle in the Nyamandlovu region. Participatory methods were used instead of postal questionnaires to reduce bias in the responses from ranchers (Rasmussen 1999). Interviews also establish personal contact with people affected by stock depredation and demonstrate the willingness of conservation practitioners to listen, which is an important step in ameliorating conflict. Following on from this ground work, there was a three-part risk assessment phase. First, the risk of a pack being extirpated was estimated from criteria such as residence status, home-range configuration, denning status, habituation to humans and likelihood of being shot by livestock managers. Second, the actual risk to livestock resulting from wild dog predation was evaluated from records of actual losses, as related to the method of livestock management and the presence of alternative (natural) prey. Availability of natural prey obviously reduces predation on livestock (Rasmussen 1999; Fritts et al. 2003). Third, and arguably most important, was the assessment of attitudes of the wider public for comparison against those of the local ranchers. Public pressure relating to the acceptability or not of killing predators can influence ranchers' attitudes.

This risk assessment phase led to the first action of circulating findings locally and nationally through newsletters, press and other media, to ensure transparency and rationality in the wild dog conservation campaign. This was effective in changing public attitudes and even generated support from ranchers, as confirmed by a reduction in the number of wild dogs being shot.

The next phase in the conflict management strategy was to determine if wild dogs were resident on a ranch, as evidenced by the location of breeding dens. If a pack was not resident, then no effective action could be taken apart from maintaining contact and relations with the land owners. If a pack was resident, then the dogs were particularly vulnerable to retaliatory killing by ranchers because the den location would become common knowledge among ranchers suffering actual or perceived livestock depredation.

Three options arise for dealing with a resident wild dog pack on ranchland: (i) do nothing, perhaps allowing the dogs to raise their pups and disperse but probably resulting in the pack being persecuted; (ii) humane culling, although this has ethical and legal implications and can undermine public support for efforts to conserve an endangered species; (iii) translocation, if there are funds for the operation and a suitable release site. In the Nyamandlovu case, translocation was the selected option and indeed there were two highly effective wild dog translocation operations in Zimbabwe, with successful breeding at the release sites followed by pack fission and long-distance dispersal (Hartwig & Rasmussen, 1999), which is essential for gene flow.

The project eased relations among ranch owners, local people and the conservation practitioners involved, generating real appreciation of attempts to find effective solutions to both livestock depredation and endangered species conservation. As a result, when new wild dogs arrived as expected to fill the vacuum caused by the translocation, ranchers tolerated the newcomers for 3 years until management options had to be revisited. There is no perfect solution when an endangered species kill livestock, and dedicated conservation practitioners are continually required at the human–wildlife interface. Nevertheless, the Zimbabwe wild dog project has provided a model for managing conflicts involving endangered carnivores in a mixed wildlife–livestock landscape.

Protecting a 'pest': dingos in Australia

The dingo (*Canis lupus dingo*) is the largest terrestrial predator in Australia and has managed to radiate into virtually all habitats across the country since its introduction from Asia around 4000–6000 years ago (Corbett 1995; Savolainen et al. 2004). It is a major predator of livestock (the primary cause of conflict with humans), causing millions of dollars of damage each year, as well as the costs associated with its control (Allen & Sparkes 2001). This includes the maintenance of one of the world's longest man-made structures, the dingo barrier fence, which is over 5000 km in length and comprises poison baiting, trapping and shooting programmes. Dingoes also carry disease, predate on pets (Dickman & Lunney 2001; Fleming et al. 2001) and, in some isolated circumstances, have attacked humans, with two instances of human death recorded.

The management of dingoes in Australia is a complex and expensive exercise and is complicated by the fact that the species is at the same time both a declared pest (legally requiring it to be controlled) and a protected native species in conservation lands (Fleming et al. 2006). Public opinion about the species is similarly divided. The situation is further complicated by the recent listing of dingoes on the International Union for Conservation of Nature and Natural Resources (IUCN) Red List as *Vulnerable*, with the main threatening processes being hybridization with domestic dogs and lethal control methods employed by humans (Johnson et al. 2007).

Dingoes are normally generalist in their approach to survival and are able to exploit most habitats and food sources. This greatly enhances the potential for conflict with humans as the number of ways in which they can interact also increases. Many 'traditional' dingo habitats have been inhabited by humans, providing competition for space and other resources. There is also a tendency for human settlement to increase the availability of some resources (such as food) for dingoes.

While human–dingo conflict is increasing dramatically in urban and semi-urban areas in Australia, large landowners in the rangelands are often the most affected (or at least, perceived to be) by these conflicts in terms of the personal, financial and psychological costs borne. Financial costs include lost livestock (mainly sheep), opportunity costs of reduced yield and control costs. Psychological costs often arise from the trauma of witnessing predation impacts upon stock and pets, as well as stress arising from financial strain. The dingo has been persecuted as a result, and in many cases, this persecution has led to an increase in the impacts that the control measures have tried to address. Increased poison baiting campaigns have, in many cases, led to an increase in predation rates on livestock. It is believed that dispersing individuals, not part of stable packs, tend to attack livestock more frequently, and the removal of local stable packs allows these dispersers to invade areas they would not normally be able to (Allen 2000; Allen & Sparkes 2001). A potential secondary effect of local dingo removal is an increase in the numbers of mesopredator and competing native herbivore (e.g. macropods) species, further impacting the livestock industry through loss of stock and reduced yield.

For the Australian livestock industry to remain viable, there needs to be a change in the way dingoes are viewed and the focus of control programmes.

To some extent, this is already happening with many abandoning the traditional sheep industry in favour of cattle, which are less likely to be impacted by dingoes. Many are also reducing baiting campaigns and using alternative non-lethal methods such as guard animals to reduce the impacts of dingoes on their stock. As the benefits of these alternative approaches become more apparent, their uptake is expected to increase. The question that remains to be answered is whether it is, as some have suggested (e.g. Daniels & Corbett 2003), too late to ensure the continued survival of the dingo as we know it today.

Culture and conflict: jaguars in South America

Similar to the dingo in Australia, the jaguar (*Panthera onca*) is the largest terrestrial predator of Central and South America. Renowned for its power, the jaguar has always been feared and hunted, yet also admired for its exceptional strength, elegance and beauty, which explains the strong cultural significance of the species. However, an additional dimension evolved in the relationship between humans and jaguars when domestic livestock was introduced to South America (Arnold 1968) and jaguar predation on cattle resulted in hostility of ranchers toward jaguars.

Attacks by jaguars on humans are extremely rare, but jaguar predation on cattle has long been documented in the Brazilian Pantanal (Roosevelt 1914) and since the late 1970s, scientific studies conducted in that region, and later in the Venezuelan Llanos and other sites, have contributed to our understanding of jaguar feeding ecology and the relationship of jaguars to livestock in the rangelands of South America (Mondolfi & Hoogesteijn 1986; Hoogesteijn & Mondolfi 1992; Crawshaw & Quigley 2002; Dalponte 2002). More than 85 species have been recorded in the jaguar's diet (Seymour 1989) from frogs to tapirs, but in some areas where cattle are ranched on prime jaguar habitat, such as the Pantanal and the Llanos, cattle can become the most frequent prey species for jaguars (Hoogesteijn & Mondolfi 1992; Crawshaw & Quigley 2002; Dalponte 2002).

Some factors increase the likelihood of cattle depredation, including scarcity of wild prey (Aranda 2002; Nunez et al. 2002; Polisar et al. 2003), persecution of jaguars, which can result in wounded individuals whose injuries make it difficult for them to capture wild prey (Rabinowitz 1986), and poor livestock husbandry practices (Schaller & Crawshaw 1980; Schaller 1983). In the Pantanal and Llanos cattle are typically left unattended, often near or even inside forested areas, which makes them more vulnerable to predation (Crawshaw & Quigley 2002; Hoogesteijn 2003). This leads to livestock losses that are unrelated to jaguars, but often blamed on them. Indeed, studies have shown that in the large ranches of the Pantanal and the Llanos predation by jaguars is a minor cause of cattle mortality compared to other causes, such as disease, abortion, malnutrition, attacks by vultures on newborn calves and puma depredation (Schaller 1983; Hoogesteijn et al. 1993). The relative economic damage associated with jaguar predation also varies. However, livestock losses to felids are generally low and less than 1–3% of total stock per year (Jackson et al. 1994; Farrell 1999), reaching 6% in the worst cases (Hoogesteijn et al. 1993).

These findings have led to a number of recommendations for resolving the conflict by attempting to decrease and compensate for economic damage. Electric fences (Saenz & Carillo 2002; Schiaffino et al. 2002; Scognamillo et al. 2002), translocation of 'problem jaguars' (Rabinowitz 1986; Linnell et al. 1997; Vaughan & Temple 2002), the introduction of water buffaloes (Hoogesteijn 2003) and improved cattle husbandry and management practices (Weber & Rabinowitz 1996; Crawshaw & Quigley 2002; Hoogesteijn et al. 2002) are examples of interventions. Compensation payments for the cattle killed by jaguars (Nowell & Jackson 1996; Hoogesteijn et al. 2002; Perovic 2002; Vaughan & Temple 2002; Conforti & de Azevedo 2003), the implementation of wildlife-based tourism (Weber & Rabinowitz 1996; Dalponte 2002; Miller 2002; Conforti & de Azevedo 2003) and trophy hunting of problem jaguars (Swank & Teer 1992) have also been tried or proposed. However, so far the evidence that any of these interventions can effectively reduce hostility towards jaguars is scarce and conservation efforts focused solely on the ecological and economic dimensions of human–jaguar conflicts have still not resulted in noticeable change throughout much of the jaguar's range.

Indeed, the above interventions will have little effect if jaguar persecution is found to be socially and culturally ingrained. Recent research on ranches and farms in Brazil suggest that rural attitudes to jaguars are not dictated by material loss caused by the animal (Conforti & de Azevedo 2003; Zimmermann et al. 2005) and negative attitudes to the species are also found among farmers who do not raise cattle (Marchini 2006). Therefore, factors other than the economics also determine negative attitudes, and the resulting hostility, to jaguars. Fear, prejudice, the social significance or simply the excitement of hunting a large predator, among other socio-cultural phenomena, also explain why people persecute jaguars. It is imperative then that we broaden the scope of our current approaches to resolve human–jaguar conflicts. Future research should turn to the human side of the conflict in order to (i) understand people's behaviour towards jaguars and identify the factors that determine hostility towards the species, (ii) develop education and communication interventions to engage landowners in the effort to resolve the conflict, improve their perceptions of the issue and increase their tolerance to jaguars, taking advantage of the exceptional socio-cultural significance of the species and (iii) compare approaches for human–jaguar conflict mitigation in a broader spectrum of ecological, economic, political, social and cultural circumstances, from large cattle ranches in the Pantanal and Llanos to small farms in other parts of the continent. Our coexistence with jaguars in the rangelands of South America depends upon the recognition that the origin – and therefore the resolution – of the conflicts between humans and jaguars lies on the human side of the conflict.

When people become prey: tigers in Asia

The tiger (*Panthera tigris*) has suffered a severe decline in its population distribution over the last decade (Biswas & Sankar 2002). It is estimated that between 3400 and 5140 individuals remain in the wild (Chundawat et al. 2008), inhabiting only 7% of the species' original range (Sanderson et al. 2006). Although now predominantly associated with tropical and temperate broadleaf forests, tigers also inhabit grassland and shrubland habitats, such as the Terai–Duar landscape, which includes the Chitwan and Royal Bardia National Parks (Nepal) and Jim Corbett National Park (India).

Much like the jaguar, the tiger is both feared and admired across its range and conflict with people has undoubtedly contributed to its population decline. Unlike its Latin American counterpart, however, tigers also occasionally attack people, a dimension that significantly complicates this case of human–carnivore conflict. In areas where attacks on people occur, fear inevitably exacerbates communities' animosity towards tigers. The resultant retaliatory killing of tigers is widespread in Asia and occurs even within protected areas. As most of Asia's protected areas are relatively small (Wikramanayake et al. 1999), tigers often range into adjacent, human-dominated habitats, where conflicts also arise. Human–tiger conflict is therefore often most severe in moderately disturbed habitat areas, such as buffer zones around protected areas (Nyhus & Tilson 2004a,b).

The scale of human–tiger conflict in rangelands, in particular, the number of attacks on humans per location, varies. Attacks on humans have been reported in both the area surrounding Chitwan and Royal Bardia National Parks where, in 1998–9, 11 deaths were reported, marking a sudden departure from the Park's prior average of 1.3 attacks per year (McDougal 1999). Attack rates also vary throughout the rest of the tigers' range. In areas of low human and tiger density, attacks on people are rare (Nowell & Jackson 1996). In the Russian Far East, for example, Miquelle et al. (1999) reported only six unprovoked attacks on humans since 1970. Conversely, in India's Dudhwa National Park, over 200 people were killed between 1978 and 1996 (Nowell & Jackson 1996), and in Sumatra over 170 people were killed or injured in tiger attacks over a 20-year period (Nyhus & Tilson 2004a). The Sundarbans of India and Bangladesh are most notorious for 'man-eating' tigers, with an official figure of an average of 24 (and an unofficial estimate of approximately 100) people killed per year (Reza et al. 2002). Attacks are generally attributed to people entering reserves to harvest natural resources (Nowell & Jackson 1996; Reza et al. 2000; Mukherjee 2003), cultivation of land around reserves (Nowell & Jackson 1996) or to hunters wounding tigers (Miquelle et al. 1999).

More often than attacking people, tigers kill livestock and a number of studies have quantified these predation losses. For example, livestock constitutes 0.45–12% of tigers' diets in four Indian protected areas (Biswas & Sankar 2002; Bagchi et al. 2003; Reddy et al. 2004; Andheria et al. 2007). In the Russian Far East, it has been estimated that up to 100 heads of livestock are killed each year by tigers (Miquelle et al. 2005), while in Sumatra at least 870 head of livestock were reportedly killed over a 20-year period (Nyhus & Tilson 2004a) and across Indochina livestock losses to tigers are common (Johnson et al. 2006). Livestock depredation tends to be particularly severe in those areas where tigers' wild prey base has been significantly reduced (Johnson et al. 2006; TIGRIS Foundation 2007).

Various methods have been implemented or suggested to reduce human–tiger conflicts, including compensation schemes (Madhusudan 2003; TIGRIS Foundation 2007), changes to cattle husbandry practices (Johnson et al. 2006), village eco-development (Bagchi et al. 2003) and the use of face masks worn on the back of the head (Nowell & Jackson 1996). As with jaguars, few quantitative data are available on the effectiveness of such mitigation techniques, although observations have been recorded. While in the Sundarbans no attacks on people wearing face masks have been reported

(Mukherjee 2003), compensation schemes have generally been found to be inefficient and ineffective (Madhusudan 2003).

More detailed and up-to-date data describing the frequency, distribution and determinants of attacks on humans by tigers in rangelands and in other habitat types are urgently required if we are to understand the dynamics of human–tiger conflicts, range wise. The extent of livestock depredation must also be examined extensively. Monitoring and evaluation of existing mitigation strategies is required to complement this research, as it is fundamental to the development of effective mitigation techniques. The applicability and feasibility of new mitigation and land management techniques must also be explored for reducing human–tiger conflict. For example, the potential of community-based work and alternative livelihood schemes to reduce the number of people entering reserves should be investigated, as should the effectiveness of economic incentive and insurance schemes in comparison to existing compensation schemes. Buffer zones, in particular, are associated with a high probability of conflict, yet, if managed specifically with this in mind, they can be used to reduce human–tiger conflict in a wider area, and the effectiveness of multiple land use practices in reducing conflict must therefore also be investigated (Nyhus & Tilson 2004b). Habitat corridors and meta-population management of tiger habitats is becoming increasingly important as habitat becomes progressively fragmented (Wikramanayake et al. 2004), and the maintenance of suitable habitat and wild prey base must remain a priority for the conservation of tigers in human-dominated landscapes (Reza et al. 2000). Finally, understanding the human dimension of human–tiger conflict is essential.

Re-emerging conflict: lynx in Europe

The previous case studies highlight that although conflicts between large carnivores and people and their livestock are a global phenomenon, the extent of the conflict varies greatly. In this final case study, we discuss what is possibly the greatest depredation conflict that occurs in the rangelands of Norway. Norway is a rugged country with the lowest human population density in Europe and where less than 5% of the land area is suitable for cultivation. The rest of the area consists of boreal forest (almost entirely used for intensive commercial forestry) and alpine tundra above the tree line.

With so little farm land it has been the tradition for centuries to graze livestock in the forests and alpine tundra habitats. In the past, grazing animals were accompanied by shepherds to protect them from predators. However, by the early twentieth century, large carnivore populations had been so reduced (wolves were actually exterminated) that it became possible to adopt a form of free-grazing where livestock (mainly sheep) were released in early summer and gathered in autumn. The animals were released into unfenced forest and alpine tundra habitat and only received occasional supervision. In the absence of large carnivores, losses for the approximate 4-month grazing season were remarkably low, in the region of 1–2% for ewes and 5% for lambs.

However, since the 1980s the populations of all four large carnivore species in Norway [brown bear (*Ursos arctos*), wolf, Eurasian lynx (*Lynx lynx*) and wolverine (*Gulo gulo*)] have begun to recover due to conservation-orientated legislation. This recovery has been associated with a dramatic increase in depredation losses of livestock to the large carnivores. Eurasian lynx are the most widespread of the four large carnivore species and are, together with wolverine, responsible for the greatest losses of livestock. Lynx have been the subject of intensive radio-telemetry based studies during the last decade (Scandlynx 2007). Between 1996 and 2005 lynx numbers have fluctuated between 300 and 500 individuals (the fluctuations are caused by hunter harvest) and compensation has been paid for between 6000 and 9000 sheep, mainly lambs killed each year. Not all losses are documented, but a combination of documented losses combined with a range of studies where sheep have been radio-collared to identify the cause of loss has confirmed that these numbers are realistic. This totals approximately 20 sheep per lynx per year. The results of an ongoing study in which lynx have been radio-collared and intensively monitored have confirmed as realistic the calculated kill rates. The project has also shown that virtually all lynx will kill sheep at some stage, but that adult males and juvenile lynx kill far more sheep than adult female lynx and that while the diet of lynx is dominated by wild ungulates, sheep do constitute a significant proportion of a summer diet (26%), which, given that unguarded livestock far outnumbered wild ungulates in our study area, was a surprisingly low proportion. In contrast to depredation by tiger and hunting dog, the data also indicated that the probability of a lynx killing sheep increased with the density of wild prey in the area. These results combined to build a picture of a conflict situation, where lynx are only incidentally killing livestock that they encounter while pursuing their wild prey. However, because unguarded sheep are found at high density throughout the natural habitats exploited by lynx and their preferred prey, lynx–sheep encounters occur frequently, resulting in depredation. The data further indicated that this is not due to specific problem individuals, but rather that all lynx at some stage kill livestock.

It is therefore not surprising that the only way lethal control reduces depredation is in circumstances where it reduces the lynx population. The modern Norwegian 'tradition', of placing free-ranging and unguarded sheep directly into forest habitats where large carnivores occur is a recipe for maximum conflict. Unfortunately, high labour costs in Norway make the restoration of the original shepherding tradition virtually impossible and even fencing sheep requires an expensive and radical restructuring of an industry that is already highly subsidized. The only alternatives for lynx are to accept losses or phase out sheep. However, in areas where other large carnivores such as bears, wolves and wolverines also occur, the combined depredation pressure on livestock holdings will probably force change. Comparative studies from France, Switzerland and Eastern Europe have shown that confining sheep to fenced fields or alpine pastures (out of the forest) dramatically reduces depredation losses per lynx and that depredation becomes more associated with specific individuals. Adopting mitigation measures such as electric fences, night-time enclosures or livestock-guarding dogs can basically reduce the problem to zero (Odden et al. 2002; Linnell & Brøseth 2003; Herfindal et al. 2005; Moa et al. 2006; Odden et al. 2006).

Conclusion

Our set of case studies illustrates that although human–carnivore conflicts occur in a great range of geographic and socio-economic contexts, on a global scale the challenges and experiences are remarkably similar. Compromises to the ecological needs of carnivores predispose them to livestock killing and the severity of each conflict is inextricably linked to the attitudes and beliefs of the communities affected. Perceptions of conflict are often formed by a few memorable or catastrophic events and reflect not only those events experienced first-hand, but historical events and often the stories from other people and communities (Naughton-Treves & Treves 2005; Treves 2007). As such, perceptions of conflict severity may not accurately reflect real conflict severity.

Similarly, the results of scientific research may not provide an entirely accurate picture. Conflict incidents tend to be randomly distributed within and between communities and the few individuals, households or communities that suffer the most devastating losses may be masked by the regional or community averages commonly used in scientific literature to describe losses. Scientific research is also rarely able to capture more extensive geographic and historic perspectives of a conflict situation due to restricted study areas and short time frames (Treves 2007). Consequently, when devising mitigation strategies, both the 'perceived' and the 'scientific' views of conflict must be considered, as when viewed together they provide a more comprehensive insight into conflict severity.

Globally, conflict mitigation techniques are plentiful, but efforts to systematically evaluate these are lacking and mechanisms for easy exchange of lessons learnt are only beginning to be established. To manage conflicts with greater success worldwide, we need to understand better the spatial and ecological dynamics of human–wild interfaces, focus on the importance of the human dimension of these conflicts, compare mitigation results and tailor mitigation approaches to the characteristics of individual cases. Although there are species-by-species differences among these conflicts, the core issues as well as the principles outlined in this chapter apply widely and are crucial to the successful management of human–wildlife conflicts on wild rangelands.

References

Allen, L. (2000) Measuring predator control effectiveness: reducing numbers may not reduce predator impact. In: Salmon, T.P. & Crabb, A.C. (eds.) *Proceedings of the 19th Vertebrate Pest Conference.* University of California Davis, San Diego, pp. 284–289.

Allen, L. & Sparkes, E.C. (2001) The effect of dingo control on sheep and beef cattle in Queensland. *Journal of Applied Ecology* 38, 76–87.

Andelt, W.F. & Gipson, P.S. (1979) Domestic turkey losses to radio-tagged coyotes. *Journal of Wildlife Management* 43, 673–679.

Andheria, A.P., Karanth, K.U. & Kumar, N.S. (2007) Diet and prey profiles of three sympatric large carnivores in Bandipur Tiger Reserve, India. *Journal of Zoology* 273, 169–175.

Aranda, M. (2002) Importancia de los pecaries para la conservacion del jaguar en Mexico. In: Medellin, R.A., Equihua, C., Chetkiewicz, C.L.B. et al. (eds.) *El Jaguar en el Nuevo Milenio.* Fondo de Cultura Económica USA, San Diego, pp. 101–107.

Arnold, O. (1968) *The Story of Cattle Ranching.* Harvey House Inc., Irving-on-Hudson.

Bagchi, S., Goyal, S.P. & Sankar, K. (2003) Prey abundance and prey selection by tigers (*Panthera tigris*) in a semi-arid, dry deciduous forest in western India. *Journal of the Zoological Society of London* 260, 285–290.

Bere, R.M. (1956) The African wild dog. *Oryx* 3, 180–182.

Biswas, S. & Sankar, K. (2002) Prey abundance and food habit of tigers (*Panthera tigris tigris*) in Pench National Park, Madhya Pradesh, India. *Journal of the Zoological Society of London* 256, 411–420.

Childes, S.L. (1988) The past history, present status and distribution of the hunting dog *Lycaon pictus* in Zimbabwe. *Biological Conservation* 44, 301–316.

Chundawat, R.S., Habib, B., Karanth, U. et al. International Union for Conservation of Nature and Natural Resources (IUCN) (2008) Panthera tigris. 2008 IUCN Red List of Threatened Species. www.iucnredlist.org. (accessed May 2, 2009).

Conforti, V.A. & de Azevedo, F.C.C. (2003) Local perceptions of jaguars (*Panthera onca*) and pumas (*Puma concolor*) in the Iguacu National Park, Southern Brazil. *Biological Conservation* 111, 215–221.

Conover, M. (2002) *Resolving Human-Wildlife Conflicts: The Science of Wildlife Damage Management*. Lewis Publishers, New York.

Corbett, L.K. (1995) *The Dingo in Australia and Asia*. University of New South Wales Press, Sydney.

Crawshaw, P.G. Jr. & Quigley, H.B. (2002) Habitos alimentarios del jaguar y el puma en el Pantanal, Brasil, con implicaciones para su manejo y conservacion. In: Medellin, R.A., Equihua, C., Chetkiewicz, C.L.B. et al. (eds.) *El Jaguar en el Nuevo Milenio*. Fondo de Cultura Económica USA, San Diego, pp. 223–236.

Dalponte, J.C. (2002) Dieta del jaguar y depredacion de ganado en el norte del Pantanal, Brasil. In: Medellin, R.A., Equihua, C., Chetkiewicz, C.L.B. et al. (eds.) *El Jaguar en el Nuevo Milenio*. Fondo de Cultura Económica USA, San Diego, pp. 209–222.

Daniels, M.J. & Corbett, L.K. (2003) Redefining introgressed protected mammals: when is a wildcat a wild cat and a dingo a wild dog? *Wildlife Research* 30, 213–218.

Dickman, C.R. & Lunney, D. (eds.) (2001) *A Symposium on the Dingo*. Royal Zoological Society of New South Wales, Mosman.

Farrell, L.E. (1999) *The Ecology of the Puma and the Jaguar in the Venezuelan Llanos*. Master of Science Thesis, University of Florida, FL.

Fleming, P.J.S., Allen, L.R., Lapidge, S.J., Robley, A., Saunders, G.R. & Thomson, P.C. (2006) A strategic approach to mitigating the impacts of wild canids: proposed activities of the Invasive Animals Cooperative Research Centre. *Australian Journal of Experimental Agriculture* 46, 753–762.

Fleming, P.J.S., Corbett, L.K., Harden, R. & Thomson, P.C. (2001) *Managing the Impacts of Dingoes and Other Wild Dogs*. Bureau of Rural Sciences, Canberra.

Fritts, S.H., Paul, W.J., Mech, L.D. & Scott, D.P. (1992) *Trends and Management of Wolf-Livestock Conflicts in Minnesota*, Resource Publication 181. United States Fish and Wildlife Service, Washington, D.C.

Fritts, S.H., Stephenson, R.O., Hayes, R.H. & Boitani, L. (2003) Wolves and humans. In: Mech, D.L. & Boitani, L. (eds.) *Wolves: Behavior, Ecology, and Conservation*. University of Chicago Press, Chicago, pp. 289–316.

Fuller, T.K., Mills, M.G.L., Borner, M., Laurenson, M.K. & Kat, P.W. (1992) Long distance dispersal by African wild dogs in East and South Africa. *Journal of African Zoology* 106, 535–537.

Hammill, J. (2005) *Is a Hunter Backlash Likely?* The Midwest Wolf Stewards Meeting, Hinckley.

Hartwig, S. & Rasmussen, G.S.A. (1999) Observations on the integration of a captive-raised African wild dog (*Lycaon pictus*) into a wild-caught pack and their adaptation to the wilderness. *Zoologische Garten* 69, 324–334.

Herfindal, I., Linnell, J.D.C., Moa, P.F., Odden, J., Austmo, L.B. & Andersen, R. (2005) Does recreational hunting of lynx reduce depredation losses of domestic sheep? *Journal of Wildlife Management* 69, 1034–1042.

Hoogesteijn, R. (2003) *Manual on the Problem of Depredation Caused by Jaguars and Pumas on Cattle Ranches*. Wildlife Conservation Society, New York.

Hoogesteijn, R., Boede, E.O. & Mondolfi, E. (2002) Observaciones de la depredacion de bovinos por jaguares en Venezuela. In: Medellin, R.A., Equihua, C., Chetkiewicz, C.L.B. et al. (eds.) *El Jaguar en el Nuevo Milenio*. Fondo de Cultura Económica USA, San Diego, pp. 183–198.

Hoogesteijn, R., Hoogesteijn, A. & Mondolfi, E. (1993) Jaguar predation and conservation: cattle mortality caused by felines on three ranches in the Venezuelan Llanos. In: Dunstone, N. & Gorman, M.L. (eds.) *Mammals as Predators*. Symposia of the Zoological Society of London. Oxford University Press, Oxford, pp. 391–407.

Hoogesteijn, R. & Mondolfi, E. (1992) *El Jaguar, Tigre Americano*. Ediciones Armitano, Caracas.

Hussain, S. (2003) The status of the snow leopard in Pakistan and its conflict with local farmers. *Oryx* 37, 26–33.

Jackson, R., Wang, Z.Y., Lu, X.D. & Chen, Y. (1994) Snow leopards in the Qomolangma nature reserve of the Tibet autonomous region. In: Fox, J.L. & Jizeng, D. (eds.) *Proceedings of the Seventh International Snow Leopard Symposium*. Snow Leopard Trust, Seattle.

Johnson, A., Vongkhamheng, C., Hedemark, M. & Saithongdam, T. (2006) Effects of human-carnivore conflict on tiger (*Panthera tigris*) and prey populations in Lao PDR. *Animal Conservation* 9, 421–430.

Johnson, C.N., Isaac, J.L. & Fisher, D.O. (2007) Rarity of a top predator triggers continent-wide collapse of mammal prey: dingoes and marsupials in Australia. *Proceedings of the Royal Society B: Biological Sciences* 274, 341–346.

Jorgensen, C.J. (1979) Bear-sheep interactions, Targhee National Forest. *International Conference on Bear Research and Management* 5, 191–200.

Linnell, J.D.C., Aanes, R., Swenson, J.E. & Smith, M.E. (1997) Translocation of carnivores as a method for managing problem animals: a review. *Biodiversity and Conservation* 6, 1245–1257.

Linnell, J.D.C. & Bjerke, T. (2002) Frykten for ulven. En tverrfaglig utredning. (Fear of wolves: an interdisciplinary study.) *NINA oppdragsmelding* 722, 1–110.

Linnell, J.D.C. & Brøseth, H. (2003) Compensation for large carnivore depredation on domestic sheep in Norway, 1996–2002. *Carnivore Damage Prevention News* 6, 11–13.

Linnell, J.D.C., Odden, J., Smith, M.E., Aanes, R. & Swenson, J.E. (1999) Large carnivores that kill livestock: do "problem individuals" really exist? *Wildlife Society Bulletin* 27, 698–705.

Macdonald, D.W. (2001) *The New Encyclopedia of Mammals*. Oxford University Press, Oxford.

Madhusudan, M.D. (2003) Living amidst large wildlife: livestock and crop depredation by large mammals in the interior villages of Bhadra Tiger Reserve, South India. *Environmental Management* 31, 466–475.

Marchini, S. (2006) Dimensões humanas dos conflitos entre gente e grandes felinos na fronteira agrícola da Amazônia: resultados preliminares. 7th International Conference for the Management of Wildlife in Amazonia and Latin America, Ilheus, Brazil.

McDougal, C. (1999) Tiger attacks on people in Nepal. *Cat News* 30, 9–10.

Miller, C.M. (2002) Jaguares, ganado y humanos: un ejemplo de coexistencia pacifica en el noroeste de Belice. In: Medellin, R.A., Equihua, C., Chetkiewicz, C.L.B. et al. (eds.)*El Jaguar en el Nuevo Milenio*. Fondo de Cultura Económica USA, San Diego, pp. 477–492.

Mills, M.G.L., Ellis, S., Woodroffe, R. et al. (eds.) (1998) *Population and Habitat Viability Assessment for the African Wild Dog (*Lycaon pictus*) in Southern Africa*. International Union for Conservation of Nature and Natural Resources/ Species Survival Commission Conservation Breeding Specialist Group, Apple Valley.

Miquelle, D.G., Merrill, T.W., Dunishenko, Y.M. et al. (1999) Hierarchical spatial analysis of Amur tiger relationships to habitat and prey. In: Seidensticker, J., Christie, S. & Jackson, P. (eds.) *Riding the Tiger: Tiger Conservation in Human-Dominated Landscapes*. Cambridge University Press, Cambridge, pp. 273–295.

Miquelle, D., Nikolaev, I., Goodrich, J., Litvinov, B., Smirnov, E. & Suvorov, E. (2005) Searching for the coexistence recipe: a case study of conflicts between people and tigers in the Russian Far East. In: Woodroffe, R., Thirgood, S. & Rabinowitz, A. (eds.) *People and Wildlife: Conflict or Coexistence?*. Cambridge University Press, Cambridge, pp. 305–322.

Moa, P.F., Herfindal, I., Linnell, J.D.C., Overskaug, K., Kvam, T. & Andersen, R. (2006) Does the spatiotemporal distribution of livestock influence forage patch selection in Eurasian lynx? *Wildlife Biology* 12, 63–70.

Mondolfi, E. & Hoogesteijn, R. (1986) Notes on the biology and status of the jaguar in Venezuela. In: Miller, S.D. & Everett, D.D. (eds.) *Cats of the World: Biology, Conservation and Management.* National Wildlife Federation, Washington, D.C., pp. 85–124.

Mukherjee, S. (2003) Tiger human conflicts in the Sundurban Tiger Reserve, West Bengal, India. *Tigerpaper* 30(2), 3–6.

Naughton-Treves, L., Grossberg, R. & Treves, A. (2003) Paying for tolerance: rural citizens' attitudes toward wolf depredation and compensation. *Conservation Biology* 17, 1500–1511.

Naughton-Treves, L. & Treves, A. (2005) Socioecological factors shaping local support for wildlife in Africa. In: Woodroffe, R., Thirgood, S. & Rabinowitz, A. (eds.) *People and Wildlife: Conflict or Coexistence?* Cambridge University Press, Cambridge, pp. 253–277.

Nowell, K. & Jackson, P. (1996) *Wild Cats: Status Survey and Conservation Action Plan.* International Union for Conservation of Nature and Natural Resources/Species Survival Commission Cat Specialist Group, Gland.

Nunez, M., Miller, B. & Lindzey, F. (2002) Ecologia del jaguar en la reserva de la biosfera Chamela-Cuixmala, Jalisco, Mexico. In: Medellin, R.A., Equihua, C., Chetkiewicz, C.L.B. et al. (eds.) *El Jaguar en el Nuevo Milenio.* Fondo de Cultura Económica USA, San Diego, pp. 107–127.

Nyhus, P.J. & Tilson, R. (2004a) Characterising human-tiger conflict in Sumatra, Indonesia: implications for conservation. *Oryx* 38, 68–74.

Nyhus, P.J. & Tilson, R. (2004b) Agroforestry, elephants and tigers: balancing conservation theory and practice in human-dominated landscapes of Southeast Asia. *Agriculture, Ecosystems and Environment* 104, 87–97.

Odden, J., Linnell, J.D.C. & Andersen, R. (2006) Diet of Eurasian lynx, *Lynx lynx,* in the boreal forest of southeastern Norway: the relative importance of livestock and hares at low roe deer density. *European Journal of Wildlife Research* 52, 237–244.

Odden, J., Linnell, J.D.C., Moa, P.F., Herfindal, I., Kvam, T. & Andersen, R. (2002) Lynx depredation on domestic sheep in Norway. *Journal of Wildlife Management* 66, 98–105.

Ogada, M.O., Woodroffe, R., Oguge, N.O. & Frank, L.G. (2003) Limiting depredation by African carnivores: the role of livestock husbandry. *Conservation Biology* 17, 1521–1530.

Perovic, P.G. (2002) Conservacion del jaguar en el noreste de Argentina. In: Medellin, R.A., Equihua, C., Chetkiewicz, C.L.B. et al. (eds.)*El Jaguar en el Nuevo Milenio.* Fondo de Cultura Económica USA, San Diego, pp. 465–476.

Polisar, J., Maxit, I. & Scognamillo, D. (2003) Jaguars, pumas, their prey base and cattle ranching: ecological interpretation of a management problem. *Biological Conservation* 109, 297–310.

Rabinowitz, A. (1986) Jaguar predation on domestic livestock in Belize. *Wildlife Society Bulletin* 14, 170–174.

Rajpurohit, K.S. (1998) Child-lifting: wolves in Hazaribagh, India. *Ambio* 28, 163–166.

Rasmussen, G. (1999) Livestock predation by the painted hunting dog *Lycaon pictus* in a cattle ranching region of Zimbabwe: a case study. *Biological Conservation* 88, 133–139.

Reddy, H.S., Srinivasulu, C. & Rao, K.T. (2004) Prey selection by the Indian tiger (*Panthera tigris tigris*) in Nagarjunasagar Srisailam Tiger Reserve, India. *Mammalian Biology* 69, 384–391.

Reza, A.H.M.A., Chowdhury, M. & Santiapillai, C. (2000) Tiger Conservation in Bangladesh. *Tigerpaper* 27, 1–5.

Reza, A.H.M.A., Feeroz, M.M. & Islam, M.A. (2002) Man-tiger interaction in the Bangladesh Sundarbans. *Bangladesh Journal of Life Science* 14(1&2), 75–82.

Roosevelt, T. (1914) *Through the Brazilian Wilderness*. Charles Scribner's Sons, New York.

Saenz, J.C. & Carrillo, E. (2002) Jaguares depredadores de ganado en Costa Rica. In: Medellin, R.A., Equihua, C., Chetkiewicz, C.L.B. et al. *El Jaguar en el Nuevo Milenio*. Fondo de Cultura Económica USA, San Diego, pp. 127–138.

Sanderson, E., Forrest, J., Loucks, C. et al. (2006) *Setting Priorities for the Conservation and Recovery of Wild Tigers: 2005–2015. The Technical Assessment*. Wildlife Conservation Society, World Wildlife Fund, Smithsonian Institution and National Fish and Wildlife Foundation-Save the Tiger Fund, New York, Washington, D.C.

Savolainen, P., Leitner, T., Wilton, A.N., Matisoo-Smith, E. & Lundeberg, J. (2004) A detailed picture of the origin of the Australian dingo, obtained from the study of mitochondrial DNA. *Proceedings of the National Academy of Sciences of the United States of America* 101, 12387–12390.

Scandlynx. (2007) Fangs tog marking av gaupe (Lynx capture). http://scandlynx. nina.no [accessed April 1, 2007].

Schaller, G.B. (1983) Mammals and their biomass on a Brazilian ranch. *Arquivos de Zoologia* 31, 1–36.

Schaller, G.B. & Crawshaw, P.G. (1980) Movement patterns of jaguar. *Biotropica* 12, 161–168.

Schiaffino, K., Malmierca, L. & Perovic, P.G. (2002) Depredacion de cerdos domesticos por jaguar en un area rural vecina a un parque nacional en el noreste de Argentina. In: Medellin, R.A., Equihua, C., Chetkiewicz, C.L.B. et al. (eds.) *El Jaguar en el Nuevo Milenio*. Fondo de Cultura Económica USA, San Diego, pp. 251–264.

Scognamillo, D., Maxit, I.E., Sunquist, M. & Farrell, L. (2002) Ecologia del jaguar y el problema de la depredacion de ganado. In: Medellin, R.A., Equihua, C.,

Chetkiewicz, C.L.B. et al. (eds.) *El Jaguar en el Nuevo Milenio.* Fondo de Cultura Económica USA, San Diego, pp. 139–150.

Seymour, K.L. (1989) *Panthera onca. Mammalian Species* 340, 1–9.

Sillero-Zubiri, C. & Laurenson, M.K. (2001) Interactions between carnivores and local communities: conflict or coexistence? In: Gittleman, J.L., Funk, S.M., Macdonald, D.W. & Wayne, R.K. (eds.) *Carnivore Conservation.* Cambridge University Press, Cambridge, pp. 282–312.

Swank, W. & Teer, J. Fundación para desarollo de las ciencias físicas, matemáticas y naturales [FUDECI] (1992) A proposed program for sustained jaguar populations. In: *Felinos de Venezuela.* Raúl Clemente Editores, Caracas, pp. 95–107.

TIGRIS Foundation (2007) Compensation for livestock predation. http://www.tigrisfoundation.nl/cms/publish/content/showpage.asp?pageID=40#top (accessed May 2, 2009).

Treves, A. (2007) *Balancing the Needs of People and Wildlife: When Wildlife Damage Crops and Prey on Livestock.* Unpublished report. Land Tenure Centre, University of Wisconsin, USA.

Treves, A., Jurewicz, R.R., Naughton-Treves, L., Rose, R.A., Willging, R.C. & Wydeven, A.P. (2002) Wolf depredation on domestic animals: control and compensation in Wisconsin, 1976–2000. *Wildlife Society Bulletin* 30, 231–241.

Treves, A. & Karanth, U. (2003) Human-carnivore conflict and perspectives on carnivore management worldwide. *Conservation Biology* 17, 1491–1499.

Treves, A. & Naughton-Treves, L. (2005) Evaluating lethal control in the management of human-wildlife conflict. In: Woodroffe, R., Thirgood, S. & Rabinowitz, A. *People and Wildlife: Conflict or Coexistence?* Cambridge University Press, Cambridge, pp. 86–106.

Treves, A., Naughton-Treves, L., Schanning, K. & Wydeven, A. (2007) Appendix H2: public opinion of wolf management in Wisconsin, 2001–2005. In: Wydeven, A.P. (ed.) *Wisconsin Wolf Management Plan.* Wisconsin Department of Natural Resources, Madison. pp. 36–47. www.dnr.state.wi.us/org/land/er/publications/wolfplan/pdfs/WIWolfManagementPlanAdd.pdf

United States Fish and Wildlife Service (2007) Endangered and threatened wildlife and plants; Final rule designating the Western Great Lakes populations of Gray Wolves as a distinct population segment; Removing the Western Great Lakes distinct population segment of the Gray Wolf from the list of endangered and threatened wildlife. *Federal Register* 72, 6051–6103.

Vaughan, C. & Temple, S. (2002) Conservacion del jaguar en Centroamerica. In: Medellin, R.A., Equihua, C., Chetkiewicz, C.L.B. et al. (eds.) *El Jaguar en el Nuevo Milenio.* Fondo de Cultura Económica USA, San Diego, pp. 355–366.

Weber, W. & Rabinowitz, A. (1996) A global perspective on large carnivore conservation. *Conservation Biology* 10, 1046–1055.

Wikramanayake, E.D., Dinerstein, E., Robinson, J.G. et al. (1999) Where can tigers live in the future? A framework for identifying high-priority for the conservation of tigers in the wild. In: Seidensticker, J., Christie, S. & Jackson, P. *Riding the Tiger: Tiger Conservation in Human-Dominated Landscapes.* Cambridge University Press, Cambridge, pp. 255–272.

Wikramanayake, E., McKnight, M., Dinerstein, E. & Joshi, A. (2004) Designing a conservation landscape for tigers in human-dominated environments. *Conservation Biology* 18, 839–844.

Woodroffe, R. & Ginsberg, J.R. (1998) Edge effects and the extinction of populations inside protected areas. *Science* 280, 2126–2128.

Woodroffe, R., Thirgood, S. & Rabinowitz, A. (2005) The impact of human-wildlife conflicts on human lives and livelihoods. In: Woodroffe, R., Thirgood, S. & Rabinowitz, A. (eds.) *People and Wildlife: Conflict and Coexistence.* Cambridge University Press, Cambridge, pp. 13–26.

Wydeven, A.P., Mladenoff, D.J., Sickley, T.A., Kohn, B.E., Thiel, R.P. & Hansen, J.L. (2001) Road density as a factor in habitat selection by wolves and other carnivores in the Great Lakes Region. *Endangered Species Update* 18, 110–114.

Wydeven, A.P., Treves, A., Brost, B. & Wiedenhoeft, J.E. (2004) Characteristics of wolf packs in Wisconsin: identification of traits influencing depredation. In: Fascione, N., Delach, A. & Smith, M.E. (eds.) *People and Predators: From Conflict to Coexistence.* Island Press, Washington, D.C., pp. 28–50.

Zimmermann, A., Walpole, M.J. & Leader-Williams, N. (2005) Cattle ranchers' attitudes to conflicts with jaguar *Panthera onca* in the Pantanal of Brazil. *Oryx* 39, 406–412.

7

Financial Incentives for Rangeland Conservation: Addressing the 'Show-Us-the-Money' Challenge

Ray Victurine[1] and Charles Curtin[2,3]

[1]Conservation Finance Program, Wildlife Conservation Society, NY, USA
[2]MIT, MA, USA
[3]Department of Environmental Studies, Antioch University, NH, USA

Introduction

Human activities have given rise to a decline in extent and condition of rangelands around the world, with the preservation of rangelands becoming a growing priority for biodiversity and ecosystem service conservation (White et al. 2000). The subdivision of individual ranches, and rangelands in general, is shown to be a threat to both biodiversity and function of grassland ecosystems (Western & Gichohi 1993; Odell & Knight 2001; Curtin et al. 2002; Knight et al. 2002; Western & Manzolillo Nightingale 2005). Yet, the growing threats to rangeland ecosystems are likely to increase until the underlying economic issues affecting landowners and their decisions regarding land use are addressed (Curtin & Western 2008). Evidence from around the world indicates that the conservation of rangeland ecosystems can be achieved through a variety of mechanisms that recognize or assign property rights

Wild Rangelands: Conserving Wildlife While Maintaining Livestock in Semi-Arid Ecosystems, 1st edition. Edited by J.T. du Toit, R. Kock, and J.C. Deutsch.

to wildlife and create markets for biodiversity and ecosystem services. This chapter examines the role of financial incentives and markets in maintaining ranches and other pastoral uses of rangelands as a strategy for preserving open space and ecological function.

Rangeland preservation can imply significant landowner costs. These costs are both direct, such as those resulting from wildlife predation or (where applicable) crop damage, or indirect through the opportunity costs incurred from not converting land to a higher value use (e.g. subdivision). As land values and alternative economic opportunities increase, so will the pressures on landowners to convert land for uses other than traditional pastoral practices that indirectly favour biodiversity and wildlife conservation (Western & Gichohi 1993; Curtin et al. 2002).

Economic incentives provide powerful tools for maintaining rangelands as open space. Private landowner choices are especially important because a large proportion of the world's biodiversity exists on private land, in a 'semi-natural matrix' outside the boundaries of formally protected areas (Brown et al. 2003; Curtin & Western 2008). These lands are extremely important for wildlife, serving as corridors and seasonal habitat and increasingly proving to be as good or better a habitat than parks and other formal conservation areas (Western 1989; Western & Gichohi 1993; Homewood 2004; Sachedina 2006). Influencing decisions in favour of conservation will require development of incentives that motivate landowners to maintain the ecological integrity of rangelands.

Rangeland conversion is subject to various economic pressures. Conversion occurs because landowners can make more money from it, and there are few conservation-related choices. Adopting conservation practices often does not enter the choice equation because opting to conserve provides landowners with insufficient income and few tangible benefits (Curtin et al. 2002). In Africa, landowners may suffer economic hardship from conservation practices, losing access to needed resources when protected areas are created or are suffering crop damage or livestock losses from migrating animals and predators (Western 1989; Homewood 2004; Sachedina 2006). On both African and North American continents, wildlife (and therefore wildlife conservation) is frequently perceived as a liability rather than an asset (Curtin 2005). In the United States, the Endangered Species Act is often considered a significant threat with the discovery of threatened or endangered species frequently considered a problem to be addressed, rather than the mark of good stewardship (Sayre 2002). In East Africa, government and donor policies frequently encourage subdivision of open landscapes, seeing it as the only means to protect individual and group landholdings (Curtin & Western 2008). As a consequence, open landscapes become fragmented, biodiversity and wildlife habitats decrease and important land connectivity disappears (Curtin et al. 2002; Homewood 2004; Curtin & Western 2008).

Conservation should be a viable land use decision, but the economics of conservation frequently rule out the option. What would happen if sustainable land management benefited landowners? What if they derived financial rewards from conservation actions? Would the incentive to convert land diminish? What levels of profitability would be required to achieve conservation results? And how would social and cultural factors affect both land management and economic decisions? Analysis of these questions will foster better understanding of landowner motivation, an essential element for devising incentives to directly address landowner needs and concerns. If these are met, the potential for achieving sustainable rangeland management objectives and biodiversity conservation on private land will increase.

Research and experience from the field indicates landowners will manage rangelands sustainably only when they have clear private incentives to do so (Nkedianye 2004; Sayre 2004; Department of Sustainability and Environment 2006). Such incentives may include benefits other than financial returns from land management. Landowners do not necessarily seek to maximize expected profits. They also care about non-financial values related to cultural and lifestyle preferences, as well as the risk associated with the financial and non-financial returns from different land use decisions. In the jargon of economists, ranchers maximize expected utility, not profits. Thus, the optimal design of an incentive for rangeland management will not necessarily require full compensation for foregone profits, nor will it necessarily be composed solely of financial incentives. Non-monetary incentives may be useful because landowners may be willing to absorb high opportunity costs. Key challenges are to devise the right level and mix of incentives and to mobilize the resources to ensure that landowners receive acceptable compensation for their conservation investments.

A century of subdivision in the American West is instructive for the African and Australian contexts where these processes are more recent (Curtin & Western 2008). In the last decades, working ranches have not proven to be effective businesses. Cattle prices have steadily declined in real terms, while costs of labour, fuel, insurance and other inputs have increased. As a result, the return-on-investment for ranching has dwindled to well-below-market norms

(Martin & Jefferies 1966; Starrs 1998). A recent survey of ranchers found that 50.4% of public land ranchers, those ranchers who run their livestock primarily on public lands, depend on non-ranching income (Gentner 1999). Meanwhile, the market value of land has increased dramatically. With developers viewing ranches as speculative investments, ranchers find much of their equity derives from the possibility of subdivision and development, putting them in a category referred to as 'land rich and money poor' (Curtin et al. 2002).

The challenge is how to transfer the incentives that drive subdivision and land fragmentation into engines of landscape preservation. The decline of rangeland and subsequent loss of biodiversity has stimulated a search for effective mechanisms and incentives that address financial needs, recognize the cultural factors that influence decision-making and provide landowners some compensation for the conservation use of their land. These mechanisms are based on the premise that people are willing to manage land sustainably and even protect wildlife if they receive fair compensation for their land as well as for their efforts, and clear title or a perception of safe land tenure.

This chapter discusses how appropriate incentives can be established through three approaches: (i) direct payment options, (ii) conservation easements and purchase of development rights (PDRs) and (iii) the development of new retail markets to increase demand for conservation. All three approaches employ economic incentives while addressing social and cultural issues that influence decision-making. They are designed to maintain land use that supports biodiversity and wildlife conservation by ensuring that conservation is self-sustaining. The implementation of these and other innovative market-based approaches offer some of the best opportunities for conserving rangeland around the world, while gaining a broader constituency for conservation efforts. For conservation to be sustainable, it must be both economically and ecologically viable.

Direct payment options

Paying landowners to conserve rangeland may be one of the most effective ways to obtain desired land use results. The payment represents the landowner incentive, or compensation, to adopt certain desired land use practices. This direct payment concept to more efficiently achieve conservation results has been discussed extensively by Paul Ferraro. Ferraro (2001a,b), Ferraro and Simpson (2001), Ferraro and Kiss (2002) and Ferraro and Simpson (2002).

He argued that targeted action would contribute to conservation results more effectively than indirect approaches, outlining a variety of direct payment mechanisms developed around the world including direct purchases, leases, easements and direct performance payments.

Direct payment schemes work because they are clear, targeted and can be designed to respond to local economic and cultural realities. They establish payments for specific, concrete actions and create a contractual relationship between the buyer and seller, who in this case is the provider of the biodiversity. This direct payment approach differs from the common practice of using development interventions as indirect incentives to achieve conservation outcomes. The indirect development model approach, where small investments in community programmes and infrastructure try to earn support for conservation from community members, generates ambiguous conservation incentives ('conservation by distraction') (Ferraro 2001a). The approach is too complex and objectives are muddled. Direct payments or contracts, on the other hand, prevent confusion and give land managers an active stake in the management of ecosystems. Land managers deliver conservation through their actions and are compensated in return.

Although the approaches and structure of the incentives or payments may differ across regions, the basic concept remains the same – direct payments pay for a conservation product, putting needed money into the pockets of landowners in exchange for a desired service or outcome, making biodiversity conservation and sustainable land management a realistic financial option.

The following cases provide examples of how direct payment options have been employed in the field. In Kenya and Tanzania, lease payments to landowners were used to ensure conservation and allow the free movement of animals through the landscape. The major difference between the countries' schemes was the source of funding. In Australia, a state government employed a bidding process through an auction, paying farmers to manage their land in an effort to find a cost-effective approach to conserve biodiversity on private land.

Direct lease payments in Kenya

Around the world, there is growing recognition of the benefits to be achieved by directly paying for conservation. In devising such schemes, however, it is important to ensure that financing is secured to ensure the desired long-term benefits. In Kenya, the Wildlife Foundation and Friends of Nairobi National Park launched an innovative programme in the Kitengela plains to provide incentives for Maasai pastoralists to maintain rangelands as wildlife migration corridors adjacent to Nairobi National Park (Gichohi 2003; Radeny et al. 2007).

Government policies and economic changes that favour agriculture over ranching have increased pressure for private landowners to convert rangeland to agricultural production. Such conversion normally leads to fencing, which in turn impedes wildlife movement. The trend towards greater conversion for agriculture, coupled with the loss of nearly 50% of the wildlife in Nairobi National Park and the adjacent ecosystem, led to fears by conservationists and the state institution in charge of wildlife, the Kenya Wildlife Service, that the integrity of the greater Nairobi National Park ecosystem and wildlife numbers in general could not be maintained (Rodriquez et al. 2006). This, in turn, led to an analysis of non-traditional options to maintain the ecosystem.

The option developed in Kitengela in 2000 involved a direct payment scheme that compensated landowners for leaving their land unfenced and in pasture. This programme, the Wildlife Conservation Lease Initiative (WCLI), negotiated short-term leases, or easements, with landowners. In exchange for foregoing fencing, not subdividing and not converting to agriculture, each landowner received a guaranteed lease payment of $4.00 per acre per year. Moreover, landowners kept control of the use of their land and could continue to keep livestock, thereby enjoying a double benefit – a direct financial benefit from the leases and a significant livestock benefit that addressed both economic and social concerns.

The lease payment scheme appears successful for two reasons. In addition to the direct financial benefits enjoyed by landowners, the payments produced significant social and conservation benefits. Lease payments were made in three installments, all of which coincided with the school terms and the precise time when families needed cash to pay school fees – a major burden on the household economy. The timing of the payments proved popular with participants. The timely cash flow facilitated an important social goal – school attendance – and relieved a significant financial burden for families in the region. And they provided direct conservation benefits.

Nkedianye (2004) reports that the lease payments stimulated a change in the attitude of residents towards wildlife. There were fewer incidences of retaliation against wildlife and an overall reduction in human–wildlife conflict. People appear to have felt adequately compensated for living with wildlife. According to Rodriquez et al. (2006), the programme achieved the participation of 117 households with a total of 8500 acres. Owners of at least another 15,000 acres were enrolled on a waiting list. At the established lease payment levels, the programme would require annual outlays of approximately $34,000 per year to maintain payments on the 8500 acres and almost $100,000 per year to lease land from all interested parties (those already signed up and on the waiting list).

The extension of the Kitengela corridor required to ensure the migration of wildlife to and from Nairobi National Park covers an area of approximately 60,000 acres. It would require annual outlays of more than $240,000, under existing practices and price levels, to lease the entire targeted range. This cost assumption needs to be tested. The excess in demand, as demonstrated by the waiting list, could indicate that landowners might be willing to accept a lower lease price. If the price could be lowered, more land could be enrolled within the current budget. At the same time, the value of products from alternative land uses, especially agricultural production, could increase landowner opportunity costs and result in a demand for higher lease payments.

The programme clearly demonstrates that many landowners are willing to keep their land as unfenced rangeland if the compensation is adequate. By ensuring an adequate return to private landowners, multiple benefits are achieved. The payments offer farmers several important advantages. They reduce the cost of losses caused by wildlife on the land, thereby compensating landowners for conservation activities. This compensation, along with the returns to livestock management, satisfies many people who wish to maintain a pastoral lifestyle. Payments also compensate for risk, or at least allow landowners to forego risk. Landowners apparently prefer the combined value of the guaranteed annual lease payment and livestock returns to the more risky returns from converting the land to annual agricultural crop production (Nkedianye 2004).

Unfortunately, not all aspects of the programme proved as successful. Despite the popularity and success of the intervention, the annual payment scheme faced serious cash shortages as early as September 2006, calling into question the sustainability of the enterprise. Some in the region attribute the reduction and loss of available funding to 'donor fatigue' (O. Petenya Yusuf, personal communication, 2006).

Gichohi (2003) reports that the existing fee payments are sufficient for people who are committed to a pastoralist lifestyle and own larger tracts of land, but may fall short for new landowners, without such traditions. With land values increasing, many landowners, especially younger ones with weaker connections to traditional lifestyles, may find the alternative financial returns too tempting to accept the conservation leases when they can earn higher returns from converting land to agricultural production.

Some solutions are available. Working with landowners and offering guaranteed leases over longer time periods might address some issues. The longer the land can be held under lease, the greater the time period to manage the land for conservation. It also allows time to generate additional resources to cover future lease payments. In many parts of East Africa, both 49- and 99-year leases are common.

Creating a long-term financing mechanism to ensure the sustainability of lease payments represents another solution that was missing from the design of the lease programme. The sustainable financing challenge is common to all direct payment schemes; the ability to make annual payments or even pay multi-year leases depends on some level of guaranteed annual cash flow. Counting on continued donor or government support is risky, although both donors and government have important roles to play in providing funding for such ventures. Diversifying funding sources and exploring innovative mechanisms to generate the funds required to make payments are essential.

Direct payment schemes are particularly well suited to the programme portfolios of endowed environmental funds that generate annual income from their investments and have the capacity to guarantee annual payments. Environmental funds traditionally have not included direct payment programmes in their portfolios, but the potential for long-term conservation benefits such as those demonstrated by Kitengela may serve to motivate some conservation endowments to dedicate a portion of their programme funds to direct payment schemes.

Creating new endowed environmental funds is an option for direct payment programmes such as Kitengela. Donors, however, have been reluctant to provide funding to capitalize endowments, they consider the opportunity costs of tying up money in an investment account too high, when conservation needs are so great in real time. Moreover, there is concern about capital losses during market downturns. Convincing donors to dedicate part of their portfolios for endowments, such as what was achieved under the Global Environment Facility in the 1990s, would provide greater flexibility and opportunity for testing and implementing direct payment schemes aimed at maintaining land use for conservation on private land.

Recent actions by donors indicate a shift in thinking on endowment funding. Donors from various European countries are beginning to look favourably on the use of endowed funds as a way to secure their investments in biodiversity conservation. In 2007, newly established conservation endowments received infusions of capital from donors new to endowments. If the trend continues and these endowments embrace direct payment schemes, lease payment programmes, such as the one developed in Kitengela, could offer an important solution to the land conversion challenge and will create social and economic value from conserving wildlife and habitat.

Direct lease payments in Tanzania

Box 7.1 describes a programme in which private sector tourism operators generate revenue for direct payments to rural Tanzanian landowners. Tourism operators and lodge owners pay landowners to maintain land for elephant dispersal outside Tarangire National Park. In exchange, the operators bring tourists to view wildlife when animals are moving outside the boundaries of the park. This arrangement gives recognition to the value of the land, and respects both private and community land ownership rights.

Box 7.1 **Developing a conservation easement: the Tarangire ecosystem in northern Tanzania**

Charles Foley, Wildlife Conservation Society

Tarangire National Park in northern Tanzania is a nationally important wildlife area. It supports the largest population of elephants in the north of Tanzania, and, till the late 1980s, regularly experienced an annual migration of over 55,000 animals (TWCM 1990), making this the third largest wildlife migration in East Africa (surpassed only by the movements of the wildebeest in the Serengeti and the white-eared kob in southern Sudan). The park has seen a steady increase in the number of visitors during the past decade and is now a staple on the northern Tanzania safari circuit, bringing in significant revenue to the park system.

Yet, the large wildlife herds of Tarangire face an uncertain future. The Tarangire ecosystem is an area of more than 20,000 km^2 (Borner 1985), stretching from the central Maasai steppe all the way to Lake Natron on the Kenyan border. Only 10% of this land is protected by the National Park, but variations in mineral content in the soil, and the dictates of rainfall, bring about the widespread dispersal of large ungulates throughout the 90% that is not protected. Herds leave the park at the onset of the rains and move to mineral-rich calving grounds, especially on the Simanjiro plains, rich in phosphate and crude protein (Voeten 1999). When the ephemeral water dries up in the dispersal area, the herds move back to the park and the permanent water the Tarangire River provides (Lamprey 1963). During very wet years, as in the El Niño year of 1998, the wildlife does not return to the park but remains on the Simanjiro plains the year round. Access to dispersal areas outside the park is therefore essential; if the large ungulate species were restricted to Tarangire's less nutritious grasslands for any lengthy period, their populations would eventually be severely reduced.

There are two major threats to this large-scale movement of wildlife: agricultural expansion and poaching. To the north and west of Tarangire, expansion by small-scale agriculturalists has brought about the disappearance of five of the nine major wildlife corridors connecting Tarangire to surrounding dispersal areas in the past 40 years. To the east and south the land is owned by Maasai pastoralists, traditionally highly tolerant of wildlife on their lands. However, in recent years they too have been increasingly cultivating or leasing their land to immigrant cultivators, putting great pressure on the long-term viability of the Simanjiro plains, the main ungulate calving grounds. The other major threat is an increase in illegal hunting driven by a large demand for game meat in nearby urban centres. During the past decade, there have been big reductions in the population of plains game (with the exception of elephant and buffalo). During this period, the wildebeest population has dropped from over 30,000 to less than 5000 today, and populations of oryx and gerenuk are facing local extinction (Nelson 2005).

The reasons for the land use change are varied. In part, it is a reflection of increasing pressures on a pastoralist society, where the poorest members are resorting to cultivation for lack of alternative forms of livelihood. Yet political factors are also at play. Land tenure insecurity stemming from past losses of village lands to the National Park have created incentives for villagers to 'individualize' land and cultivate, which is seen as providing an indelible stamp of land ownership. A lack of vested interest in wildlife resources means that local communities have few incentives to prevent either exploitation of wildlife on their lands or the adoption of wildlife-incompatible land use practices. These factors driving the over-exploitation of Simanjiro's wildlife have to be addressed if management strategies are to be effective.

The most effective conservation models in Maasailand have been agreements negotiated by tour companies that provide a fee per client in exchange for access to village land. In two villages, where the village is part owner of a safari lodge, such agreements have provided the communities with substantial revenue. As a result, both villages have set aside large areas of village land for wildlife conservation and tourism, combined with limited livestock grazing. However, most villages in the key wet season dispersal areas do not attract significant tourism investments and have limited potential to earn revenue in this way. The main calving grounds in Sukuro and Terat villages are far from the park boundary (over 50 km in some areas), where wildlife densities are highly variable, both seasonally and spatially, and access during the wet season is problematic. Creating value from wildlife in these areas that will encourage conservation by the villages is a challenge.

To address this problem, concerned safari tour operators formed a group called ELAND (Enterprise Linkage to Conservation and Development), a collaborative partnership among private sector interests, to whom the long-term sustainability of the ecosystem is of key economic importance. They decided that the issue of protecting the Tarangire dispersal areas was a business problem. The ELAND group need the calving areas to be protected so that wildlife populations in the park remain healthy, which their enterprises depend on. The local villagers deliver a service by providing land, but receive no compensation, and instead suffer costs through animal attacks on their livestock and crops. The tour operators decided to offer to pay local villages in the key dispersal areas an annual remuneration to establish a conservation easement, where land use practices detrimental to wildlife use would be prohibited. What makes this initiative different from other private partnership initiatives in Tanzania between tour operators and villages is that none of the tour operators involved in ELAND had any direct business interests in these areas; they were in effect proposing to pay for protection of land that their clients would never visit.

ELAND representatives presented this idea to leaders of Terat village, an important village in the Simanjiro plains, which cover a substantial portion of the ungulate calving grounds. After comprehensive consultation with their village constituency, the village leadership agreed to set up a conservation easement on approximately $120 \, \text{km}^2$ of village land. The agreement was signed in December 2005 for an initial period of 5 years. Under the terms of the agreement, the land will be set aside for cattle grazing and wildlife use only; all cultivation and permanent settlements are prohibited. ELAND will pay the village an annual sum of $5000 with an additional $5000 provided by conservation organizations to hire four local villagers that will act as game scouts to protect and monitor wildlife populations. A key aspect of the agreement is that the village will retain full ownership of the land and will be able to reverse their decision in the future when the current contract expires.

ELAND is now looking to consolidate its conservation efforts in the area and apply the model to neighbouring village lands. Critical in this process will be to encourage the two adjacent villages, Sukuro and Emboret, to develop a similar approach of easement development so that a corridor of protected land can be established linking Tarangire National Park to the calving grounds on the Simanjiro. There are currently signs of growing appreciation for this type of arrangement in both villages, though for this to succeed would require more companies to join the ELAND initiative to ensure that funds are available to finance any further agreements.

To a certain degree, the approach is similar to efforts in southern Africa where community ownership rights to wildlife and land were codified, allowing people to earn money from wildlife management and establish economic opportunities through conservancies. Although not always successful, nor equally effective across individual conservancies and communities, the establishment of property rights and ownership has led to improved wildlife and land management in many communities. Because income depended on the existence of wildlife for hunting and tourism, land use decisions favoured conservation and people sought diversification of income through practices that maintained conservation objectives. As a result, ventures involving wildlife tourism and sustainable harvesting of plant species from rangelands developed, creating new sources of income and incentives against land conversion, as examples in southern Africa demonstrate.

Although the Tarangire lease payments do not create wildlife ownership, they do establish a direct link, based on tourism, between income and conservation land use. The tourism-derived payments serve as an effective incentive for people to keep their land in rangeland for the benefit of wildlife populations.

Auctions for direct payments in Australia

Auction systems provide another opportunity to implement direct payments to protect biodiversity and maintain rangeland. While the East African cases illustrate direct payments from non-governmental organizations (NGOs) and the private sector, an Australian case offers an innovative example of direct payments by government for biodiversity conservation on private land, setting the conservation price via auction.

With over 1 million hectares of native vegetation on private land, and with 35% of threatened vegetation occurring primarily on private land, the State Government of Victoria recognized that biodiversity conservation would require development of an effective private land management strategy to actively involve land managers (Stoneham et al. 2002a). In seeking a solution, the State Government of Victoria instituted a novel approach to direct payments, introducing an auction system called *BushTender* to pay landowners to protect biodiversity and maintain sustainability of its rangelands. This performance-based approach directs payments to landowners for specific biodiversity outputs from land management activities proposed by the landowners. The design ensures that landowners are engaged in the management decisions that affect their property, with direct budget support provided by the government. In addition to the payments, the government offers technical assistance to support best conservation practices.

The auction system involves the participation of individual bidders, each making a decision about his or her cost for delivering a service – in this case, improved land management for biodiversity conservation. In the *BushTender* auction system, the government expects a large number of bids with a variety of price offers. With a large number of bids the likelihood of collusion among participants to push up the price paid for services is low, while price variety offers the government the ability to select sites at the best possible price. This process reduces what are called information rents, that is, the profits accruing to someone with inside information that could affect results and skew pay-out levels (Stoneham et al. 2002a; Ha et al. 2004). As a result, the buyer of environmental services, in this case the State of Victoria, can maximize its conservation returns from its limited budget.

In the *BushTender* case, the auction system worked by matching the buyer (the State Government of Victoria) with the sellers (individual landowners), allowing the landowners to voluntarily offer land management and conservation services on their land to the government, at a price the landowners felt was fair and reasonable. The government would decide whether or not to take the offer. On the basis of a call for proposals, landowners submitted sealed bids to the government offering to undertake specific land management activities aimed at protecting biodiversity at a set price. The government evaluated each bid to determine the feasible amount of biodiversity conservation per dollar and selected the most desirable sites to achieve cost-effective conservation.

Once each bid was accepted, the government and the landowner signed a contract whereby the landowner agreed to certain actions and outputs, and the government paid based on compliance. The effort required landowner reporting and government monitoring to ensure success.

The auction procurement process effectively established bidding rules and market competition to reduce the incentive for sellers to inflate their contract prices. In the process, it reduced the cost of discovering the correct price to pay landowners, which can often be time-consuming and expensive when done through strategic, bilateral bargaining between conservation buyers and landowners (Stoneham et al. 2002a).

BushTender began as a trial in two phases between 2001 and 2003 in northern Victoria and Gippsland. A 2006 report commissioned by the government revealed a high compliance rate among participating landowners, with between 84 and 94% of landholders satisfactorily completing their management commitments each year. The report indicated high levels of landowner satisfaction with the programme and demonstrated 97% of participants opting for long-term contracts of at least 6 years (Department of Sustainability and Environment 2006; Stoneham et al. 2002b).

The *BushTender* agreements stipulated inputs and actions for implementation over a 3-year period. The government made an initial payment to successful bidders to cover necessary capital costs (e.g. fencing) with annual payments made on the basis of performance progress. Progress and performance were measured annually, with both parties having responsibility for monitoring. The budget for the pilot phase was set at $400,000 for the initial 3-year period and the auction process allowed the government to ensure conservation and management of 3200 hectares of native vegetation in priority areas.

The auction process yields a diversity of bids, allowing the government to select those which offer the best biodiversity value. The range of bids, and especially the fact that the government received many low ones, demonstrates that many landowners show a willingness to share the cost of land management and, in this case, the cost of biodiversity conservation on their land with the government. As a result, the auction provides a mechanism to achieve more biodiversity conservation at a lower overall cost than more traditional fixed price contracting methods. The analysis indicates that achieving the equivalent biodiversity benefits on the same number of hectares under a fixed price system would require a sevenfold increase in budget, or an investment of around $2.7 million (Stoneham et al. 2002a).

The auction system allowed the Victoria Government to take advantage of a wide range of bids by landholders to provide biodiversity and land management services. And most landowners did not bid profit-maximizing market rates but implicitly entered into a cost-sharing arrangement with the government for sustainable land management. This allowed more conservation to be achieved with the limited funds available. Farmers benefited from contract payments that provided them with resources to manage their own land, and the state achieved more conservation for its investment.

The *BushTender* programme demonstrates the important role government can play in providing performance payments to private landowners to protect biodiversity and ensure sustainable land use practices. The positive outcomes resulted from the government's willingness to explore market options and provide sufficient budgetary allocations to fund the programme. The State Government of Victoria has committed $3.2 million to undertake targeted *BushTender* projects in Northeast and Central Victoria, and a further

$2.7 million to spread the *BushTender* programme to other parts of Victoria over a 3-year period.

It is difficult to speculate about the future of *BushTender*. With the maturation of the programme, the level of cost-effectiveness could decline as bidders gain more price information and market knowledge. As farmers become aware of the value of the government's payments to all participants, they could begin to bid up the price of their participation in order to capture greater rent, thereby increasing the cost of biodiversity protection (Stoneham et al. 2002a). However, in this type of auction, the price for biodiversity conservation may be difficult to infer. Even if bid prices were to increase, cost sharing between government and private landowners is likely to continue, thereby mitigating against significant price increases and allowing the programme to deliver the desired conservation benefits at a favourable price.

Popularity of direct payment systems

The auction and other direct payment systems give important recognition to the role of private landowners in the management of rangelands and biodiversity. This recognition can begin to foster better understanding and collaboration between conservationists and private landowners.

Experiences in North America demonstrate that landowners are more likely to respond positively to conservation measures when their ownership rights are recognized and payments are based on an established contract. In a study in Texas on landowner perceptions regarding ecosystem services and cost sharing for land management, Olenick et al. (2005) found that on average potential participating landowners found performance contracts the only type of mechanism to which they were positively disposed. They were willing to enter into land management agreements as part of specific contracts with clear requirements and payments. Most respondents considered the ideal performance contract period to be between 6 and 10 years.

The direct payment approach is popular because it recognizes property rights and offers direct compensation for conservation practices, addressing landowners' concerns that conservation is costly. In direct payment schemes, the buyer and seller reach an agreement on the value of conservation, and the seller (provider of the conservation benefit) receives an acceptable compensation. As one US landowner stated when asked about conservation, 'it does not trump the bank', indicating the importance of the conservation rental payments to the landowners (Kowal 2006) and the importance they place on recognition that they need to be compensated for adopting alternative land uses. No matter what the model is or how it is applied, paying farmers to conserve land assigns economic importance to conservation outcomes and engages private landowners as partners in their achievement.

Conservation easements and purchase of development rights programmes

Conservation easements and PDR programmes are legal tools that remove the development potential of an owner's private lands in exchange for money or other considerations (such as access to forage, tax deductions, etc.). The advantage of this approach is that it allows landowners to capitalize on the speculative portion of their property value while retaining the right to conduct and benefit from non-development activities (such as agricultural production, ranching and conservation) (Curtin et al. 2002). The value of an easement is generally the difference between the market value of the land and the value of the land after removal of its development rights. For ranches, this value typically ranges from 20 to 90% of the property's market value (Veslany 2002). Once the right to subdivide, build or convert to another use is gone, the ranch's value rests on its livestock productivity and potentially other ecological and amenity values. This, in turn, provides the rancher incentives to preserve the land's ecological function and range conditions, thus providing a legal framework for restoring the incentive structures of the pre-land speculation era prior to World War II (Curtin et al. 2002).

A disadvantage of these approaches is that they work only on lands under private ownership, that is, on deeded acres. In North America, land managers utilize a matrix of private land holdings (deeded acres) and public lands, with their own mandates, rules and regulations. As a result, much of the land managed by North American ranchers is not easily subject to easement or PDRs. Ranchers are often highly dependent on the use of public lands for their economic well-being and thus are unlikely to sell or donate conservation easements on their deeded property if there is the potential, due to land use regulations, of losing access to adjacent public lands. For the rancher, the loss of any grazing use on the deeded land would result in a loss of most or all of its value for livestock production and perhaps for conservation purposes as well, because the landscape becomes too fragmented to be viable economically or ecologically. In many cases, the only solution to this dilemma is to include an escape clause in the conservation easement under which the easement terminates immediately if there is a loss of associated public lands grazing leases. This overall approach provides a way to avoid fragmentation and stimulate the conservation potential and integrity of many landscapes. In addition, scientific research by groups such as the Malpai Borderlands Group (MBG) of southeastern Arizona and southwestern New Mexico documents that ranching and conservation are compatible and leads to new approaches to achieve a balance between land use for ranching and conservation (Box 7.2).

Box 7.2 **Generating financial support for conservation through science and economics: Malpai Borderlands Group**

Charles Curtin

'We wanted the best and most credible scientists in the U.S. working with us... If the information and research is honest and unbiased, we'll let the chips fall where they may.' – Bill McDonald, Rancher, MacArthur 'Genius' Award Recipient, and Executive Director of the Malpai Borderlands Group (MBG).

The MBG of southeastern Arizona and southwestern New Mexico has pioneered links between collaborative conservation and science with conservation easements. The MBG arose from a concern for the need to restore fire to the landscape to sustain and restore ecosystem function (Curtin 2002, 2005; Curtin & Western 2008). A hallmark of MBG's approach is a focus on large-scale experimental science. Especially significant are two large-scale (over 1000 acres) experiments examining climate, fire and grazing. These experiments can be, in turn, linked to hundreds of monitoring plots to get a view of how management plays out across the rest of the project area. Of the more than 400 collaborative groups involved in the 2005 White House Conference on Collaborative Conservation, the MBG is perhaps the only one to invest in large-scale experimental science to test the group's practices. Through an investment in experimental science in the peer-reviewed literature, the MBG and collaborators have been able to document that ranching and conservation are compatible, a point that is key to sustaining the local economy and culture.

Experimental science helps ensure that conservation targets are met. When conservation is undertaken without systematic accounting for successes and failures (through the necessary collection of quantitative data), it is hard to determine whether conservation efforts are effective or warrant the cost. Research entailing replicated experiments to test hypotheses is a departure from conventional monitoring approaches in that they test underlying assumptions, in the process generating deeper understanding of the ecological processes to be conserved or managed. For example, in the borderlands the conservation efforts are predicated on the assumption that fire and grazing are important for sustaining 'ecological health', which is the core goal of the Group's mission statement. The experimental science programme, coupled with monitoring, has proved that both are key for sustaining ecosystem processes such as maintaining diversity at local and landscape levels. The results have been communicated through numerous peer-reviewed publications. Experimental research has the potential to fundamentally change understanding and management built upon the confluence of local and science-based knowledge (Curtin 2005). The importance of peer-reviewed literature, coupled with external science oversight through yearly meetings of external science advisors, is recognized as additional important parts of generating credibility (Curtin 2002).

From an economic perspective, science is significant because it provides the credibility that supports other facets of conservation. For the MBG, being able to demonstrate the ecological viability of ranching and how it is key in sustaining open landscape creates the ability to secure the capital to purchase development rights that are locally called 'conservation easements'. Early in its existence, the MBG generated interest in conservation easements through 'grassbanking', a process by which participating ranches trade a development right on their deeded acres for the right to graze cattle on the grassbank. This allows the rancher to remove cattle to another ranch and rest and restore their own land. This programme achieved the dual goals of improving the ecological integrity and function of the participating ranches while securing the development rights to that land.

Grassbanking proved so popular that all the MBG's available spaces filled, so the group switched to a programme of cash payments for development rights (e.g. 'conservation easements'). Since the mid-1990s, the group, in concert with other organizations working in the region, has conserved 60% of the private land within MBG's million-acre planning area. The revenue from the sale of development rights has enabled ranchers to refinance, erase existing debt or allow individuals to buy out other family members and consolidate their holdings, with the result that ranching has become more viable in the region.

In the long run, the foundation in experimental science may even prove key to niche marketing of beef that allows ranchers to obtain higher and more consistent prices for their product if they can document that production and land management meet certain standards. In this case, the documentation through science that the ranching is ecologically sustainable and actually contributes to land health establishes the proof of compliance with existing standards. Moreover, a foundation in science gives groups a competitive edge; being able to document results can set an organization apart in a crowded field of community-based groups competing for scarce conservation dollars.

Grassbanking: leveraging a natural resource market to achieve conservation

As conservation organizations begin to embrace market solutions, they often find common ground with private landowners in devising markets for natural resources and ecosystem services. Matador Ranch (Box 7.3) presents an example of a successful grassbank, a conservation tool that leases forage land to ranchers at a discount in exchange for practising conservation on their home ranches. The grassbank creates a market for high-quality forage and provides an effective incentive for better ecological stewardship by keeping ranchers responsible for the management of the land. As a result, there is greater local support for conservation than would arise from locking up the land through an NGO purchase. But, perhaps most importantly, the grassbank enables ranches to become better integrated with local ecological and economic realities (Curtin 2005).

Box 7.3 **Hedging for conservation and a ranching future: Matador Ranch Grassbank**

Bob McCready, The Nature Conservancy

'I don't want the grassbank held up as the cure-all for all problems. But it's the best tool we've found for reducing conflict, and for working constructively with different interests.' – Dale Veseth

Dale Veseth is a rancher in Phillips County, Montana, an area The Nature Conservancy and other conservation organizations call the Glaciated Plains. Veseth's ranch and those surrounding it are in the middle of North America's largest remaining expanse of mixed-grass prairie – a veritable sea of grass, but not one without threats. Although it is unlikely that the next decade will bring the kind of ex-urban development seen along the Rocky Mountain Front of Colorado, there is increasing pressure to convert these grasslands into dry land wheat farms.

The Glaciated Plains is one of the most ecologically significant grasslands in North America. This area encompasses 3 million acres of grasslands and represents one of the last large wild expanses of the Great Plains. Due to its size, intact nature and mosaic of mixed-grass prairie, big sagebrush steppe and other diverse habitats, the Glaciated Plains hosts the most intact assemblage of wildlife in the Great Plains of North America. Among the rare and endangered species that occur here are the black-footed ferret, black-tailed prairie dog, swift fox, burrowing owl, sage grouse, ferruginous hawk, mountain plover, long-billed curlew, Baird's sparrow, Sprague's pipit, chestnut-collared longspur, McCown's longspur and lark bunting. In fact, the Glaciated Plains is home to North America's highest number of endemic grassland bird species – an ecological group of birds that is declining more rapidly than any other on the continent.

The Nature Conservancy purchased the 60,000-acre Matador Ranch in 2000. The ranch comprises 31,000 acres of private land and 29,000 acres of grazing leases on state and federal land. The ranch abuts the Fort Belknap Reservation to the west, is 40 miles south of the Montana Hi-Line town of Malta, and just north of the C.M. Russell

National Wildlife Refuge and the Missouri River National Monument. The purchase permits realization of a conservation vision that spans far beyond its boundaries.

Phillips County is one of the more lightly populated counties in the United States, with the population declining from 30,000 in the 1930s to 4500 today. This is a landscape beset by long winters, unpredictable weather and frequent drought, factors leading to very low profit margins for ranchers, a conservative outlook and a suspicion of outsiders and new ideas. In response, The Nature Conservancy convened an advisory committee of local ranchers to develop the conservation vision and strategies for the Matador Ranch and Glaciated Plains. Phillips County ranches are typically large and have been managed by the same families for three, four and even five generations. Following multiple advisory committee meetings, it became clear that there was a need to adapt the typical ranching business model to become more flexible to deal with the vagaries of the cattle market and periodic drought. The overriding goal was to achieve conservation outcomes and create a more stable economic situation, in order to maintain ranching operations and a rural lifestyle for generations to come.

The parties agreed on a strategy to develop a Matador Ranch Grassbank. A grassbank can be defined as a partner-based collaborative process where forage is traded between landowners for conservation benefits. The grassbank operates as follows: ranchers pay a lease rate to graze their cattle on the Matador Ranch, but as a requirement for participation, agree to maintain their native grassland and enroll in a weed-prevention programme. For additional discounts, they can select from a menu of short-term habitat protection and management actions directed at rare or declining species. During periods of drought, access to the abundant grass on the Matador allows ranchers to move their cattle off their own ranch, preventing the need to overgraze their own land or to sell off part of their herd. If ranchers receive all available discounts, they will receive a 40% discount off the full market value of a grazing lease. Since 2002, 11–14 ranches that together manage between 166,000 and 292,500 acres enrolled in the grassbank each year. Combined with the 60,000-acre Matador, the grassbank has led to the protection and improved management of nearly 350,000 acres of this biologically important landscape.

The grazing lease discount system has positively impacted a huge swath of the Glaciated Plains, but there are less tangible benefits as well. Since working with The Nature Conservancy, the ranching community has become more receptive to other conservation opportunities. Whereas little interest in conservation easements previously existed, area ranchers now view easements as an increasingly attractive option. In addition, Matador Grassbank participants and other area ranchers formed the non-profit Rancher Stewardship Alliance in 2003. The goal of this group is to develop cost-effective conservation practices which, when implemented by prairie ranchers, will result in sustained health and vitality of native wildlife, ranchers and rural communities (see www.ranchersstewardshipalliance.org).

Extensive stakeholder discussion and collaboration on management of both grassbank and home ranch lands are necessary to make the grassbank market function. Equally important are a clear set of management goals, because resting rangelands alone does not intrinsically lead to landscape recovery (Curtin 2002). Rested lands can actually degrade, compared with managed lands, highlighting the need for active management such as monitoring and restoration, as well as viewing management in its larger climatic and environmental context (Curtin & Brown 2001). This includes preventing the processes that led to landscape degradation in the first place. Livestock managers must therefore be willing and able to change their management practices. Without regard for these factors, ranchers can find themselves having made a considerable investment but not having realized a commensurate return in range productivity.

Grassbanking draws on land use practised by nomadic cultures throughout history. The concept has parallels with Spanish land grants and grazing associations that once typified much of American ranching (Remley 2000). Grassbanking also resembles the communal grassing practices of the Maasai and other herding cultures (e.g. Western & Gichohi 1993; Homewood 2004). But the ability to overcome ecological and economic constraints has limits. Grassbanks can provide a false sense of security in times of drought, an important consideration given the still-unknown ramifications of climate change. Livelihood diversification, deriving from multiple forms of consumptive and non-consumptive ranch income, may be the only long-term solution for ranchers. In addition, grassbanks themselves are subject to drought and other environmental degradation, creating a need for more dynamic approaches to grass reserves. The East African experience is instructive in this respect. East African pastoralists designate temporary or smaller localized grass refuges called *ol-opololi* or *ol-okeri* in response to seasonal rainfall or pasture conditions. By varying the location and size of grass refuges from season to season, the pastoralist attempt to more closely reflect the ecological realities of seasonality and microclimate conditions. In this way, East African pastoralists avoid the constraints of small land holdings by tailoring livestock foraging to ecological flux, much as free-ranging wildlife does (Curtin & Western 2008).

Emerging payments for ecosystem services

Emerging markets for ecosystem services may also play an important future role in creating demand for rangelands conservation. Government regulation has created market opportunities for ecosystem services such as carbon and biodiversity conservation. The growth in the newly emerging carbon market is a result of cap-and-trade systems developed within the framework of the Kyoto Protocol. Grasslands are well suited for carbon sequestration because the rate of sequestration is high and carbon is securely stored as root mass (Browne et al. 2007). Improved land use management, including reducing deforestation and soil disturbance, can decrease carbon dioxide emissions.

Although international agreements and national programmes to establish payments for avoided emissions from large-scale land management remain under negotiation, some mechanisms for emission reduction payments have emerged through the voluntary carbon market. In North Dakota, Ducks Unlimited, an NGO, collaborates with landowners and the U.S. Fish and Wildlife Service to enroll native prairie and Conservation Reserve Program Land in the government's Grassland Easement Program (Nicholson 2008). Through the programme, landowners agree to manage the land in accordance with predetermined conservation practices. In addition to the government's easement payments, farmers receive payment from Ducks Unlimited, negotiated through the Chicago Climate Exchange, for carbon stored in the soil. The easement and carbon payments produce enough income to motivate farmers to leave the land in conservation, thereby protecting biodiversity and preventing carbon emissions. As carbon markets develop and countries search for ways to reduce their greenhouse gas emissions, rangelands conservation will be better linked to markets, and landowners will enjoy greater economic benefit from implementing conservation practices.

Creating demand for sustainable products

The development of products that are linked to sustainable land use and conservation represents another approach to create conservation incentives. Commercializing nature's diversity is not straightforward. Market creation occurs in many ways and involves different actors, including governmental, the business, and the non-governmental sector.

NGOs can also play an important role in helping create new markets for conservation. They have expertise in convening parties, brokering deals and providing important technical expertise. They also contribute to the public education process, working with producers and consumers alike to help address market failure and improve opportunities for producers, such as farmers and ranchers, to participate in the marketplace.

For example, in Zambia, the Wildlife Conservation Society launched its Community Markets for Conservation (COMACO) project to create income opportunities for farmers while also building support for conservation. Farmers who affiliate with one of COMACO's three trading centres can sell their produce for higher prices than offered by middlemen operating in the region. In exchange, the farmers agree to sign contracts that commit them to removing snares and protecting wildlife and habitat, thereby linking economic benefit directly to improved land and conservation management practices. A 2007 business plan completed for COMACO reported that through an effective outreach training approach, COMACO's farmers are now practising composting and zero-tillage farming practices, creating improved conditions that reduce soil and water loss while reducing the need to clear new land for farming (COMACO 2007). As a result, more habitat is available for wildlife and forest and overall land degradation is reduced. Over the period 2003–6, the value-added processing and marketing to larger buyers in the capital and regionally has resulted in a 74% increase in the annual household income of participating farmers. This corresponds with COMACO's wildlife monitoring results, which show positive trends in wildlife populations between 2002 and

2006, and demonstrate wildlife numbers that are significantly higher around the areas where COMACO operates in comparison to non-COMACO sites. COMACO hopes to expand markets regionally and take advantage of the consumer desire for organic as well as *green* products and by doing so continue to gain local support for sustainable land management and conservation.

Demand is increasing worldwide for food and non-food products that meet *sustainable* or *green* attributes (e.g. organic, locally produced, wildlife friendly, etc.). This is especially the case among higher income consumers in wealthier countries. The *Phoenix Business Journal* (2008) reports the value of the market for green and healthy products at $209 billion with estimates of growth to $400 billion in 2010. In the United States alone, the number of consumers who identify themselves as regular buyers of green products rose from 12 to 36% between 2007 and 2008, with the estimates for 2009 remaining at the latter figure despite the economic downturn (Environmental Leader 2009). Green products fulfil consumers' desire for health, quality and activism, and as demand for quality and purposeful consumption grows, so does the demand for sustainable products (Jensen 2006). As a result, retail outlets featuring these products are introducing them to a broader range of consumer groups.

This consumer demand fuels the private sector's burgeoning role in biodiversity conservation. Investment in production approaches that provide conservation benefits has increased. Commitments by high-profile companies to base their sourcing decisions on certain practices (e.g. fair trade, organic cotton, sustainably harvested wood) allow producers to profit from land uses that support resource conservation.

Sustainable ranching for the speciality beef market

Conventional beef producers in the western United States are struggling. The cattle industry is fraught with economic uncertainty, causing producers to face significant risk. This uncertainty is compounded by changes in land valuation and urbanites' increased demand for a rural lifestyle that drive up the price of rural land. Developers offering huge sums tempt ranchers to sell their land – and many do, realizing they can get significantly more money from the sale of their land than from raising cattle (Curtin et al. 2002). Many ranches sold for this reason are subdivided into recreationally oriented 'ranchettes' (Curtin et al. 2002; Sengupta & Osgood 2002) and large expanses of land become subject to fragmentation as a result of differing land uses. Fencing off parcels and siting buildings without attention to the overall landscape can disrupt wildlife and cattle-grazing corridors and put more intense pressure on grazing resources. While zoning and other policy measures can address some of these impacts, incentives are more likely to elicit positive responses from landowners and result in conservation-oriented management decisions.

Work conducted by the Corporation for the Northern Rockies (CNR) (Lil Erickson, personal communication, 2005) indicates that ranchers, like many of their counterparts around the world, value their lifestyle and culture. These values, however, are being lost as land is developed. Many ranchers are committed to staying on their land even if that choice appears irrational, preferring suboptimal financial returns to a radical change in lifestyle, as long as income earned allows them to provide for their families. This view is in part a lifestyle choice and in other part the rancher's perceived role of himself as a provider of food and fibre to the country. The latter is a keen source of pride for many ranchers. As a result, ranchers are likely to assume the opportunity cost of lower returns if they can keep and work their land, with the implication that they do not need to capture the total value of development rights to adopt sustainable management practices.

Developing markets for meat raised sustainably may be a key factor in curbing pressure for conversion and enabling ranchers to keep and continue working their land. New markets depend on changing demand and supply conditions (Landell-Mills & Porras 2002). Consumer demand for healthy and safe beef and other meat products is growing in North America. Consumption of meat in the United States has grown by 12% since 1990, and by 50% in Mexico over the same period (Jensen 2006). Rising incomes – especially in richer countries but also in developing countries – will fuel demand for high-quality niche meat products, marketed for specific attributes, including health, safety, sustainability and conservation.

Recognizing the market opportunities, CNR took a lead on sustainably produced, natural beef (produced with no hormones or antibiotics). The corporation operates on both sides of the market chain, working with ranchers to change agricultural practices and negotiating with companies and buyers to stimulate product demand. CNR's aim is to capture a price premium over traditional beef, high enough to support and keep ranchers working their land.

Converting to sustainable production requires effort. Sustainable practices produce healthier animals than conventional ranching, resulting in lower veterinary costs and higher quality beef. Lower input costs from improved practices, however, have been shown to be insufficient to change rancher behaviour (Lil Erickson, personal communication, 2005). This is especially true in a conservative culture, where early adopters face potential ridicule from colleagues. CNR found ranchers needed a clear market signal that they would be rewarded for adopting sustainable practices. In this case, the niche market for natural beef responded. CNR linked up with Montana Legend, a local retailer of sustainable beef, which offered a price premium of 12 cents per pound over normal commodity prices. The premium provided two important benefits: motivating farmers to stay on the land and leading ranchers to adopt sustainable practices.

Building a market for high-quality natural beef is critical to creating a price premium. Quality is the primary market-driving factor. Sustainability, and the 'green' label, earns that market premium which businesses like Montana Legend can afford to pay to the producers. The green label may open the door to niche markets, but companies need to ensure product quality to gain and maintain their customer base. Montana Legend has realized success in this, with demand for its products doubling every year over the period from 2003 to 2006. With such growth, Montana Legend's new challenge is to source quality natural beef. CNR thus plays an important broker role – working with ranchers to provide technical assistance and advice to improve the quality of beef production so that more Montana ranchers can gain access to the growing sustainable beef market.

The collaboration between an NGO, working on behalf of producers, and a retail company with access to speciality markets delivered both economic and conservation benefits. As Montana Legend discovered, one of the greatest challenges to developing a market for conservation-oriented products is reliable supply. Often, producers are small scale and their ranches are located in isolated areas, making it difficult to organize production to respond to growing market demand. CNR's ability to organize and gain the trust of ranchers helped address a key supply constraint for the company while Montana Legend's ability to market sustainably produced beef increased economic benefits to ranchers, with the result being opportunities for improved management of ranchlands in the Northern Rockies.

Building demand for sustainable cuisine

The lack of supply of sustainable products is a real market constraint – a type of market failure, which Xanterra Parks and Resorts discovered in its efforts to introduce sustainable cuisine at its national park concessions. This U.S. National Park Concessionaire operates in several US parks, including Yellowstone National Park. As part of its commitment to sustainability, Xanterra created a sustainable cuisine programme, buying organic food and sustainably produced and fair trade products from suppliers such as Montana Legend. Xanterra's goal is, by 2015, to source 50% of all food served in its hotels and restaurants from sustainable producers, with sustainable defined as natural (organic, sustainable land use practices, no hormones or antibiotics) and local.

Xanterra has met success in its strategy. In 2004, Xanterra purchased $1.4 million worth of sustainable cuisine products. By 2006 that figure had increased by 120% to $3.1 million across a variety of product categories (http://www.xanterra.com/Sustainable-Cuisine-385.html). Xanterra finds that many consumers purchase products just because they are local. Xanterra sells out local products first; for example, their Yellowstone Lodge menu choices from Montana are exhausted more quickly than sustainable choices from Oregon. Quality is another factor in this success. Xanterra can sell local products, especially sustainable beef, at higher prices because the quality is so high and consumer response positive. Most visitors are willing to pay higher prices for sustainable, locally produced products, although some customers complain about the regional bias.

Xanterra's most serious challenge is product sourcing. The company faces both quality and quantity constraints in achieving its sustainable cuisine goal. Xanterra originally considered sourcing from US states connected geographically to Yellowstone National Park. That strategy changed to a broader regional approach when local producers were unable to meet demand. This broadening of geographic scope allowed Xanterra to source sustainable products from the Northwest region and even California to meet consumer demand and maintain its commitment to selling sustainable products.

Efforts, like Xanterra's, to introduce consumers to more sustainable and healthier products has stimulated the market for sustainable beef and met with a positive response from ranchers, as discussed in the case of the CNR. This case demonstrates that an early commitment to sustainability and demand for high quality, local products can pose challenges: demand can significantly exceed available supply, especially in early years. However, as companies remain committed to sourcing sustainably and demand grows and is sustained, the supply gap should disappear as producers respond both in terms of quantity and quality. The growing demand for sustainable products may be the stimulus needed to not only keep land in range use but also to foster broader conservation-friendly agricultural practices.

Summary and conclusions

Sustainable management and biodiversity conservation on rangelands may ultimately depend on providing appropriate incentives to private landowners to make conservation a viable financial option. This chapter has outlined three broad approaches to creating incentives: (i) direct payments, (ii) conservation easements and PDR programmes and (iii) creation of new markets. These are summarized in Table 7.1.

The direct payment approach pays landowners to manage land for conservation, while easements and PDRs use legal tools to allow an exchange of land value for money or other benefits. The sustainable market approach is less direct, but succeeds by creating a demand for sustainably produced products and ecosystem services and by ensuring a price that motivates market participation. What they all have in common is the creation of economic value around conservation.

Performance-based contracts and direct payments stimulate biodiversity conservation and sustainable land management by establishing mechanisms for monitoring conservation outcomes to ensure compliance and to create partnerships between government, NGOs, private sector entities and landowners. They also recruit landowner participation and cost sharing in biodiversity conservation, thereby achieving cost-effective conservation. Performance payments can be effective and efficient, but their success depends on having sufficient and consistent resources to ensure commitments to relatively long-term contracts. The need to commit funds over time is a constraint that buyers of services need to address during the planning process. Payments for ecosystem services, such as payments over time to ensure carbon dioxide sequestration in soil over a long-time horizon, offer one possibility of resolving the lack of financing to create incentives for conservation on private land. Some form of market or payment system for land-based carbon sequestration will likely emerge from climate change negotiations. Scientists generally agree that effective land management can reduce greenhouse gas emissions significantly at a relatively low price. Society's desire to reduce emissions globally will

Table 7.1   **Financial instruments to conserve rangelands.**

Instrument	Mechanisms	Essential ingredients

Direct payment options for conservation
 	
Examples:
 	
Kitengela, Kenya
Tarangire, Tanzania
Victoria, Australia
North Dakota, USA
	• Lease payments
	• Land rentals
	• Land management contracts
	• Auctions (*BushTender*)
	• Payments for ecosystem services (soil carbon payments)
		1. Clear and enforceable contracts
		2. Sustainable source of financing including, *inter alia*:
		  • Long-term government commitments
		  • Private sector annual payments
		  • Payments from endowed funds
		  • Long-term donor funds
		  • Long-term climate mitigation payments
		3. Recognized/respected property rights including private sector support

Legal and policy instruments that create value for conservation *Malpai Borderlands, Arizona* *Matador Ranch, Montana* *Ducks Unlimited, North Dakota*	• Conservation easements • Purchase of development rights • Grassbanking • Carbon trading	1. Private deeded land, cannot be effected on public lands 2. Funds available to purchase development rights 3. Legal and policy instruments exist, or can be introduced to provide stimulus
Development of markets for sustainable products *Montana Legend Sustainable Beef* *Xanterra Sustainable Cuisine* *COMACO*	Demand side: companies increase demand for sustainable and conservation-friendly products Supply side: price premiums paid to producers to make sustainable land management feasible	1. Robust market 2. Price premiums 3. Product of sufficient quality 4. Ability to meet demand once it is created 5. Willingness of ranchers to change land management practices

COMACO, Community Markets for Conservation.

provide opportunities for public–private investments in improved rangelands management and conservation.

Environmental funds offer a workable mechanism to secure long-term performance payments, but for some reason these funds have been slow to adopt performance payments. As environmental funds and the stakeholders they serve seek answers to growing conservation challenges, direct payment options will likely be seen as a viable alternative and employed where most effective. Payments that are well-directed and require measurable performance will likely be the most successful and taken most seriously by participating landowners.

The market growth for products that support biodiversity and provide healthy and sustainable alternatives provides producers with an incentive for conservation. Demand for a variety of different products – such as forage, water, other environmental services, ecotourism and quality food – offers a stimulus for farmers and ranchers to adopt improved agricultural and land management practices. The challenge lies in making these markets work effectively. The market for ecosystem services and conservation benefits from resource use is not mature, but is expected to expand as a result of dwindling resources, growing consumer demand and government regulation. Natural and sustainable food and fibre products face supply and quality-control constraints that market participants are working to address. As consumers seek healthier lifestyles, however, producers will respond and embrace opportunities that support sustainable land use. The challenge lies in accessing these markets – having both the information about markets and the capability of producing the quality they demand.

In developing financial incentives for conservation, there is no one solution that works everywhere. Government, NGOs and the private sector must be creative and innovative in devising solutions. Experiences from East Africa, Australia and the United States demonstrate options for achieving successful conservation-based land management. Many other options will develop as people explore the use of incentives to achieve conservation objectives. The key is the understanding of local, cultural, social and economic drivers, as well as the motivation of landowners, and then designing programmes and policies that stimulate the development of innovative approaches and market mechanisms to best respond to existing conditions. This chapter indicates that conservation can be achieved if people are compensated fairly. Showing landowners the money places a value on that conservation and delivers results.

References

Borner, M. (1985) The increasing isolation of Tarangire National Park. *Oryx* 19, 91–96.

Brown, J.H., Curtin, C.G. & Brithwaite, R.W. (2003) Management of the semi-natural matrix. In: Bradshaw, G.A. & Marquet, P.A. (eds.) *How Landscapes Change: Human Disturbance and Ecosystem Fragmentation in the Americas.* Springer-Verlag, Berlin, pp. 327–343.

Browne, D.M., Ringelman, J. & Dell, R. (2007) Case study – carbon research in Prairie Wetlands and Grassland Systems. In: Browne, D.M. & Dell, R. (eds.) *Conserving Waterfowl and Wetlands Amid Climate Change.* Ducks Unlimited, Inc., Memphis, pp. 37–39.

COMACO (2007) Business Plan. Wildlife Conservation Society, NY.

Curtin, C.G. (2002) Cattle grazing, rest, and restoration in arid lands. *Conservation Biology* 16, 840–842.

Curtin, C.G. (2005) Linking complexity, conservation, and culture in the Mexico/US borderlands. In: Lyman, M.W. & Child, B. (eds.) *Natural Resources as Community Assets: Lessons from Two Continents.* Sand County Foundation & Aspen Institute, Monona, Washington , D.C., pp. 235–258.

Curtin, C.G. & Brown, J.H. (2001) Climate and herbivory in structuring the vegetation of the Malpai Borderlands. In: Bahre, C.J. & Webster, G.L. (eds.) *Changing Plant Life of La Frontera: Observations on Vegetation in the United States/Mexico Borderlands.* University of New Mexico Press, Albuquerque, pp. 84–94.

Curtin, C.G., Sayre, N.F. & Lane, B.D. (2002) Transformations of the Chihuahuan Borderlands: grazing, fragmentation, and biodiversity conservation in desert grasslands. *Environmental Science and Technology* 218, 55–68.

Curtin, C.G. & Western, D. (2008) Rangelands as a global conservation resource: lessons from cross-cultural exchange between pastoral cultures in East Africa and North America. *Conservation Biology* 22, 870–877.

Department of Sustainability and Environment (2006) *BushTender – The Landholder Perspective: A Report on Landholder Responses to the BushTender Trial.* State of Victoria, Department of Sustainability and Environment, East Melbourne.

Environmental Leader (2009) Consumer Survey: Growth of 'Green' Consumption on Hold. http://www.environmentalleader.com/2009/03/06/consumer-survey-growth-of-green-consumption-flounders/, 6 March, 2009 (accessed May 2009).

Ferraro, P.J. (2001a) Global habitat protection: limitations of development interventions and role for conservation performance payments. *Conservation Biology* 15, 1523–1739.

Ferraro, P.J. (2001b) Reconciling the long-term conservation needs of large mammals and local communities: conservation performance payments. In: Wikramanayake, E., Dinerstein, E. & Loucks, C.J. (eds.) *Terrestrial Ecoregions of the Indo-Pacific: A Conservation Assessment.* Island Press, Washington, D.C., pp. 185–187.

Ferraro, P.J. & Kiss, A. (2002) Direct payments to conserve biodiversity. *Science* 298, 1718–1719.

Ferraro, P.J. & Simpson, R.D. (2001) Cost-effective conservation: a review of what works to preserve biodiversity. *Resources* 143, 17–20.

Ferraro, P.J. & Simpson, R.D. (2002) The cost-effectiveness of conservation performance payments. *Land Economics* 78, 339–353.

Gentner, B. (1999) *Characteristics of Public Lands Grazing Permittees.* Masters Thesis, Oregon State University.

Gichohi, H.W. (2003) Direct payments as a mechanism for conserving important wildlife corridor links between Nairobi National Park and its wider ecosystem: The Wildlife Conservation Lease Program. V World Parks Congress, Durban, South Africa.

Ha, A., Strappazzon, L., Crowe, M. & Todd, J. (2004) *Contract Design and Multiple Environmental Outcomes: An Economic Perspective.* State of Victoria, Department of Primary Industries, Melbourne, Victoria, Australia.

Homewood, K.M. (2004) Policy, environment and development in African rangelands. *Environmental Science and Policy* 7, 125–143.

Jensen, H.H. (2006) Consumer issues and demand. *Choices* 21, 165–169.

Knight, R.L., Gilgert, W.C. & Marston, E. (eds.) (2002) *Ranching West of the 100th Meridian: Culture, Ecology, and Economics.* Island Press, Washington , D.C.

Kowal, J. (2006) Farmers and Conservationists Form a Rare Alliance. New York Times, December 27, 2006.

Lamprey, H. (1963) Ecological separation of the large mammal species in the Tarangire Game Reserve, Tanganika. *East African Wildlife Journal* 1, 63–92.

Landell-Mills, N. & Porras, I.T. (2002) *Silver Bullet or Fools' Gold?: A Global Review of Markets for Forest Environmental Services and their Impact on the Poor.* International Institute for Environment and Development, London.

Martin, W.E. & Jefferies, G.L. (1966) Relating ranch prices and grazing permit values to ranch productivity. *Journal of Farm Economics* 48, 233–242.

Nelson, F. (ed.) (2005) Social and Ecological Dynamics and Complexity in the Simanjiro Plains: A Roundtable Discussion. Tanzania National Resource Forum, Workshop Report No. 8. Tanzania National Resource Forum, Arusha, Tanzania.

Nicholson, B. (2008) New ND carbon credit program aims to help ducks. International Business Times. August 8, 2008. http://www.ibtimes.com/articles/20080808/new-nd-carbon-credit-program-aims-to-help-ducks_all.htm (accessed May 2009)

Nkedianye, D. (2004) *Testing the Attitudinal Impact of a Conservation Tool Outside a Protected Area: The Case for the Kitengela Wildlife Conservation Lease Programme for Nairobi National Park*. Masters thesis, University of Nairobi, Nairobi, Kenya.

Odell, E.A. & Knight, R.L. (2001) Songbird and medium-sized mammal communities associated with ex-urban development in Pitkin County, Colorado. *Conservation Biology* 15, 1143–1150.

Olenick, K.L., Kreuter, U.P. & Conner, J.R. (2005) Texas landowner perceptions regarding ecosystem services and cost-sharing land management programs. *Ecological Economics* 53, 247–260.

Phoenix Business Journal (2008) Nielsen Co. event pegs growth of green products to $400 billion by 2010. June 5, 2008. http://www.bizjournals.com/phoenix/stories/2008/06/02/daily52.html (accessed May 2009).

Radeny, M., Nkedianye, D., Kristjanson, P. & Herrero, M. (2007) Livelihood choices and returns among pastoralists: evidence from Southern Kenya. *Nomadic Peoples* 11(2): 31–55.

Remley, D. (2000) *Bell Ranch: Cattle Ranching in the Southwest, 1824–1947*. Yucca Tree Press, Los Cruces.

Rodriquez, L.C., Henson, D., Moran, D., Nkedianye, D., Reid, R. & Herrero, M. (2006) Private farmers' compensation and viability of protected areas: The Case of Nairobi National Park and Kitengela Dispersal Corridor. Paper delivered at the 8th Annual BIOECON Conference on the Economic Analysis of Ecology and Biodiversity, Kings College, Cambridge, 29–30 August, 2006.

Sachedina, H. (2006) Conservation, land rights and livelihoods in the Tarangire Ecosystem of Tanzania: increasing incentives for non-conservation compatible land use change through conservation policy. Conference paper presented to Pastoralism and Poverty Reduction in East Africa: A Policy Research Conference, International Livestock Research Institute, Symposium: Wildlife and Pastoralists, Nairobi, Kenya, 27–28 June, 2006.

Sayre, N.F. (2002) *Species of Capital: Ranching, Endangered Species, and the Urbanization of the Southwest*. University of Arizona Press, Tucson.

Sayre, N.F. (2004) Viewpoint: the need for qualitative research to understand ranch management. *Journal of Range Management* 57, 668–674.

Sengupta, S. & Osgood, D.E. (2002) The value of remoteness: a hedonic estimation of ranchette prices. *Ecological Economics* 44, 91–103.

Starrs, P.F. (1998) *Let the Cowboy Ride: Cattle Ranching in the American West*. The Johns Hopkins University Press, Baltimore.

Stoneham, G., Chaurdri, V., Ha, A. & Strappazzon, L. (2002a) *Auctions for Conservation Contracts: An Empirical Examination of Victoria's BushTender Trial*. Department of Natural Resources and Environment, Melbourne, Victoria, Australia.

Stoneham, G., Chaurdri, V., Ha, A. & Strappazzon, L. (2002b) Victoria's *Bush-Tender* Trial: a cost sharing approach to biodiversity. Sheep and Wool Industry Conference, Hamilton, Victoria.

Tanzania Wildlife Conservation Monitoring [TWCM] (1990) *Wildlife Census: Tarangire 1990*. Tanzania Wildlife Conservation Monitoring, Arusha.

Veslany, K. (ed.) (2002) *Purchase of Development Rights: Conserving Lands, Preserving Western Livelihoods*. Western Governors' Association, The Trust for Public Land & National Cattlemen's Beef Association, Denver, San Francisco & Centennial.

Voeten, M. (1999) *Living with Wildlife: Coexistence of Wildlife and Livestock in an East African Savanna System*. PhD Thesis, Wageningen University.

Western, D. (1989) Conservation without parks: wildlife in the rural landscape. In: Western, D. & Pearl, M. (eds.) *Conservation for the Twenty-First Century*. Oxford University Press, New York, pp. 158–165.

Western, D. & Gichohi, H. (1993) Segregation effect and the impoverishment of savanna parks: the case for ecosystem viability analysis. *African Journal of Ecology* 31, 269–281.

Western, D. & Manzolillo Nightingale, D.L. (2005) *Keeping East African Rangelands Open and Productive*. African Conservation Centre, Nairobi.

White, R., Murray, A. & Rohweder, M. (2000) *Pilot Analysis of Global Ecosystems: Grassland Ecosystems*. World Resources Institute, Washington, D.C.

Part II
Case Studies

8

Biodiversity Conservation in Australian Tropical Rangelands

Stephen T. Garnett[1], John C.Z. Woinarski[1,2],
Gabriel M. Crowley[3] and Alex S. Kutt[4]

[1]School for Environmental Research, Charles Darwin University,
Northern Territory, Australia
[2]Department of Natural Resources, Environment, the Arts
and Sport, Northern Territory, Australia
[3]Tropical Savannas Cooperative Research Centre,
Northern Territory, Australia
[4]CSIRO Sustainable Ecosystems, Queensland, Australia

Introduction

The Australian continent has three broad biogeographic realms: the fertile fringes of temperate southern and eastern Australia, characterized by Gondwanan elements; the arid and semi-arid core and the monsoonal tropics of northern Australia, which blend the distinctive Australian stock with a substantial pantropical biota (Keast 1981). The fertile lands of southern and eastern Australia have endured substantial environmental transformation since European settlement 220 years ago and their biodiversity is now compromised by processes that beset densely populated and highly modified lands worldwide (Beeton et al. 2006).

By contrast, the rangelands of central and northern Australia have experienced relatively little purposeful environmental modification and remain

Wild Rangelands: Conserving Wildlife While Maintaining Livestock in Semi-Arid Ecosystems,
1st edition. Edited by J.T. du Toit, R. Kock, and J.C. Deutsch.
© 2010 Blackwell Publishing

sparsely populated. Nevertheless, the biota and environments of these range-lands have changed, sometimes drastically, in the last two centuries. The well-documented extinction of species in arid and semi-arid Australia con-stitutes one of the world's highest rates of biodiversity loss in recent history: 18 native mammal species of arid and semi-arid Australia, about a third of the endemic terrestrial mammal fauna, have become extinct during the last 150 years (Finlayson 1961; Burbidge & McKenzie 1989; Morton 1990). Unlike the development drivers in temperate Australia, this decline is associated mainly with pervasive and insidious factors, particularly predation by the feral cat and European fox and environmental change brought about by exotic herbivores, particularly rabbits and – in some areas – livestock.

Until recently, the biota of the rangelands of monsoonal northern Australia appeared more robust. Australia's tropical rangelands are seen as one of the last great regions on earth with a continuous natural environment (Garnett & Crowley 2003; Woinarski 2003a,b). And these rangelands do indeed retain secure populations of many species that have disappeared or declined drasti-cally from elsewhere in Australia: magpie goose *Anseranas semipalmata*, bush stone-curlew *Burhinus grallarius*, red-tailed black-cockatoo *Calyptorhynchus banksii*, grey-crowned babbler *Pomatostomus temporalis*, pale field-rat *Rat-tus tunneyi* and spectacled hare-wallaby *Lagorchestes conspicillatus* are some examples. However, this apparent security is now being challenged by a medley of factors and there is increasing evidence that shows that elements of the biota of the north Australian rangelands are in decline (Bowman & Panton 1993; Russell-Smith et al. 1998; Franklin 1999; Woinarski et al. 2001; Woinarski & Fisher 2003; Woinarski & Catterall 2004; Franklin et al. 2005; Woinarski et al. 2006b; Hannah et al. 2007). In this chapter, we contrast the conservation opportunity and value of the largely unmodified rangelands of tropical northern Australia with the evidence of biodiversity decline within them. We attempt to understand the causal links between these apparently contradictory perspectives and we propose some mechanisms, in practice and policy, that will better maintain and enhance the biodiversity values of Australia's tropical rangelands.

Setting

Most of Australia consists of rangeland, much of it with low and irregular rainfall. In northern Australia, however, the regular monsoonal influence changes the nature of the biota in a broad sweep of land from Broome in north-western Australia, across the northern part of the Northern Territory to Townsville on the Queensland coast (Keast 1981; Bowman 1996). Apart from a sliver of land along well-watered parts of the east coast, dominated by rainforest and intensive agriculture, these 1.9 million km^2 are dominated by a sparsely-populated tropical savanna woodland. Three main features have shaped this environment – its geological age, the seasonality of rainfall and the natural susceptibility of its vegetation to fire.

In broad terms, the land is old, weathered and infertile (Williams et al. 1985). There has been no recent uplift or glaciation and almost no recent volcanism that could have released new nutrients. There are few fertile patches, most of these being concentrated on alluvial plains and flood-out areas of the coastal fringe, in the former (Tertiary) swamplands of the Barkly Tableland and the basaltic black soil of the Victoria River Downs.

The climate consists of an extended dry season from May to November, followed by a monsoonal wet season from December to April (Dick 1975). Cyclones are a regular feature of the wet seasons and can have long-lasting effects on vegetation, particularly riparian and other rainforest patches where new breaks in the canopy can allow invasion of grasses and weeds. Away from the coast, cyclones produce flooding rains that periodically shape landscapes.

Fire is an inevitable event in such a markedly seasonal climate. Before humans arrived in Australia, fire is assumed to have occurred primarily at the end of the dry season, with lightning from early wet season storms as the only source of ignition.

Human history

The future of conservation of biodiversity in the tropical Australian rangelands cannot be understood without appreciating the region's prehistory. The first humans reached northern Australia 40–60,000 years ago from Southeast Asia (Roberts et al. 1990; Miller et al. 2005). Subsequent arrivals were few, but included, about 5000 years ago, a colonization event that included domestic Southeast Asian dogs, the genetic stock of the Australian dingo *Canis familiaris dingo* (Savolainen et al. 2004). At the same time the intensity of human land-scape use increased, possibly in association with increased climate variability (Turney & Hobbs 2006). More regular interaction with Southeast Asia began soon after 1700, when fishermen and traders started visiting from what is now Indonesia, although they did not settle (MacKnight 1972).

European explorers encountered Australia in the early seventeenth century and established permanent settlements in tropical Australia in the early nineteenth century. European diseases may have preceded the spread of settlers across the continent, depleting the Indigenous population before the European settlers arrived (Campbell 2002). In the latter part of the nineteenth century, the European influence started to dominate the landscape with the spread of pastoralism across the breadth of tropical Australia, regardless of the presence of Indigenous owners – and more quickly where there was gold. European settlement was also subsequently driven by mining for lead, zinc, copper, iron, bauxite, manganese, uranium and diamonds.

The arrival of Europeans severely disrupted Indigenous land management. Many Indigenous people were violently displaced to make way for cattle production (Reynolds 1982; Roberts 2005). Some worked later in the cattle industry until changes in employment conditions on pastoral properties forced them into towns (May 1994; Smith 2003), where they joined others who had been drawn into missions and settlements, often for administrative convenience. The European domination of land management through the twentieth century was a gradual process, however; only in the 1950s did the European population exceed that of the Indigenous people in the Northern Territory (Landrigan & Wells 2005) and there are still more Indigenous than non-Indigenous people living outside the larger settlements across most of the region. The rangelands of northern Australia are still sparsely populated, with the total human population density of 0.001 people/hectare outside the major towns (or 8.4 km^2/person) and pressures for rural depopulation continue as the cost of providing services and jobs to sustain highly dispersed, small and remote settlements increases.

Tenure and the economy

Tropical rangelands occur in the northern parts of the states of Queensland and Western Australia, which have relative autonomy over land-use decisions and the Northern Territory, for which land-use decisions are partly determined by the national government. Under each form of government, there are broadly four types of tenure that dominate land use, and through this, the fate of biota.

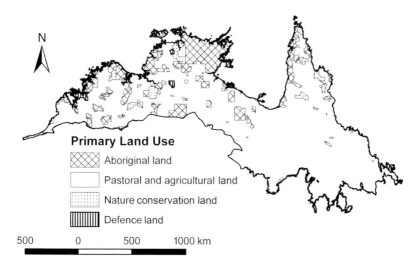

This map incorporates data that are © Commonwealth of Australia

Figure 8.1　**Land tenure across the Australian tropical rangelands (from Crowley 2006).**

Pastoral leases, almost entirely devoted to cattle production, make up about 75% of the tropical savannas (Figure 8.1). These are owned by the government but leased to private companies or individuals for periods ranging from a few decades to perpetuity. Properties can be very large, the average being 2200 km² (Fisher et al. 2004) and the largest over 30,000 km². Much of the most productive land is held by five companies that control over 200,000 km² of the rangelands (Martin & van Klinken 2006). The most profitable pastoral properties are on the natural grasslands of the Victoria River Downs, Gulf Plains, Mitchell Grass Downs and Barkly Tablelands (Stafford-Smith et al. 2000) and the least profitable enterprises occurring in the smaller holdings of central Queensland, the more rugged upland areas (such as the north Kimberley) and/or regions most remote from ports and population centres (Holmes 2000). In the last 10 years, the profitability of the beef industry in northern Australia has been at historical high levels, forcing up land prices. Pressure to gain greater returns on capital is driving intensification of grazing (Ash & Stafford-Smith 2003). The dominant and most profitable market for cattle from northern Australia is now live export, mostly to southeastern Asia, with little local value addition or employment diversification.

Until recently, a condition of a pastoral lease was property improvement, which usually meant fencing, provision of additional watering points and overall intensification of grazing. Land clearance was, until very recently, encouraged through tax laws, until constrained by legislation (Queensland and Western Australia) or government policy (Northern Territory). Leasehold conditions generally include some provision for land management 'duty of care', but there have been few, if any, instances when leases have been revoked because of overstocking, erosion or other evidence that this duty has not been met. In 2007, the Queensland Government introduced a system of tying lease renewals to land condition and management agreements, and other jurisdictions are considering measures to encourage environmentally sustainable use on production properties. The industry is at least partly subsidized by governments through provision of public infrastructure like roads, through tax deductions on production and fuel usage and through acceptance of environmental degradation, or the provision of funds to reverse degradation caused by the cattle grazing (Godden 1997). Whether this is offset by flow-on benefits from employment in the industry is unclear.

Lands held by Indigenous people under a collective title make up about 15% of the area. While mining, plantation forestry and pastoralism do occur on the land of this tenure, they do so only after extended negotiation over permit and benefit-sharing conditions. Benefits from economic activity are necessarily distributed among multiple owners, which can make individual investment of capital by residents unrewarding. Thus, not only do many Indigenous people not want broad-scale development of their lands, there is a range of economic disincentives to doing so. This contributes to the problem of high levels of unemployment and impoverishment on Indigenous lands. Almost all jobs within Indigenous communities are in the government sector, though some people supplement their income with the sale of art and increase their protein intake by hunting and gathering (Altman 2004). Policies are currently being developed that aim to move Indigenous people from welfare and into regular employment. This usually entails people leaving their land altogether, much of which is already unoccupied for extended periods. Indigenous people can also have a right of access to their traditionally owned lands that are currently under pastoral leases owned by others, but are rarely given a say in the management of this land. Where pastoral leases are in the hands of Indigenous people, lease conditions require the land to be managed for cattle. Pastoral production is once again being seen as an important source of employment for Indigenous Australia. Indeed, a Memorandum of Understanding among pastoralists, government and peak bodies representing

Indigenous people of the Northern Territory, which grants grazing licenses on Aboriginal land to a number of pastoralists in exchange for employment, infrastructure and support for the training and employment of Aboriginal cattlemen, originally signed in 2003, was re-signed and extended in August 2006 (Anonymous 2006). This agreement is likely to affect biodiversity in the tropical rangelands of the Northern Territory negatively by increasing cattle numbers, but positively by increasing resident land management capacity.

Conservation lands make up 6% of the landscape, mostly in more rugged and less fertile areas. Most conservation lands are managed by government agencies, but a small number are now being run by charitable trusts. The broad aim of conservation lands has been to keep habitats in a pre-colonial condition and maintain viable populations of all species and examples of all communities. With rare exceptions, management resources are limited and priority is typically given to infrastructure development for tourism rather than optimisation of conservation land management.

Beyond conservation reserves, most tourists are independent travellers, often intent on catching fish and spending as little money as possible (Griener et al. 2004). Although they have little influence on the economy, they can cause localized land degradation, wildfire and weed dispersal as well as reduce fish stocks in inland rivers and offshore (Griener et al. 2004).

Mining tenures occupy less than 0.5% of the landscape, most of which is encompassed by bauxite leases on Cape York Peninsula in Queensland. However, mining also occurs under permit on Indigenous land, and by right on pastoral leases. Many of the region's towns are centred around existing or former mines, with mines providing the economic underpinning for Weipa, Mt. Isa, Borroloola, Gove, Jabiru and Kununurra. Although active mining usually persists for only a few decades, a legacy of environmental change may remain, shaping local patterns of human settlement and biodiversity. Nevertheless, even if dust plumes (Griffiths 1998; Andersen et al. 2002) and river degradation (Markich et al. 2002) can extend impacts beyond the actual mine site, most mining impact on biodiversity is localized and often temporary.

Biodiversity conservation

Current state

The wildlife of Australia's tropical savannas is notably lacking in the conspic-uous congregations of large mammals that characterize rangelands in other tropical regions of the world: the largest native herbivorous mammal is the typically solitary common wallaroo *Macropus robustus* at 55 kg and the largest extant native carnivorous mammal, other than the relatively recently arrived dingo, is the northern quoll *Dasyurus hallucatus* at 1 kg. This absence is relatively recent: a distinctive, rich and varied megafauna community (of large mammals, terrestrial reptiles and birds) characterized the rangelands of northern Australia until their rapid disappearance between 20,000 and 50,000 years ago, soon after the arrival of humans (Brook & Bowman 2002, 2005; Miller et al. 2005). Some large mammals, notably the thylacine *Thylacinus cynocephalus* and the Tasmanian devil *Sarcophilus harrisi*, persisted in the rangelands of northern Australia until a few thousand years ago, their memory preserved in Indigenous rock art (Calaby & White 1967; Murray & Chaloupka 1984).

In contrast to the modern poverty of the large fauna, the rangelands of northern Australia continue to support an extraordinary richness of ants, termites and reptiles. A 1 hectare area in these tropical savannas typically contains more than 100 ant species (Andersen 1992, 2000; Andersen et al. 2000); and the herpetofauna of the 20,000 km² Kakadu National Park comprises more than 150 species (Woinarski & Gambold 1992; Press et al. 1995). Spectacular aggregations are another feature of the fauna, with particularly large concentrations of waterfowl, pythons, crocodiles, frogs and rodents associated with the unusually fertile floodplains of some large river systems (Morton et al. 1990a, 1990b, 1993; Madsen & Shine 1999; Madsen et al. 2006).

The geographical patterning of the biota is driven largely by rainfall. There are gradual species replacements and changes in richness associated with the generally simple rainfall gradients from the northern coast towards the drier interior (Bowman 1996; Woinarski et al. 1999). However, there is remarkably little change in species composition across the lowland plains stretching in a 2000 km East–West arc from Cape York Peninsula in Queensland to the Kimberley in Western Australia (Woinarski et al. 2005). In contrast, isolated sandstone ranges, especially those of western Arnhem Land and the north Kimberley, have a distinctive biota, high levels of endemism and many relict species (Woinarski et al. 2006a). Likewise, small patches of rainforests embedded in unusually fire-protected or moist pockets of the savanna, support distinctive and rich plant and animal assemblages (Russell-Smith 1991).

Extreme seasonality is a major driver of the ecology of most of the region's plant and animal species (Woinarski et al. 2005). Plant phenology is typically highly synchronized and tightly associated with rainfall patterning (Williams et al. 1999b). Many plants are annual, and many animal species are completely or largely inactive during the long dry season. Climatic seasonality drives major changes in the availability of food resources and many animal species respond to this by shifting habitat and/or by dispersing across widely differing scales (Madden & Shine 1996; Price et al. 1999; Woinarski et al. 2000, 2005). Such forced movements rely on the maintenance of landscape-scale linkages and connectivity.

Overall the modern biota of northern Australia has remained unusually intact. Unlike most other parts of Australia, there have been few extinctions since European settlement and even these have been only on the southern fringes of these rangelands: the paradise parrot *Psephotus pulcherrimus* disappeared from central Queensland during the first few decades of the twentieth century; and a few mammal species, such as the burrowing bettong *Bettongia lesueur*, whose largely inland range extended marginally into the northern rangelands, disappeared between the 1930s and 1950s (McKenzie 1981). However, recent evidence suggests both broad-scale historic decline and/or recent ongoing decline for many groups of species. Such declines are best documented for many species of granivorous birds (Figure 8.2), mammals such as bandicoots, possums, larger rodents and larger dasyurids (Figure 8.3) and fire-sensitive plants and fauna species that may be closely associated with these species (Bowman & Panton 1993; Russell-Smith et al. 1998; Franklin 1999; Woinarski et al. 2001, 2006b; Franklin et al. 2005). However, these groups may

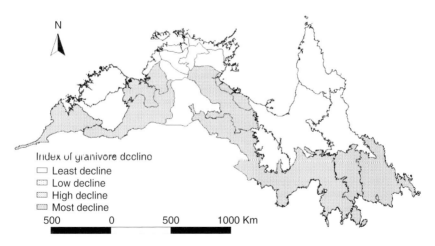

Figure 8.2   **Patterns of decline of granivorous birds across northern Australia (from Crowley 2006 based on Franklin 1999).**

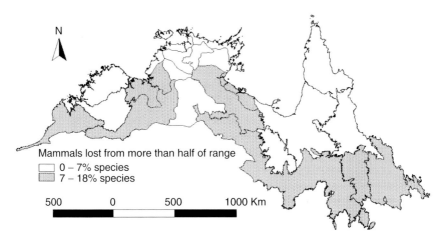

N

Mammals lost from more than half of range
☐ 0 – 7% species
▨ 7 – 18% species

500   0   500   1000 Km

Figure 8.3   **Patterns of mammal decline across northern Australia (from Crowley 2006 based on McKenzie & Burbidge 2003).**

be simply the most obvious of a larger set of species disturbed from a previous equilibrium associated with intricate land management by Indigenous peoples (Yibarbuk et al. 2001; Woinarski & Fisher 2003; Woinarski & Catterall 2004). Indicative of this pattern of change, a total of 54 plant, 11 invertebrate, 9 freshwater and/or estuarine fish, 1 amphibian, 7 terrestrial reptile, 15 bird and 14 terrestrial mammal species are listed as critically endangered, endangered or vulnerable in the tropical savanna rangelands of the Northern Territory alone.

Factors driving current situation

Weeds and structural vegetation change

European settlement brought to insular Australia numerous vigorous exotic plant species. In Australia as a whole there are at least 2700 such species, of which 370 are legally declared as noxious species by States and Territories. The top 20 individually cover between 0.2 and 12.4% of the continent, with an estimated $4 billion cost to industry (Sinden et al. 2004). The annual cost of weeds to the livestock industry (loss of production, control measures) throughout Australia has been estimated as $315–45 million, with a further $112 million cost to public authorities (Sinden et al. 2004). Data for the tropical savannas are less readily available, though as an indication of the problem, over 5% of the flora of the Kimberley in Western Australia (108 spp.), 5% of the Northern Territory flora (230 spp.) and 13% of the Queensland flora (1220 spp.) are exotic (Wheeler et al. 1992; Dunlop et al. 1995; Csurhes & Edwards 1998). Of the 622 naturalized species in the Australian rangelands as a whole, 26% are considered a threat to biodiversity (Martin et al. 2006). Most of the $25 million spent on rangeland weed control by the Australian government since 1996 and the $56 million on projects with a weed management component, was spent in the tropics (Martin & van Klinken 2006). Uncosted in these estimates are the changes that have occurred in the structure of the native vegetation which, in some places, is considered a worse problem for pastoralists and biodiversity than exotic weeds.

Various stakeholders have characterized three main types of weed in the savannas: native woody shrubs occurring at an increased density; exotic species universally condemned as having an environmental impact (exotic woody species, often garden escapees or agricultural blunders); and exotic species that continue to be spread deliberately by one stakeholder group because of a presumed, though narrow, benefit (Grice 2002). Although empirical data on the effects of weeds on biodiversity are markedly few, there is widespread agreement that lack of documentation does not acquit weeds of major impacts on many forms of biodiversity (Grice 2006).

Increased density of woody native plants in Australian tropical savannas is also a global phenomenon, perhaps accentuating the ecological dynamism in this system in response to disturbance regimes. In some places the thickening is a consequence of natural climatic fluctuation causing mass recruitment, followed some years later by drought-driven dieback (Fensham & Holman 1999). In others, it is the result of the grazing of fuel loads or active fire prevention that affects fire intensity (Butler & Fairfax 2003; Sharp & Whittaker 2003).

It is in the latter situation that the biodiversity effects appear to be greatest (see section Fire). However, carbon dioxide fertilization may also contribute (Bond & Midgley 2000), and this has been used to justify ongoing mechanical clearance of vegetation, with suggestions that the thickening is an abnormal state of nature deserving redress (Burrows et al. 1990).

The principal woody exotics affecting the savanna are rubber vine *Cryptostegia grandis* (costs $Aus 2.1 million/year), lantana *Lantana camara* ($Aus 2.2 million), prickly mimosa *Mimosa pigra* ($Aus 0.6 million), mesquite *Prosopis* spp ($Aus 1.7 million), prickly acacia *Acacia nilotica* ($Aus 2.2 million) and *Sicklepod* ($Aus 2.6 million), which collectively invade and alter riparian, dry rainforest, wetland and grassland environments. Scrambling shrubs such as rubber vine and lantana, both garden escapees, have smothered millions of hectares of mesic environment in northern Queensland. Rubber vine dominates about 700,000 hectares of riparian areas and floodplains, climbing and entwining emergent trees until heavy enough to topple them. Lantana invades dry rainforest patches in tropical savannas, allowing fire incursion and profound structural change. The distinctive and formerly unusually extensive inland dry rainforest Forty Mile Scrub in northern Queensland, largely a national park, is now mostly dominated by lantana, promoted by the combined forces of fire, pigs and mechanical disturbance (Fensham et al. 1994). Noogoora burr *Xanthium occidentale* and *X. strumarium* ($Aus 0.63) has invaded over two million hectares in Australia, including 360 km along the Fitzroy River and large sections of the Ord and Victoria Rivers. Mimosa, prickly acacia and mesquite all form dense impenetrable woody thickets in grasslands and wetlands. Again many millions of hectares have now been infested, reducing native plant diversity and abundance and changing or reducing the fauna assemblages present (Braithwaite et al. 1989; Whitehead et al. 1990), although the full effects of these weeds on biodiversity are poorly known (Grice 2006).

Nearly all introduced pasture grasses and legumes were considered desirable to the grazing industry when introduced (Eyles et al. 1985), but they have spread rapidly and aggressively with permanent irreversible effects on Australia's rangelands and its wildlife (Cook & Dias 2006). A report on the benefit–cost ratio of introducing buffel grass sums up the philosophy of those who, with idealistic vigour, drove pasture plant introductions in Australia:

'Due to the discounting processes used in the analysis, a low weighting would be placed upon future negative impacts [of introduced grasses] compared to the positive impacts that started much sooner than the negative impacts. To some extent this represents a problem of intergenerational distribution of benefits and costs'

(Chudleigh & Bramwell 1996).

There are many examples where one arm of government has promoted the use of exotic plants, while another funds the eradication or control of those same species (CRC Australian Weed Management 2003). Over the last 100 years, attempts have been made to introduce more than 8200 pasture plants to Australia, including more than 2600 grass and 2200 legume species (Cook & Dias 2006). A retrospective review of the fate of over 450 introduced pasture species found that only 4 were useful and did not become weeds whereas 60 became weeds and 17 were both useful and considered weeds (Lonsdale 1994).

In the wettest areas, *Brachiaria* spp., *Hymenachne amplexicaulis* and *Echinochloa* spp. were introduced to create ponded pastures, but have spread into natural wetlands where they replace native grass and sedge species, dominate open water patches, eliminate fish habitat and reduce bird nesting opportunities (e.g. the magpie goose) (Ferdinands et al. 2005). In semi-humid savannas, gamba grass *Andropogon gayensis*, a vigorous African perennial species, has been promoted widely for use in pastures, but readily invades intact ecosystems, creating a fuel load 10 times that typical of non-invaded environments. Not only are native understorey species smothered but woody mid-storey and canopy species may also be lost as a result of increased fire intensity (Rossiter et al. 2003). This species, already widespread, has an ecological envelope that could allow it to spread across all northern tropical Australia. Another problematic exotic pasture species is Buffel grass *Cenchrus* spp. (mainly *ciliaris*). It has been promoted actively by government agencies and pastoralists since the 1950s to improve pastoral productivity and for erosion control and is used most widely where broad-scale clearing has occurred. It is also aerially sown or spreads naturally across uncleared vegetation. Estimates of present cover range up to 50 million hectares (Hannah et al. 2007) with continental scale modelling suggesting that 60% of Australia is susceptible to its invasion (Lawson et al. 2003). Recent examination of the genetic makeup of buffel cultivars in Australia suggests that new hybrids are forming in the wild to match local climate, soils and landscapes (Friedel et al. 2006). Where buffel grass has invaded intact ecosystems, the intensity of fire has increased, killing native grass and woody species in a cycle of invasion, fire promotion and invasion (Butler & Fairfax 2003). Uncleared fragments invaded by buffel are typically further disturbed by heavy grazing (Hannah et al. 2007) and several studies identify the potential for buffel pastures to drain nitrogen from already typically nutrient-poor environments (Schmidt & Lamble 2002).

Feral animals

Many exotic animals have been introduced, deliberately or accidentally, into Australia's tropical savannas. Some of these are now amongst the region's most widespread and abundant species and have substantially changed the dynamics and composition of the landscapes they now inhabit.

There has been no stocktake of the number of exotic invertebrates in Australia's tropical savannas, but some are of particular concern. The most notable for biodiversity impacts are a series of aggressive tramp ant species, including the big-headed ant *Pheidole megacephala* and yellow crazy ant *Anoplolepis gracilipes*, both known to spread rapidly, including to otherwise 'undisturbed' native vegetation, form large aggressive colonies and substantially alter the assemblage of invertebrates, plants and vertebrates, typically towards simplification. These two species are currently relatively restricted in the rangelands of northern Australia, but are difficult to control and likely to expand substantially (Hoffman et al. 1999; Young et al. 2001).

Fourteen species of feral vertebrate are present in tropical Australia, away from human settlements including a fish (tilapia, currently being actively confined to east coast catchments), an amphibian, two bird species (both confined to the east) and 10 mammals. Those of greatest concern to native biodiversity are horse *Equus caballus*, donkey *Equus asinus*, Asian water buffalo *Bubalus bubalis*, pig *Sus scrofa*, cat *Felis catus* and cane toad *Bufo marinus* (Norris & Low 2005).

The large herbivores fill many of the ecological gaps not dominated by domestic stock and contribute substantially to total grazing pressure on tropical landscapes (Fisher et al. 2004). Together – cattle on the extensive pastoral estate and feral herbivores on other tenures – result in a remarkably small proportion of the Australian tropical rangelands being free from the impacts of recently introduced herbivorous mammals. Buffalo affect both the long-term structure of the savanna (Werner 2005) and the composition of grasslands, their wallowing channels causing salt water intrusion into large areas of seasonal freshwater swamps (Skeat et al. 1996). The elimination of buffalo from large parts of their range to control disease during the 1980s allowed for some relief from their impact, but populations are now recovering rapidly (Anonymous 1999).

The ecological impact of feral pigs is surprisingly little known, given the visibility of their activities, and some stakeholders value them as a good source of food, particularly on Indigenous lands (Norris & Low 2005). They turn over large areas of freshwater wetland each year and their varied diet includes frogs, earthworms, plant tubers and the eggs of turtles (Pavlov 1995). They may also have a substantial impact on rainforest patches and grasslands, selectively grazing species like *Alloteropsis semialata*, a grass species important to many small mammals and granivorous birds and they reduce recruitment of termite mounds (Crowley et al. 2004).

Large exotic predators provide a different perspective on Australian attitudes to feral pests. Since their relatively recent arrival, dingoes are thought to have been at least partly responsible for the elimination of the thylacine and Tasmanian devil on the mainland, through competition or through hosting of exotic diseases. Dingoes also hunt critically endangered northern hairy-nosed wombats *Lasiorhinus krefftii* and excavate the eggs of endangered marine turtles, but this predation is only thought to be significant because these species are additionally threatened by other causes. Although dingoes interbreed with domestic dogs, there are still many animals in which there has been little genetic mixing (Savolainen et al. 2004) and they are often treated as a native part of the Australian fauna. Indeed, dingoes may now have a net benefit on native biodiversity, with recent work suggesting that dingoes may regulate cat and fox populations, particularly around livestock water points (Robley et al. 2004). Nevertheless, they cost the livestock industry $66 million per annum (McLeod 2004), so government-sponsored baiting and bounty programs are widespread.

The effects of cats are more insidious. Cats first became widespread across Australia at the end of the nineteenth century (Abbott 2002) and, along with the fox, which is largely absent from the tropical rangelands, are strongly implicated in the loss of several arid zone species. Their impact in the tropics is poorly known, although they are a predator of the endangered Julia Creek Dunnart (Lundie-Jenkins & Payne 2004). Though a common predator of many native species, there has been no attempt to remove cats at a landscape scale in the rangelands of northern Australia to test the effects of their absence.

Most feral animals have been present in the Australian tropical rangelands for well over a century, a history and an extent now so substantial that it is difficult to provide a precise assessment of their ecological impact. In contrast, the impact of the cane toad is being recorded in real time as it spreads inexorably across the tropical rangelands. A global example of biological control gone wrong, it was deliberately released by the Australian Bureau of Sugar Experimental Stations in 1935 in northern Queensland to control two species of cane beetle, a task in which it completely failed. It now occupies all of tropical Queensland and most of the Top End of the Northern Territory and is on the verge of invading Western Australia. Eventually, it is predicted that toads will occupy 20 million ha of mainland Australia. Elapid snakes, varanid lizards and dasyurid marsupials are poisoned when they prey on toads. In Kakadu, the long-term resident mammal the northern quoll *D. hallucatus* could no longer be found within 12 months of the toads arriving (Watson & Woinarski 2004) and goanna *Varanus* spp. numbers declined substantially (Griffiths & McKay 2005). Though some level of population recovery appears to occur in affected species some time after toad colonization, there are inevitably evolutionary impacts yet to be played out and some species effectively disappear altogether – the extremely sensitive quolls have persisted only in fragmented pockets of broken country within the Queensland range of toads.

Pastoralism

Though mining, tourism and government services form a larger share of the regional economies, pastoralism is by far the dominant commercial use of land across northern Australia and has the greatest impact on biodiversity. Cattle numbers have been increasing in northern Australia since the 1860s. Initially, only British *Bos taurus* breeds were used, but each year a high proportion died from disease and drought stress at the end of the dry season. This meant relatively few cattle were grazing perennial grasses at the start of the wet season when they were first resprouting. From the 1970s, Brahman cattle *Bos indicus* and their hybrids were introduced and were more disease resistant and better able to survive dry periods. The survival, intake and weight gain of cattle was improved further with the introduction of treatments for botulism, as well as by dietary and hormone supplements. Better husbandry also resulted in greater survival of newly weaned calves, so inherent herd growth and turnoff also increased. At the turn of the millennium, these trends have accelerated rapidly with a series of seasons with relatively high rainfall coinciding with higher than average prices (Ash & Stafford-Smith 2003). Increasing demand, particularly from markets in south-eastern Asia, has also stimulated the industry and led to increased prices for pastoral leases. High levels of capital investment, compounded by escalating operational costs, particularly fuel, have increased pressure on pastoralists to raise herd sizes and off-take. This has been assisted by replacement or supplementation of native grasses with exotic grasses and legumes, increases in the provision of artificial water sources, clearance of native vegetation and increased fencing to reduce paddock sizes and overgrazing.

The introduction and recent intensification of grazing has had substantial effects on natural systems. Soils have been compacted, reducing water penetration and increasing runoff, causing widespread erosion in some landscapes.

Erosion of sediment onto the Great Barrier Reef is 2–6 times pre-European levels (Furnas 2003). Watering points are particularly hard hit, with sacrifice zones away from both natural and artificial watering points where little grows and erosion is rife (Ludwig et al. 1999). The aureole of grazing effects extends to the distance that cattle can travel each day, which can be over 8 km in many environments (Landsberg et al. 1997; James et al. 1999; Landsberg et al. 1999). Drainage lines have also come under heavy grazing pressure, with increasing bank erosion and sedimentation of drainage channels while the provision of supplements has had a heavy ecological impact in the tropics by increasing the appetite of cattle (Tothill & Mott 1985). Livestock grazing also reduces fuel loads, particularly in areas that remain damp well into the dry season. This changes fire regimes, in many cases then favouring shrub invasion (Sharp & Whittaker 2003). Impacts reach their apogee where exotic forage plants have been introduced. In some parts of the rangelands, legumes that spread through uncleared woodland provide extra nitrogen, increasing the appetite of cattle through the dry season (Tothill & Mott 1985). These legume–grass systems can be inherently unstable, with the elimination of native grasses, particularly if grazed in the early wet season (Jones & Jones 1977; Jones et al. 1997). Some grasses will also spread naturally into uncleared systems, particularly where the understorey is disturbed by cattle.

Many native species have been adversely affected by cattle grazing. The most palatable plants are rapidly grazed out, even under light grazing (Landsberg et al. 1999), and there is successive species loss as grazing intensity increases, starting with the softer perennial grasses through to the remaining perennial species until only annual species and weeds remain. Perennial species are particularly susceptible to grazing in the early wet season (Ash & McIvor 1998). Pastoralism also has direct or indirect adverse impacts upon a range of native invertebrate (Woinarski et al. 2002; Churchill & Ludwig 2004) and vertebrate species (Fisher 2001; Woinarski & Ash 2002). The widespread decline of granivorous bird species (Franklin 1999), while correlated with pastoral intensification, may be associated with changes in fire management, vegetation structure or seed availability, all of which are increasingly different from the ungrazed state as pastoralism is intensified.

Fire   When European explorers recorded their first impression of Australia, two things stood out, the strange flora and fauna and the horizons filled with smoke (Lee 1925; Preece 2002). Even in some of the early accounts, the link between Indigenous burning practices and biological patterns was apparent

(Jack 1922). Current understanding of traditional Indigenous management of fire is imperfect, but we do know that fires were lit for much of the year as people travelled through the landscape (Russell-Smith 2002). Each fire added to a progressive network of burns, breaking up the fuel layer, so that few fires travelled far, and leaving patches of unburnt country that served as temporary refugia for fire-sensitive species. We also know that fires were lit around some of the more fire-sensitive communities under mild conditions, maintaining their boundaries. It is thus believed that under Indigenous management, fires were moderate in intensity, canopies were protected and soils maintained, at least in comparison to contemporary regimes (Russell-Smith 2002; Williams et al. 2002).

It has proved difficult to maintain the dynamic fire management system that is assumed to have operated under Indigenous management. Significant factors have changed: population concentration (Whitehead 1999; Whitehead et al. 2003), the spread of large grazing herbivores and possibly carbon dioxide fertilization (Archer & Schimel 1995). Currently, up to one quarter of northern Australia is burnt in fires of considerable extent and intensity every year. In regions where few people manage the landscape and there is relatively little pastoralism, around 50% of the country is burnt each year. In other areas, where cattle and feral herbivores have eaten much of the grass or where pastoralists manage primarily to conserve fodder, large areas remain unburnt each year and sometimes do not burn for decades (Crowley & Thompson 2005). The impacts of the current regimes are substantial: trees and some fire-sensitive communities are being lost from significant areas (Williams et al. 1999a), while in others, such as fire-dependent grasslands and grassy woodlands, they are replaced by thicker vegetation communities (Neldner et al. 1997; Crowley & Garnett 1998; Sharp & Whittaker 2003), and species dependent on these contracting communities or that require a fine-scale fire mosaic are declining (Garnett & Crowley 1995, 2004). These changes seem inherently linked to the removal of Indigenous stewardship of the land (Bowman 1998).

There has been a concerted effort in recent years to shift the burning patterns to include more preventative fires in the early-dry season, which are presumed to be relatively cool, patchy and localized, and fewer in the late-dry season when hotter temperatures, drier fuels and stronger winds make for hotter, less patchy and more extensive fires (Press 1987). Anywhere this has been attempted, land mangers relate tales of failure; where, in the course of a week or 10 days the country had dried out, the fuel has cured from not supporting fire to a single dropped match causing fires of phenomenal scale.

There is also a reluctance to burn in the early dry season among land managers, who not only wish to preserve forage but also consider early burning as a factor contributing to vegetation thickening (Crowley 2001; Crowley et al. 2004). Recreating the mosaic is therefore fraught with difficulty. This is true even on conservation reserves. This is because the ecological problems are related to how and not by whom the country is managed. Whether designated over seemingly unproductive lands or former pastoral properties, the gazettal of protected areas in northern Australia has usually contributed to the depopulation process and has often been associated with a phase of attempted fire suppression. Also gazettal is rarely followed by adequate resourcing to ensure destocking and removal of introduced herbivores, fencing and weed control or fire management. Hence, these areas are equally in need of intensive fire management (Woinarski & Fisher 2003).

Trends and future drivers

Since European settlement, the tropical savannas of northern Australia have been largely a struggling backwater, geographically distant from the nation's engine rooms and largely quarantined from many of the main drivers that have beset rangelands elsewhere in the World. It is likely that this stasis will be increasingly challenged by the consequences of global socio-economic factors including climate change.

Climate change

Anthropogenic changes in climate will influence biodiversity in tropical Australia, through changes in sea level, temperature, rainfall, cyclones and atmospheric carbon dioxide fertilization.

Some of the most productive systems in the Australian rangelands, indeed in the world, are coastal floodplains (Madsen et al. 2006). These relatively recent geomorphological features formed after the last interglacial as terrestrial erosion filled Quaternary river valleys that were flooded by the interglacial rise in sea level (Woodroffe 1993). Global sea level rise could inundate many of these coastal plains in the next 50 years. This is likely to cause the loss of one of the key features of Australian tropical biodiversity, the concentrations of aquatic and other species on the coastal floodplains (Eliot et al. 1999).

In the last 50 years, rainfall has increased for all but the eastern portion of tropical Australia (Australian Bureau of Meteorology 2006), probably as a result of the influx of Asian aerosols and warming of water off Northwestern Australia (Rotstayn et al. 2007). There is greater certainty that temperatures will rise across the Australian tropical savannas, but few predictions of how changes in rainfall or temperature will affect the biota. Some generalist species may be able to move across the current relatively gradual climate gradients of northern Australia to reposition themselves in newly suitable climatic locations, but such re-positioning may not be possible for species with more tightly defined habitat requirements, such as those confined to the sandstone massifs where endemism is currently highest (Woinarski et al. 2006a), in situations where the vegetation continuity is interrupted by intensive development or other isolation and for species dependent upon a complex array of environmental factors that may not re-configure elsewhere with sufficient rapidity under climate change. Changes in longer term cycles of rainfall patterns may also affect the cycle of eucalypt die-off and regeneration noted on the drier fringes of the tropical rangelands (Fensham & Holman 1999), but again, models are insufficient to predict the direction of change.

Changes in cyclone activity may affect biodiversity directly and indirectly. Though cyclone frequency is thought unlikely to change, cyclones that do occur are likely to have increased intensity (Walsh et al. 2004). In 2005 and 2006, northern Australia was affected by three of the strongest cyclones, Cyclones Ingrid, Larry and Monica, ever recorded to make landfall in the region. Such intense cyclones are likely to be accompanied by severe storm surges, which can exacerbate the effects of sea level rise and may alter the dynamics of habitats that require relatively long disturbance-free intervals. Some islands have been denuded of species (Palmer et al. 2007) and there may be similar species loss in islands of habitat. Some species in monsoon rainforest may be unable to persist for long or frequent periods without the tall dense canopy structure of undisturbed rainforest. Severe cyclones are also likely to render rainforest patches more susceptible to invasion by weeds and non-rainforest taxa.

Increases in atmospheric carbon dioxide appear to increase tree growth over that of grasses (Stokes et al. 2005). Carbon dioxide fertilization may thus increase the conversion of grassland to shrubland, decreasing the profitability of rangeland grazing and disadvantaging species that rely on savanna woodlands with an open structure. Control of this shrubby growth may be possible with judicious use of fire, but this will become increasingly difficult where cattle grazing is reducing grass biomass that both competes with woody vegetation and provides the main fuel for fires. However, the other main effect of increasing carbon dioxide is on carbon and nitrogen dynamics. The nitrogen content of forage will decline as carbon dioxide increases and this will have an impact on the nutritional value of forage for native herbivores as well as for livestock (King et al. 2004), although some negative effects of carbon dioxide increase may be mitigated by temperature rise (Zvereva & Kozlov 2006).

Indirect effects of global climate change may have even greater consequences for the savanna rangelands of northern Australia. Loss in productivity and shortage of water in the prime agricultural landscapes of south-eastern and south-western Australia will create pressure to increase the intensity and productivity of land use in northern Australia. External impacts may be even more pronounced if climate change reduces productivity in the densely populated lands of Southeast Asia.

Demography

In contrast to tropical rangeland environments outside Australia, the rural human population of this continent's savannas is predicted to grow relatively slowly (Ash & Stafford-Smith 2003). The government welfare programs that currently make possible many small isolated Indigenous settlements appear likely to be wound back and the dysfunctionality and lack of endogenous economic activity in and around larger communities is likely to drive people towards tropical Australia's urban and peri-urban fringe. This is a continuation of the depopulation described above. With each removal, the motivation for and knowledge required to survive on the country has been lost. Consequently, large areas of country are not being visited on a regular basis, let alone managed for fire with the same intricacy that marked traditional usage (Preece 2007). Across the broader landscape, the remaining population is likely to be scattered between ephemeral mining and agricultural communities, many of which will be serviced by itinerant staff from elsewhere. There is also likely to be a decline in the number of people on pastoral properties as technology allows automation of cattle management, with equally small population centres servicing passing tourists, the numbers of which will be determined by the price of automotive and aviation fuel relative to personal wealth (Garnett & Lewis 1999; Prideaux et al. 2001; Fargher et al. 2003). This trend may be slightly offset by increasing access of pastoral companies to Indigenous-owned lands.

Thus, without substantial government investment, there will be a shortage of people across much of the rangeland to carry out the active land management needed to counter current and impending threats.

A small potential workforce that could be engaged in such management still exists among Indigenous communities wishing to remain in the country. The proportion of Indigenous people in the population is likely to continue to rise, with the proportion of young Indigenous people in the population being substantially higher than that of Australia's temperate zone (Taylor & Kinfu 2006). As the wider Australian workforce ages, there should be more employment opportunities for relatively young Indigenous people but it remains uncertain whether the next generation of Indigenous people will have the health, education and skills to take these opportunities.

Pastoral intensification

The past decade has seen rapid intensification of the pastoral industry (Ash & Stafford-Smith 2003; Stokes et al. 2006) Stocking rates and the extent of impact of cattle grazing are increasing as the number of watering points is increased, cattle are constrained behind fences to allow closer management that will maximize weight gains and there is continued pressure to expand the area under exotic pasture grasses. Though restrictions on vegetation clearance have been introduced, continued reduction in native vegetation cover seems inevitable with increasing grazing pressure and increased fire intensity resulting from the spread of exotic grasses and pressure to thin the forest by mechanical means. And there is an ongoing ambition to increase the size of the tropical cattle herd.

Not all intensification is necessarily bad for biodiversity. Much of the worst environmental damage due to pastoralism occurred in the first few decades following pastoral settlement (Woinarski & Catterall 2004; Franklin et al. 2005); and through experience, regulation, or the inevitability of loss of carrying capacity, many properties have pulled back from practices that laid the country bare. In some instances, there has been an increasing realization that grass-fed tropical rangeland beef has a price advantage in some markets if associated with ISO14001 environmental management systems (Ash & Stafford-Smith 2003), although such management systems aim to conserve function not biodiversity. Nevertheless, there has been investment in research to minimize environmental impacts from new technologies as well as fencing to protect riparian and other environmentally sensitive ecosystems, education of pastoral managers in the need for a more holistic approach to pastoral managements and the adoption of management systems that both maximize long-term income and reduce deleterious environmental effects.

Economic viability of pastoral properties may be increased without increasing pressure on land and biodiversity condition where it is achieved through augmenting a small number of trained staff to rural industries (who are increasingly difficult to attract) (Oxley et al. 2006) with technological management tools, such as remote sensing to assess pasture condition or the development of remotely driven technologies to manage herds and the replacement of expensive non-renewable fuels by renewable energy (Ash & Stafford-Smith 2003). There is also an increasing understanding among policymakers that tenure should not constrain land use to pastoralism alone and that hybrid income sources, including stewardship of country for conservation purposes (Morton et al. 1995; Fargher et al. 2003), can both increase profitability and sustainability.

On balance, however, while the rate of degradation may have slowed, the reach of cattle grazing into sensitive habitats continues apace. Therefore, while we may see a more sustainable industry that manages with care the natural resources on which it depends, the continuing decline of the grazing sensitive biota seems inevitable.

Other intensification

As ecological systems, the rangelands of northern Australia remain largely intact, characterized by low human population density and extensive land use, with only a few relatively small pockets of intensive modification for horticulture, forestry and mining (Woinarski 2004). The lack of intensive development is due, in part, to the limited extent of fertile soils; to economic marginality, itself partly a function of logistics and modest environmental productivity, and to current tenure arrangements, with most lands allocated to pastoralism, with most of the remainder being returned to the control of Indigenous communities that have so far had neither the inclination nor the capacity to undertake intensive land development. However, demand and market forces can overcome these constraints. For example, the fluctuations in gold price have periodically led to the re-opening of formerly abandoned mines in these rangelands and the global hunger for wood products may make it profitable to exploit the extensive eucalypt woodlands of the north Australian rangelands. Improvements in crop varieties, technological advances and enhanced access to world markets are all likely to increase the likelihood of expansion of intensive development in northern Australia. Retention of large areas of natural landscape in the north Australian rangelands may also begin to be seen by some as an unaffordable indulgence, especially so if climate change adversely affects the densely populated arable regions of southern Asia.

These drivers will result in increasing pressure for the continued development of the rangelands of north Australia over the next 20–50 years, with monopolization of fertile pockets by various horticultural industries, in the less fertile matrix, broad-scale forestry (based either in plantations of exotic tropical species or pulping of native eucalypts) and an expanded mining and associated refinery/manufacturing sector. However, for more than 150 years, northern Australia has resisted similarly rapacious development agendas (Woinarski & Dawson 2001); these lands unlock their bounty unwillingly.

Solutions

There are five proximal threats to biodiversity in the tropical Australian rangelands: that land will be cleared for pastoralism or horticulture and be replaced by exotic plants or intensive development; that existing vegetation will be replaced by exotic species; that feral animals will destroy critical components of natural systems; that fire will occur in a manner that elements of the biota will not be able to tolerate; and that climate change will cause the sea level to rise and local conditions to become intolerable at a rate faster than that to which species can adapt. Economic drivers and demographic trends impinge on how these proximal threats are managed, but all solutions will need to include answers on how the proximal threats will be countered.

The first threat, that of land clearance, requires ongoing political resolve to minimize environmental harm even if economic conditions render investment profitable, at least in the short-term. Legislation that limits land clearance is a step in this direction, and the longer the majority of Australia's north remains uncleared, the greater will be its rarity, and hence its value, in global terms. The next three threats require active investment in land management. At the moment, at a regional level, all can be countered, even if there are patches where weeds and feral animals have so transformed systems that they will never be restored. At least no species endemic to these rangelands has become extinct since European settlement and no tropical Australian ecosystem has tipped over a threshold of no return. All but the most severely overgrazed lands (Mott et al. 1979) are known to be able to recover biological function if rested for long enough. The last proximal threat, climate change, is driven by processes beyond the Australian tropics, though contributions to its mitigation and responses to immediate threat also require active engagement with the land. Here, we take a tenure-based approach to managing these five proximal threats.

Conservation lands

The Australian nation has inherited a conservation perspective, derived largely from Europe, that some vestiges of natural areas should be retained as representative showpieces in a mostly modified landscape. Thus, Australian environmental policies seek to capture 5 or 10 or 15% of every environment within conservation reserves. Implicitly – or sometimes explicitly – this renders the dominant matrix redundant to conservation and hence open to exploitation and degradation.

Australian researchers have been at the forefront of approaches that select sets of areas of high conservation value with the aim of conserving the minimal extent possible, [e.g. Margules & Pressey (2000)]. This is generally a worthy sentiment for fragmented regions, but its frame of reference is inappropriate or insufficient for extensive ecologically intact landscapes. Such reductionism will compromise the main conservation assets of the north Australian rangelands – their continuity, extent and ability to allow broad-scale ecological processes to flourish. And it is certainly inappropriate in an environment where climatic change is increasingly certain to move species' environmental envelopes beyond existing boundaries of conservation lands. An alternative approach is acceptance that broad-scale environmental values include enough continuity and connection to allow the ongoing operation of ecological and evolutionary processes (Mackey et al. 2001).

Even by the terms of the conventional conservation paradigm, the north Australian rangelands are poorly protected. While there are some outstanding conservation reserves, most notably Kakadu National Park, the 6.2% of the rangelands contained within the conservation reserve system is highly biased towards rocky infertile areas with high aesthetic appeal and away from fertile lowlands (Woinarski et al. 1996). Typically, they are managed with few resources that are spent primarily on tourism development rather than on biodiversity management.

The conservation estate has also been contested and remains fragile. Most reserves are accessible to mining, most famously the mines and lease within Kakadu National Park, and there has been little enthusiasm from governments or industry to transfer profitable pastoral leases on relatively fertile lands to conservation tenures. Most conservation reserves in the region have also been subjected, generally successfully, to Indigenous land rights claims subsequent to proclamation, with the increasing likelihood that such reserves will be managed for multiple land management objectives, of which biodiversity conservation is but one, although this can also enable more intensive management. Against this trend, the reserve system is currently being augmented by private sector philanthropists with the necessary funds to secure productive real estate. Such non-governmental conservation groups have recently transformed several pastoral properties into conservation reserves.

However, outside the sandstone massifs and relictual rainforest patches, there are few areas of high biodiversity that can be cauterized away from a surrounding low-value landscape. Even threatened species can have vast and often fluctuating distributions and vegetation communities are more likely to be threatened as a result of degradation by invasive species than through contraction to an easily definable remnant that can be rescued by a simple shift in tenure. It is thus apparent that the future of biodiversity conservation in the rangelands of Australia will not be secured by efficiently selected incremental additions of pockets of otherwise unwanted land. Rather, it will require fundamental changes in environmental and development policy and practice across the broader landscape.

The Australian tropical rangelands are defined by their extensive and intact ecological systems. Maintenance of these qualities should be the foundation of conservation and land-use planning. In the reverse of current practice, areas dedicated to intensive development should be sited to minimize their consequences to this natural matrix. The natural matrix need not be locked up nor unpopulated in order to maintain its conservation value. Rather, nurturance of biodiversity on these lands will require sympathetic stewardship by people, managing fire regimes and minimizing the extent and impacts of weeds and feral animals. At a regional level, such a conservation idyll may be uneconomic and may not match the utilization demands of some powerful interest groups. The challenge is to rise above such parochial perspectives and to recognize instead that, on a global scale, intensive development of northern Australia may never be of significant economic consequence. Instead, extensive natural environments are likely to become an increasingly precious commodity, and the maintenance of that asset should be a shared responsibility.

Pastoral lands

The question has been asked as to whether pastoralism is the most appropriate use of the tropical savanna landscape and whether the long-term costs are worth the short-term gain (Holmes 2000). While many of the small properties in Queensland whose marginal profitability drove overgrazing, even if their owners had good intentions, are now being reconsolidated (McAllister et al. 2006), many larger properties are populated so sparsely that extensive areas of low productivity are essentially unmanaged, allowing for the proliferation of feral animals and weeds and the propagation of undesirable fire regimes. Only properties owned by the largest pastoral companies have the resources to undertake comprehensive management and internal environmental policy imperatives to do so. While replacement of pastoralism is unlikely for many political reasons, the industry is currently more influential than its demographic or financial impact would suggest can be maintained in the long term (Fargher et al. 2003). While part of the pastoral industry's success rests on romantic notions of the outback, there is also a lack of viable alternative uses of the land and there remains an underlying assumption in Australian policy that all land must serve some economic purpose. However, a rigorous analysis of whether pastoralism is an economic benefit to the country has not been undertaken, at least not one in which future costs are assessed with a realistic future discount rate. Only should the costs of the environmental damage and its subsequent repair at government expense be seen to outweigh the benefits, might the emphasis on pastoral production diminish.

As it is, assuming the pastoral industry will continue, a vision of landscape stewardship is not entirely inconsistent with trends in the management of pastoral lands. However, biodiversity on pastoral lands will only be maintained where there are sufficient incentives for management to extend beyond that required for cattle alone. Current government investment in conservation funding focuses on small-scale remediation projects, such as closing off cattle access to rivers or wetlands and erection of fences to control stock movements. However, additional fencing may simply facilitate pastoral intensification, or merely deny cattle access to environments that feral pigs continue to degrade.

Some incentives are already in place for more sustainable management across entire properties, including demonstrated benefits of conservative grazing practices, but there has been little political will to institute negative incentives that punish environmentally deleterious land management. As it is, adoption of improved management is likely to be gradual. In the short term, there is good evidence that substantial parts of properties are often uneconomic to keep in production and could be set aside for conservation as part of the broader landscape matrix envisaged above (Ash & Stafford-Smith 2003). However, they will continue to lose species unless there are incentives for them to be managed actively, suppressing weeds, controlling feral animals and maintaining a finely grained fire matrix.

If stewardship is to be rewarded, it remains to be determined how management will be funded and who will undertake it. The benefits of biodiversity conservation in areas that are not producing profit for leaseholders accrue to the wider community; thus it should be the government who pays for this stewardship. A range of stewardship arrangements has now been tested in many parts of the world and methodologies are available that aim to ensure their success through monitoring and accountability (Cashore 2002). The more interesting question is who will undertake the work. At one level, labour is increasingly hard to attract to leasehold lands. On the other, unemployment among young Indigenous people is extremely high. Currently many Indigenous people who could potentially work on pastoral properties or on areas set aside for conservation lack many of the necessary skills and rely heavily on welfare payments. Yet, once they were pivotal to the pastoral industry (May 1994; Smith 2003). However, should policy settings change and education programs become more effective, conservation management of lands set aside on pastoral properties could provide an important source of employment. The recent renewal of the Northern Territory Indigenous Pastoral Program includes the training and employment of Indigenous people in weed, fire and feral animals control as well as pastoral land management and could be extended to include biodiversity conservation on pastoral lands.

There may also be a legal requirement that Indigenous people benefit from any conservation stewardship agreements on leasehold land. It has been argued that pastoral leases only allow leaseholders to keep cattle to graze grass. Other uses of leasehold land may therefore be contestable under the *Native Title Act* 1994, an Australian legislation that recognizes that Indigenous tenure can coexist with the more recently imposed tenure systems. In some parts of Australia and under some tenure types, native title is held to be extinguished, but this is unlikely to be true for most pastoral lands across tropical Australia. Thus Indigenous native title owners could have rights to any benefits that might derive from land set aside for conservation on pastoral leases, rights that could be awarded through contracts to undertake conservation management (Productivity Commission 2001). A confrontational approach, however, will impede change; the chances of improved management will be much greater by taking advantage of opportunities available under Indigenous Land Use Agreements to develop tailored benefit-sharing arrangements that provide returns to leaseholders, Indigenous native title holders and the broader community (National Native Title Tribunal 2006).

Indigenous-owned lands

Knowledge of how to maintain the natural landscapes of northern Australia persists in varying degrees among many Indigenous communities. While Indigenous ecological knowledge will need to adapt to deal with exotic weeds and herbivores, climate change and the resultant disrupted fire regimes, reinstating land management is a high priority for many Indigenous organizations. This will require a reversal of existing population trends that mean that Indigenous-owned lands of the future are likely to be even emptier than they are today (Altman 2006). Drawing on the current discussion about the direction for Indigenous land management, we propose four complementary and interdependent approaches to maintaining conservation values on Indigenous-owned land.

The first approach is to value work undertaken in land management as a genuine contribution to a sustainable Australia. To an extent, the value of this work is acknowledged through the funding of Indigenous ranger programs on national parks and the periodic support given to Indigenous ranger groups to undertake land management. However, the political will to provide government support fluctuates over time, whereas land management requires a long-term institutional commitment. An alternative is partnership between business and Indigenous land owners. Mining companies are starting to invest in social capital to ensure ongoing access to resources (Batten & Birch 2005). There is also an example from the Northern Territory where, as part of its licence to operate, a gas company has entered into a 17-year $22 million agreement with an Indigenous group to undertake intensive fire management in the early dry season that will reduce the extent of later fires

(Scrymgour 2006). This follows research demonstrating that minimizing late-dry season fires substantially reduces net annual greenhouse gas emissions (Williams et al. 2004; Cook et al. 2005), so that the fire management should offset the emissions of the gas company.

The second approach is to accept urban drift as inevitable, but then equip young Indigenous people with the skills to engage with their lands on their own, economically independent terms. Indigenous people do not necessarily lose traditional ownership or connection to the land simply by moving to the cities. As long as they can regularly return to their country, they can continue to undertake traditional obligations to the land and contribute to western priorities of land and biodiversity management. Those who can move between urban settings and traditional lands can bring new skills to combine with the old. This vision of young people going into 'orbit' and returning to leadership roles in their communities has been espoused by Indigenous leaders (Pearson 2003). It has the potential to provide greater opportunity for the independent Indigenous initiative rather than land management being a form of contractual obligation, often following non-Indigenous land management prescriptions and objectives. Such a model will only have biodiversity benefits if it results in, or is combined with, ongoing land management, particularly fire management through the year.

The third approach is to facilitate development of a local knowledge-based economy. To an extent, this is already happening. Australian Indigenous art, which is based on both traditional design and modern interpretations, is marketed profitably worldwide. In the Northern Territory, the centre of the industry, the art market, is valued at $100 million per annum (Altman 2001). Cultural tourism and music production also offer potential to generate endogenous income streams. However, the fact that this economic activity is happening, but active land management is diminishing, suggests that this is not a long-term solution for the broader landscape. Drier areas in particular have limited potential for tourism, and art and music will only support a part of the population. Certainly, the current income is insufficient to provide economic independence from government welfare and the income generated is rarely invested in increasing capacity for land management. Biodiversity benefits may come from developing Indigenous enterprises that focus on these neglected areas, particularly, the collection of fruits in areas that would have been visited only at times of seasonal abundance.

The final approach is to facilitate development of locally owned resource-based industries including pastoralism and wildlife. History, however, would suggest that the potential for such industries is limited. Many lands only remained under Indigenous control during the colonization phase of Australian history because they had low pastoral potential. Given the suggestion that the slightly better country is already being mined of natural capital by current broad-scale land uses, there would seem little base for optimism that such industries could be established profitably, let alone sustainably, on most Indigenous land. However, the return of Indigenous control over land alienated from its traditional owners by the granting of pastoral leases is likely to see a hybrid approach to management of Indigenous pastoral properties.

Wildlife-based industries, including the use of native plant resources, may become profitable if markets can be established for high value, readily transportable products. But current policy settings favour the capture of any profitable plant product in horticulture while wild animal-based products are likely to face the same problems with productivity as pastoralism. Also, attempts to introduce selected hunting of wild crocodiles were recently stopped because they were considered cruel (Campbell 2005). Nevertheless, the full economic analysis of potential productivity under favourable policy settings has not been undertaken.

There is no one solution to ensuring biodiversity conservation on Indigenous-owned land and many approaches will be tried in the coming years. The most important factor in their success will probably be the extent to which Indigenous people themselves set agendas and benefit from conservation programs.

Conclusion

The savanna rangelands of tropical Australia offer an intriguing mix of conservation and socio-economic issues, in part similar and in part contrasting to the major rangeland biomes elsewhere in the World. They have the virtue of falling entirely within a single relatively prosperous nation. The relative intactness of the natural systems has allowed persistence, in modern times, of nearly all of the biota, in contrast to many other parts of Australia. But, it is becoming increasingly clear that important elements of that biota are in decline and that there is an accelerating rate of detrimental change brought about by invasive elements.

The natural values of Australia's tropical rangelands have been retained, partly by default, through the repeated historical failure of intensive development rather than by conscious choice. Their infertility, remoteness and extreme climate have largely repelled a century and a half of development ambition. However, they will not do so in future without proactive policy development and implementation, particularly if climate change radically reduces agricultural production and water availability elsewhere in the broader region. As it is, their ecological integrity is at least subtly modified and almost universally controlled by a hegemonic land use, pastoralism, and is partly compromised by the chronic impoverishment of many of the resident communities. Although the current formal conservation reserve network is generally meagre and unrepresentative, the region's extraordinarily low population density and predicted low rate of population growth present, on a world scale, an unrivalled opportunity for retention of a vast, relatively undisturbed natural landscape; but the low population and production potential also present challenges to land and biodiversity management that requires both funds and people.

This paper stresses that the Australian tropical rangelands need active management in order to control exotic pests, changes in vegetation structure and limit the extent of wildfires if their biodiversity is not only to remain intact but also to continue providing ongoing ecosystem services. We propose that the answer to biodiversity conservation in such circumstances is not the creation, and subsequent neglect, of small 'representative' reserves. Nor is it the aggregation of its population in tiny settlements and coastal towns with an empty landscape in between. Rather, the future provides a powerful opportunity for renewal of Indigenous cultural engagement and the development of a pastoral industry in which environmental conservation can take precedence over production, and where both contribute to enterprise profitability. The key to this is allocating appropriate economic value to environmental services delivered by people who understand the landscape and are dedicated to living in it.

References

Abbott, I. (2002) Origin and spread of the cat, *Felis catus*, on mainland Australia, with a discussion of the magnitude of its early impact on native fauna. *Wildlife Research* 29, 51–74.

Altman, J. (2001) *Sustainable Development Options on Aboriginal Land: The Hybrid Economy in the Twenty-First Century*, Discussion Paper No. 226. Centre for Aboriginal Economic Policy Research, Canberra.

Altman, J. (2004) Economic development and indigenous Australia: contestations over property, institutions and ideology. *Australian Journal of Agricultural and Resource Economics* 48, 513–534.

Altman, J. (2006) *In Search of an Outstations Policy for Indigenous Australians*, Working Paper 34. Centre for Aboriginal Economic Policy Research, Canberra.

Andersen, A.N. (1992) Regulation of 'momentary' diversity by dominant species in exceptionally rich ant communities in the Australian seasonal tropics. *American Naturalist* 140, 401–420.

Andersen, A.N. (2000) *The Ants of Northern Australia: A Guide to the Monsoonal Fauna*. CSIRO Publishing, Melbourne.

Andersen, A.N., Azcárate, F.M. & Cowie, I.D. (2000) Seed selection by an exceptionally rich community of harvester ants in the Australian seasonal tropics, *Journal of Animal Ecology* 69, 975–984.

Andersen, A.N., Hoffmann, B.D., Müller, W.J. & Griffiths, A.D. (2002) Using ants as bioindicators in land management: simplifying assessment of ant community responses. *Journal of Applied Ecology* 39, 8–16.

Anonymous. (1999) *Kakadu National Park Plan of Management*. Kakadu Board of Management and Parks Australia, Commonwealth of Australia, Jabiru.

Anonymous. (2006) Breaking new ground: Indigenous Pastoral Program, Northern Territory. Memorandum of Understanding between Central Land Council, Indigenous Land Corporation, Northern Territory Government, Northern Land Council and the Northern Territory Cattlemen's Association. www.nt.gov.au/dpifm/Primary_Industry/Content/File/publications/books_reports/indigenous_pastoral_program.pdf

Ash, A. & Stafford-Smith, D.M. (2003) Pastoralism in tropical rangelands: seizing the opportunity to change. *Rangeland Journal* 25, 113–127.

Ash, A.J. & McIvor, J.G. (1998) How season of grazing and herbivore selectivity influence monsoon tall-grass communities of Northern Australia. *Journal of Vegetation Science* 9, 123–132.

Archer, S. & Schimel, D.S. (1995) Mechanisms of shrubland expansion: land use, climate or CO_2. *Climatic Change* 29, 91–99.

Australian Bureau of Meteorology (2006) Trend maps – Australian climate variability and change 1950–2000 (mm/10 years). http://www.bom.gov.au/cgi-bin/silo/reg/cli_chg/trendmaps.cgi.

Batten, J.A. & Birch, D. (2005) Defining corporate citizenship: evidence from Australia. *Asia Pacific Business Review* 11, 293–208.

Beeton, R.J.S., Buckley, K.I., Jones, G.J., Morgan, D., Reichelt, R.E. & Trewin, D. (2006) *Australia State of the Environment 2006, Independent Report to the Australian Government Minister for the Environment and Heritage*. Department of Environment and Heritage, Canberra.

Bond, W.J. & Midgley, G.F. (2000) A proposed CO_2-controlled mechanism of woody plant invasion in grasslands and savannas. *Global Change Biology* 6, 865–869.

Bowman, D.M.J.S. (1996) Diversity patterns of woody species on a latitudinal transect from the monsoon tropics to desert in the Northern Territory, Australia. *Australian Journal of Botany* 44, 571–580.

Bowman, D.M.J.S. (1998) Tansley Review no. 101, The impact of Aboriginal landscape burning on the Australian biota. *New Phytologist* 140, 385–410.

Bowman, D.M.J.S. & Panton, W.J. (1993) Decline of *Callitris intratropica* R.T. Baker & H.G. Smith in the Northern Territory: implications for pre- and post-European colonization fire regimes. *Journal of Biogeography* 20, 373–381.

Braithwaite, R.W., Lonsdale, W.M. & Estbergs, J.A. (1989) Alien vegetation and native biota in tropical Australia: the impact of *Mimosa pigra*. *Biological Conservation* 48, 189–210.

Brook, B.W. & Bowman, D.M.J.S. (2002) Explaining the *Pleistocene megafaunal* extinctions: models, chronologies and assumptions. *Proceedings of the National Academy of Sciences of the United States of America* 99, 14624–14763.

Brook, B.W. & Bowman, D.M.J.S. (2005) One equation fits overkill: why allometry underpins both prehistoric and modern body sized-biased extinctions. *Population Ecology* 47, 137–141.

Burbidge, A.A. & McKenzie, N.L. (1989) Patterns in the modern decline of Western Australia's vertebrate fauna: causes and conservation implications. *Biological Conservation* 50, 143–198.

Burrows, W.H., Carter, J.O., Scanlan, J.C. & Anderson, E.R. (1990) Management of savanna for livestock production in north-east Australia: contrasts across the tree-grass continuum. *Journal of Biogeography* 17, 503–512.

Butler, D.W. & Fairfax, R.J. (2003) Buffel grass and fire in a gidgee and brigalow woodland: a case study from central Queensland. *Ecological Management and Restoration* 4, 120–125.

Calaby, J.H. & White, C. (1967) The Tasmanian Devil (*Sarcophilus harrisi*) in northern Australia in recent times. *Australian Journal of Science* 29, 473–475.

Campbell, I. (2005) Crocodile safari hunting proposal rejected – crocodile culls to continue. http://www.deh.gov.au/minister/env/2005/mr06oct05.html. (accessed 28 August 2006).

Campbell, J. (2002) *Invisible Invaders: Smallpox and Other Diseases in Aboriginal Australia, 1780–1880.* Melbourne University Press, Melbourne.

Cashore, B. (2002) Legitimacy and the privatization of environmental governance: how non-state market-driven (NSMD) governance systems gain rule-making authority. *Governance* 15, 503–529.

Chudleigh, P. & Bramwell, T. (1996) *Assessing the Impact of Introduced Tropical Pasture Plants in Northern Australia.* Commonwealth Scientific and Industrial Research Organisation Task Force Interested in Introduced Pasture Plants, Brisbane.

Churchill, T.G. & Ludwig, J.A. (2004) Changes in spider assemblages along grassland and savanna grazing gradients in northern Australia. *Rangeland Journal* 26, 3–16.

Cook, G.D. & Dias, L. (2006) Turner Review No. 12: it was no accident: deliberate plant introductions by Australian government agencies during the 20th century. *Australian Journal of Botany* 54, 601–625.

Cook, G.D., Liedloff, A.C., Eager, R.W. et al. (2005) The estimation of carbon budgets of frequently burnt tree stands in savannas of northern Australia, using allometric analysis and isotopic discrimination. *Australian Journal of Botany* 53, 621–630.

Cooperative Research Centre for Australian Weed Management (2003) *Weed Management Guide: Hymenachne or Olive Hymenachne Hymenachne amplexicaulis*. Department of Environment and Heritage, Canberra.

Crowley, G.M. (2001) Grasslands of Cape York Peninsula: a fire-dependent habitat. In: Dyer, R., Jacklyn, P., Partridge, I., Russell-Smith, J. & Williams, R. (eds.) *Savanna Burning: Understanding and Using Fire in Northern Australia*. Cooperative Research Centre for Tropical Savannas Management, Darwin, p. 34.

Crowley, G.M. (2006) *Developing a Community Driven Knowledge system for Biodiversity Management in the Australian Tropical Savannas Region*. Cooperative Research Centre for Tropical Savannas Management, Darwin.

Crowley, G.M. & Garnett, S.T. (1998) Vegetation change in the grasslands and grassy woodlands of east-central Cape York Peninsula, Australia. *Pacific Conservation Biology* 4, 132–48.

Crowley, G.M., Garnett, S.T. & Shephard, S. (2004) *Management Guidelines for Golden-Shouldered Parrot Conservation*. Queensland Parks and Wildlife Service, Brisbane.

Crowley, G.M. & Thompson, P. (2005) Managing perceptions: can a change in attitude towards fire and its management ameliorate environmental problems in Australia's north? *Wingspan* 15, S12–S14.

Csurhes, S. & Edwards, R. (1998) *Potential Environmental Weeds in Australia: Candidate Species for Preventative Control*. Department of Environment and Heritage, Canberra.

Dick, R.S. (1975) A map of the climates of Australia according to Köppen's principles of definition. *Queensland Geographic Journal* 3, 33–69.

Dunlop, C.R., Leach, G.J. & Cowie, I.D. (1995) *Flora of the Darwin region*. Conservation Commision of the Northern Territory, Darwin.

Eliot, I., Finlayson, C.M. & Waterman, P. (1999) Predicted climate change, sea-level rise and wetland management in the Australian wet-dry tropics. *Wetlands Ecology and Management* 7, 63–81.

Eyles, A.G., Cameron, D.G. and Hacker, J.B. (1985) *Pasture research in northern Australia - its history, achievements and future emphasis*. Commonwealth Scientific and Industrial Research Organisation Division of Tropical Crops and Pastures, Brisbane.

Fargher, J.D., Howard, B.M., Burnside, D.G. & Andrew, M.H. (2003) The economy of Australian rangelands – myth or mystery? *Rangeland Journal* 25, 140–156.

Fensham, R.J., Fairfax, R.J. & Cannell, R.J. (1994) The invasion of *Lantana camara* L. in Forty Mile Scrub National Park, north Queensland. *Australian Journal of Ecology* 19, 297–305.

Fensham, R.J. & Holman, J.E. (1999) Temporal and spatial patterns in drought-related tree dieback in Australian savanna. *Journal of Applied Ecology* 36, 1035–1050.

Ferdinands, K., Beggs, K. & Whitehead, P. (2005) Biodiversity and invasive grass species: multiple-use or monoculture? *Wildlife Research*, 32, 447–457.

Finlayson, H. H. (1961) On central Australian mammals, Part IV: the distribution and status of central Australian species. *Records of the South Australian Museum* 41, 141–191.

Fisher, A. (2001) *Biogeography and Conservation of Mitchell Grasslands in Northern Australia*. Northern Territory University, Australia.

Fisher, A., Hunt, L.P., James, C. et al. (2004) *Review of Total Grazing Pressure Management Issues and Priorities for Biodiversity Conservation in Rangelands: A Resource to Aid NRM Planning*. Desert Knowledge Cooperative Research Centre, Alice Springs and Cooperative Research Centre for Tropical Savannas Management, Darwin.

Franklin, D.C. (1999) Evidence of disarray amongst granivorous bird assemblages in the savannas of northern Australia, a region of sparse human settlement. *Biological Conservation* 90, 53–68.

Franklin, D.C., Whitehead, P.J., Pardon, G., Matthews, J., McMahon, P. & McIntyre, D. (2005) Geographic patterns and correlates of the decline of granivorous birds in northern Australia. *Wildlife Research* 32, 399–408.

Friedel, M., Puckey, H., O'Malley, C., Waycott, M., Smyth, A. & Miller, G. (2006) *Buffel Grass: Friend or Foe?* Desert Knowledge Cooperative Research Centre, Alice Springs.

Furnas, M. (2003) *Catchments and Corals: Terrestrial Runoff to the Great Barrier Reef*. Australian Institute of Marine Science, Townsville.

Garnett, A.M. & Lewis, P.E.T. (1999) *Trends in Rural Labour Markets*. Centre for Labour Market Research, Perth.

Garnett, S.T. & Crowley, G.M. (1995) The decline of the Black Treecreeper *Climacteris picumnus melanotaon Cape York Peninsula, Emu* 95, 66–68.

Garnett, S.T. & Crowley, G.M. (2003) Cape York Peninsula. In: Mittermeier, R.A., Mittermeier, C.G., Gil, P.R. et al. (eds.) *Wilderness: Earth's Last Wild Places*. Cemex, Mexico City, pp. 220–229.

Garnett, S.T. & Crowley, G.M. (2004) *Recovery Plan for the Golden-Shouldered Parrot (Psephotus chrysopterygius) 2003–2007*. Queensland Parks and Wildlife Service, Brisbane.

Godden, D. (1997) *Agricultural and Resource Policy*. Oxford University Press, Melbourne.

Grice, A.C. (2002) *Weeds of Significance to the Grazing Industries of Australia. Final Report (Project Number COMP.045)*. Meat and Livestock Australia, North Sydney.

Grice, A.C. (2006) The impacts of invasive plant species on the biodiversity of Australian rangelands. *Rangeland Journal* 28, 27–36.

Griener, R., Mayocchi, C., Larson, S., Stoeckl, N. & Schweigert, R. (2004) *Benefits and Costs of Tourism for Remote Communities – Case Study for the Carpentaria Shire in North-West Queensland*, Commonwealth Scientific and Industrial Research Organisation Sustainable Ecosystems and Tropical Savannas Cooperative Research Centre, Darwin.

Griffiths, A.D. (1998) *Impact of Sulphur Dioxide Emissions on Savanna Biodiversity at Mt Isa, Queensland. Unpublished final report to Mt Isa Mines*. Commonwealth Scientific and Industrial Research Organisation and Tropical Savannas Cooperative Research Centre, Darwin.

Griffiths, A.D. & McKay, J.L. (2005) *Monitoring the Impact of Cane Toads on Goanna Species in Kakadu National Park*. Kakadu National Park, Department of Environment and Heritage, Jabiru.

Hannah, D., Woinarski, J.C.Z., Catterall, C.P., McCosker, J.C., Thurgate, N.Y. & Fensham, R.J. (2007) Impacts of clearing, fragmentation and disturbance on the bird fauna of Eucalypt savanna woodlands in central Queensland, Australia. *Austral Ecology* 32, 261–276.

Hoffman, D.H., Andersen, A.N. & Hill, G.G.E. (1999) Impact of an introduced ant on native rainforest invertebrates: *Pheidole megacephala* in monsoonal Australia. *Oecologia* 120, 595–604.

Holmes, J. (2000) Land tenure and land administration in northern Australia: needed future directions. In: *Land Administration and Land Management in the Tropical Savannas: A Better Way – Proceedings of the Tropical Savannas CRC Forum*. Tropical Savannas Cooperative Research Centre, Darwin.

Jack, R.L. (1922) *Northmost Australia*. George Robertson and Co., Melbourne.

James, C.D., Landsberg, J. & Morton, S.R. (1999) Provision of watering points in the Australian arid zone: a review of effects on biota. *Journal of Arid Environments* 41, 87–121.

Jones, R.J. & Jones, R.M. (1977) The ecology of Siratro-based pastures, In: Wilson, J.R. (ed.) *Plant Relations in Pastures*. Commonwealth Scientific and Industrial Research Organisation, Melbourne, pp. 353–367.

Jones, R.J., McIvor, J.G., Middleton, C.H., Burrows, W.H., Orr, D.M. & Coates, D.B. (1997) Stability and productivity of *Stylosanthes* pastures in Australia. I. Long-term botanical changes and their implications in grazed *Stylosanthes* pastures. *Tropical Grasslands* 31, 482–493.

Keast, A. (ed.) (1981) *Ecological Biogeography of Australia*. Dr W. Junk Publishers, The Hague.

King, J.Y., Mosier, A.R., Morgan, J.A., LeCain, D.R., Milchunas, D.G. & Parton, W.J. (2004) Plant nitrogen dynamics in shortgrass steppe under elevated atmospheric carbon dioxide. *Ecosystems* 7, 147–160.

Landrigan, M. & Wells, J.V. (2005) Populating the northern territory. In: Wells, J.T., Dewar, M. & Parry, S. (eds.) *Modern Frontier: Aspects of the 1950s in Australia's Northern Territory*. Charles Darwin University Press, Darwin, pp. 15–22.

Landsberg, J., James, C.D., Morton, S.R. et al. (1997) *The Effects of Artificial Sources of Water on Rangeland Biodiversity*. Environment Australia. Canberra.

Landsberg, J., O'Connor, T.G. & Freudenberger, D. (1999) The impacts of livestock grazing on biodiversity in natural systems. In: Jung, H.J.G. & Fahey, G.C. (eds.) *Vth International Symposium on the Nutrition of Herbivores*. American Society of Animal Science, Savoy.

Lawson, B.E., Bryant, M.J. & Franks, A.J. (2003) Assessing the potential distribution of buffel grass (*Cenchris ciliaris* L.) in Australia using a climate soil model. *Plant Protection Quarterly* 19, 155–163.

Lee, I. (1925) *Early Explorers in Australia: From the Log Books and Journals*. Methuen and Co. Ltd., London.

Lonsdale, W.M. (1994) Inviting trouble: introduced pasture species in northern Australia. *Australian Journal of Ecology* 19, 345–354.

Ludwig, J.A., Eager, R.W., Williams, R.J. & Lowe, L.M. (1999) Declines in vegetation patches, plant diversity and grasshopper diversity near cattle watering-points in the Victoria River District, Northern Territory. *Rangeland Journal* 21, 135–149.

Lundie-Jenkins, G. & Payne, A.L. (2004) *Recovery Plan for the Julia Creek Dunnart (Sminthopsis douglasi) 2000–2004*. Queensland Parks and Wildlife Service, Brisbane.

Mackey, B., Nix, H. & Hitchcock, P. (2001) *The Natural Heritage Significance of Cape York Peninsula*. ANUTECH, Canberra and Queensland Environmental Protection Agency, Brisbane.

MacKnight, C.C. (1972) Macassans and Aborigines. *Oceania* 47, 283–319.

Madden, E. & Shine, R. (1996) Seasonal migration of predators and prey: a study of pythons and rats in tropical Australia. *Ecology* 77, 149–156.

Madsen, T. & Shine, R. (1999) The adjustment of reproductive threshold to prey abundance in a capital breeder. *Journal of Animal Ecology* 68, 571–580.

Madsen, T., Ujvari, B., Shine, R., Buttemer, W. & Olsson, M. (2006) Size matters: extraordinary rodent abundance on an Australian tropical floodplain. *Austral Ecology* 31, 361–365.

Margules, C. & Pressey, R. (2000) Systematic conservation planning. *Science* 405, 243–253.

Markich, S.J., Jeffree, R.A. & Burke, P.T. (2002) Freshwater bivalve shells as archival indicators of metal pollution from a copper-uranium mine in tropical northern Australia. *Environmental Science and Technology* 36, 821–832.

Martin, T.G., Campbell, S.D. & Grounds, S. (2006) Weeds of Australian rangelands. *Rangeland Journal* 28, 3–26.

Martin, T.G. & van Klinken, R.D. (2006) Value for money? Investment in weed management in Australian rangelands. *Rangeland Journal* 28, 63–75.

May, D. (1994) *Aboriginal Labour and the Cattle Industry: Queensland from White Settlement to the Present.* Cambridge University Press, Cambridge.

McAllister, R., Abel, R.J.N., Stokes, C.J. & Gordon, I.J. (2006) Australian pastoralists in time and space: the evolution of a complex adaptive system. *Ecology and Society* 11(2), 41. http://www.ecologyandsociety.org/vol11/iss2/art41/ (accessed 6 May 2007).

McKenzie, N.L. (1981) Mammals of the Phanerozoic south–west Kimberley, Western Australia: biogeography and recent changes. *Journal of Biogeography* 8, 263–280.

McKenzie, N.L. & Burbidge, A.A. (2003) Mammals. In: *Australian Terrestrial Biodiversity Assessment 2002.* National Land and Water Resources Audit, Department of Primary Industries and Energy and Australian Government Publishing Service, Canberra, pp. 83–96.

McLeod, R. (2004) *Counting the Cost: Impacts of Invasive Animals in Australia, 2004.* Pest Animal Control Cooperative Research Centre, Canberra.

Miller, G.H., Fogel, M.L., Magee, J.W., Gagan, M.K., Clarke, S.J. & Johnson, B.J. (2005) Ecosystem collapse in pleistocene Australia and a human role in megafaunal extinction. *Science* 309, 287–291.

Morton, S.R. (1990) The impact of European settlement on the vertebrate animals of arid Australia: a conceptual model. *Proceedings of the Ecological Society of Australia* 16, 210–213.

Morton, S.R., Brennan, K.G. & Armstrong, M.D. (1990a) Distribution and abundance of ducks in the Alligator Rivers region, Northern Territory. *Australian Wildlife Research* 17, 573–590.

Morton, S.R., Brennan, K.G. & Armstrong, M.D. (1990b) Distribution and abundance of magpie geese, *Anseranas semipalmata*, in the Alligator Rivers Region, Northern Territory *Australian Journal of Ecology* 15, 307–320.

Morton, S.R., Brennan, K.G. & Armstrong, M.D. (1993) Distribution and abundance of Brolgas and Black-necked Storks in the Alligator Rivers Region, Northern Territory. *The Emu* 93, 88–92.

Morton, S.R., Stafford-Smith, D., Friedel, M., Griffin, G.F. & Pickup, G. (1995) The stewardship of arid Australia: ecology and landscape management. *Journal of Environmental Management* 43, 195–217.

Mott, J.J., Bridge, B.J. & Arndt, W. (1979) Soil seals in tropical tall grass pastures of Northern Australia. *Australian Journal of Soil Research* 30, 483–494.

Murray, P. & Chaloupka, G. (1984) The Dreamtime animals: extinct megafauna in Arnhem Land rock art. *Archaeology in Oceania* 19, 105–116.

National Native Title Tribunal (2006) What is an indigenous land use agreement (ILUA)? Native Title Fact Sheets No. 2b, www.nntt.gov.au/publications/data/files/NTF_2b.pdf (accessed 11 September, 2006).

Neldner, V.J., Fensham, R.J., Clarkson, J.R. & Stanton, J.P. (1997) The natural grasslands of Cape York Peninsula, Australia: description, distribution and conservation status. *Biological Conservation* 81, 121–136.

Norris, A. & Low, T. (2005) *Review of the Management of Feral Animals and their Impact on Biodiversity in the Rangelands: A Resource to Aid NRM Planning*. Pest Animal Control Cooperative Research Centre, Canberra.

Oxley, T., Leigo, S., Hausler, P., Bubb, A. & MacDonald, N. (2006) *NT Wide Pastoral Survey 2006*. Department of Primary Industry, Fisheries and Mines, Darwin.

Palmer, C., Brennan, K. & Morrison, S. (2007) Short-term effects of a category 5 cyclone on terrestrial bird populations on Marchinbar Island, Northern Territory. *Northern Territory Naturalist* 19, 15–24.

Pavlov, P.M. (1995) Pig, *Sus scrofa*. In: Strahan, R. (ed.) *The Mammals of Australia*. Reed Books, Sydney, pp. 715–717.

Pearson, G. (2003) Man cannot live by service delivery alone. In: *Opportunity and Prosperity Conference*. Melbourne Institute, Melbourne.

Preece, N. (2002) Aboriginal fires in monsoonal Australia from historical accounts. *Journal of Biogeography* 29, 321–336.

Preece, N. (2007) Traditional and ecological fires and effects of bushfire laws in north Australian savannas. *International Journal of Wildland Fire* 16, 378–389.

Press, A.J. (1987) Fire management in Kakadu National Park: the ecological basis for the active use of fire. *Search* 18, 244–248.

Press, T., Lea, D., Webb, A. & Graham, A. (eds.) (1995) *Kakadu. Natural and Cultural Heritage and Management*. Australian Nature Conservation Agency and Australian National University, Darwin.

Price, O.F., Woinarski, J.C.Z. & Robinson, D. (1999) Very large area requirements for frugivorous birds in monsoon rainforests of the Northern Territory, Australia. *Biological Conservation* 91, 169–180.

Prideaux, B., Wei, S. & Ruys, H. (2001) The senior drive tour market in Australia. *Journal of Vacation Marketing* 7, 209–219.

Productivity Commission (2001) Harnessing private sector conservation of biodiversity. Commission Research Paper, Ausinfo, Canberra.

Reynolds, H. (1982) *The Other Side of the Frontier*. Penguin, Ringwood, Victoria.

Roberts, T. (2005) *Frontier Justice: a history of the Gulf Country to 1900*. University of Queensland Press, Brisbane.

Roberts, R.G., Jones, R. & Smith, M.A. (1990) Thermoluminescence dating of a 50,000-year old human occupation site in northern Australia. *Nature* 345, 153–156.

Robley, A., Reddiex, B., Arthur, T., Pech, R. & Forsyth, D. (2004) *Interactions Between Feral Cats, Foxes, Native Carnivores and Rabbits in Australia.* Arthur Rylah Institute for Environmental Research, Department of Sustainability and Environment, Melbourne.

Rossiter, N.A., Setterfield, S.A., Douglas, M.M. & Hutley, L.B. (2003) Testing the grass-fire cycle: alien grass invasion in the tropical savannas of northern Australia. *Diversity and Distributions* 9, 169–176.

Rotstayn, L.D., Cai, W., Dix, M.R. et al. (2007) Have Australian rainfall and cloudiness increased due to the remote effects of Asian anthropogenic aerosols? *Journal of Geophysical Research* 112, D09202.

Russell-Smith, J. (1991) Classification, species richness and environmental relations of monsoon rain forest in northern Australia. *Journal of Vegetation Science* 2, 259–278.

Russell-Smith, J. (2002) *Pre-Contact Aboriginal and Contemporary Fire Regimes of the Savanna Landscapes of Northern Australia: Patterns, Changes and Ecological Processes.* Department of Environment and Heritage, Canberra.

Russell-Smith, J., Ryan, P.G., Klessa, D., Waight, G. & Harwood, R. (1998) Fire regimes, fire-sensitive vegetation and fire management of the sandstone Arnhem Plateau, monsoonal northern Australia. *Journal of Applied Ecology* 35, 829–846.

Savolainen, P., Leitner, T., Wilton, A.N., Matisoo-Smith, E. & Lundeberg, J. (2004) A detailed picture of the origin of the Australian dingo, obtained from the study of mitochondrial DNA. *Proceedings of the National Academy of Sciences of the United States of America* 101, 12387–12390.

Schmidt, S. & Lamble, R.E. (2002) Nutrient dynamics in Queensland savannas: implications for the sustainability of land clearing for pasture production. *Rangeland Journal* 24, 96–111.

Scrymgour, M. (2006) Multi-million dollar Arnhem Land Greenhouse gas fire sale. Minister for Environment Press Release, www.nt.gov.au/nreta/publications/mediareleases/pdf/2006/08/scryfire20060824.pdf (accessed 25 August 2006).

Sharp, B.R. & Whittaker, R.J. (2003) The irreversible cattle-driven transformation of a seasonally flooded Australian savanna. *Journal of Biogeography* 30, 783–802.

Sinden, J., Jones, R., Hester, S. et al. (2004) *The Economic Impact of Weeds in Australia,* Technical Series #8. Cooperative Research Centre for Australian Weed Management, Sidney.

Skeat, A.J., East, T.J. & Corbett, L.K. (1996) Impact of feral water buffalo. In: Finlayson, C.M. & von Oertzen, I. (eds.) *Landscape and Vegetation Ecology of the Kakadu Region, Northern Australia.* Kluwer Academic Publishers, Dordrecht, pp. 155–177.

Smith, T. (2003) Aboriginal labour and the pastoral industry in the Kimberley Division of Western Australia: 1960–1975. *Journal of Agrarian Change* 3, 552–570.

Stafford-Smith, M., Morton, S. & Ash, A. (2000) Towards sustainable pastoralism in Australia's rangelands. *Australian Journal of Environmental Management* 7, 190–203.

Stokes, C., Ash, A., Tibbett, M. & Holtum, J. (2005) OzFACE: the Australian savanna free air CO_2 enrichment facility and its relevance to carbon-cycling issues in a tropical savanna. *Australian Journal of Botany* 53, 677–687.

Stokes, C.J., McAllister, R.R.J. & Ash, A.J. (2006) Fragmentation of Australian rangelands: processes, benefits and risks of changing patterns of land use. *Rangeland Journal* 28, 83–96.

Taylor, J. & Kinfu, Y. (2006) Differentials and determinants of Indigenous population mobility. In: Hunter, B.H. (ed.) *Assessing the Evidence on Indigenous Socio-economic Outcomes: A Focus on the 2002 NATSISS*, CAEPR Research Monograph No. 26. The Australian National University, Canberra, pp. 57–67.

Tothill, J.C. & Mott, J.J. (1985) Australian savannas and their stability under grazing. *Proceedings of the Ecological Society of Australia* 13, 317–322.

Turney, C.S.M. & Hobbs, D. (2006) ENSO influence on Holocene Aboriginal populations in Queensland, Australia. *Journal of Archaeological Science* 33, 1744–1748.

Walsh, K.J.E., Nguyen, K.C. & McGregor, J.L. (2004) Fine-resolution regional climate model simulations of the impact of climate change on tropical cyclones near Australia. *Climate Dynamics* 22, 47–56.

Watson, M. & Woinarski, J.C.Z. (2004) *Vertebrate Monitoring and Resampling in Kakadu National Park, 2003.* Parks Australia North and Parks and Wildlife Commission of the Northern Territory, Darwin.

Werner, P.A. (2005) Impact of feral water buffalo and fire on growth and survival of mature savanna trees: an experimental field study in Kakadu National Park, northern Australia. *Austral Ecology* 30, 625–647.

Wheeler, J.R., Rye, B.L., Koch, B.L. & Wilson, A.J.G. (eds.) (1992) *Flora of the Kimberley Region of Western Australia.* Department of Conservation and Land Management, Perth.

Whitehead, P.J. (1999) Is it time to fill the north's empty landscapes? *Savanna Links* September/October 6–7.

Whitehead, P.J., Wilson, B.A. & Bowman, D.M.J.S. (1990) Conservation of coastal wetlands of the Northern Territory of Australia: The Mary River Floodplain. *Biological Conservation* 52, 85–111.

Whitehead, P.J., Woinarski, J.C.Z., Franklin, D. & Price, O. (2003) Landscape ecology, wildlife management and conservation in northern Australia: linking policy, practice and capability in regional planning. In: Storch, I. & Bissonette, J.A. (eds.) *Landscape Ecology and Resource Management: Linking Theory with Practice.* Island Press, Washington, D.C., pp. 227–259.

Williams, R.J., Cook, G.D., Gill, A.M. & Moore, P.H.R. (1999a) Fire regime, fire intensity and tree survival in a tropical savanna in northern Australia. *Austral Ecology* 24, 50–59.

Williams, J., Day, K.J., Isbell, R.F. & Reddy, S.J. (1985) Soils and climate. In: Muchow, R.C. (ed.) *Agro-Research for the Semi-Arid Tropics: North-West Australia.* University of Queensland Press, St Lucia, pp. 31–69.

Williams, R.J., Griffin, A.J. & Allen, G.E. (2002) Fire regimes and biodiversity in the savannas of north Australia. In: Bradstock, R.A., Williams, J. & Gill, A.M. (eds.) *Flammable Australia: Fire Regimes and Biodiversity of a Continent.* Cambridge University Press, Cambridge, pp. 281–304.

Williams, R.J., Hutley, L.B., Cook, G.D., Russell-Smith, J., Edwards, A. & Chen, X. (2004) Assessing the carbon sequestration potential of mesic savannas in the Northern Territory, Australia: approaches, uncertainties and potential. *Functional Plant Biology* 31, 415–422.

Williams, R.J., Myers, B.A., Eamus, D. & Duff, G.A. (1999b) Reproductive phenology of woody species in a north Australian tropical savanna. *Biotropica* 31, 626–636.

Woinarski, J.C.Z. (2003a) Arnhem Land. In: Mittermeier, R.A., Mittermeier, C.G., Gil, P.R. et al. (eds.) *Wilderness: Earth's Last Wild Places.* CEMEX, Mexico City, pp. 230–237.

Woinarski, J.C.Z. (2003b) Kimberley. In: Mittermeier, R.A., Mittermeier, C.G., Gil, P.R. et al. (eds.) *Wilderness: Earth's Last Wild Places.* CEMEX, Mexico City, pp. 238–245.

Woinarski, J.C.Z. (2004) The forest fauna of the Northern Territory: knowledge, conservation and management In: Lunney, D. (ed.) *Conservation of Australia's Forest Fauna.* Royal Zoological Society of New South Wales, Sydney, pp. 36–55.

Woinarski, J.Z.C., Andersen, A.N., Churchill, T.B. & Ash, A.J. (2002) Response of ant and terrestrial spider assemblages to pastoral and military land use and to landscape position, in a tropical savanna woodland in northern Australia. *Austral Ecology* 27, 324–333.

Woinarski, J.C.Z. & Ash, A.J. (2002) Responses of vertebrates to pastoralism, military land use and landscape position in an Australian tropical savanna. *Austral Ecology* 27, 311–323.

Woinarski, J.C.Z. & Catterall, C.P. (2004) Historical changes in the bird fauna at Coomooboolaroo, northeastern Australia, from the early years of pastoral settlement (1873) to 1999. *Biological Conservation* 116, 379–401.

Woinarski, J.C.Z., Connors, G. & Franklin, D.C. (2000) Thinking honeyeater: nectar maps for the Northern Territory, Australia. *Pacific Conservation Biology* 6, 61–80.

Woinarski, J.C.Z., Connors, G. & Oliver, B. (1996) The reservation status of plant species and vegetation types in the Northern Territory. *Australian Journal of Botany* 44, 673–689.

Woinarski, J.C.Z. & Dawson, F. (2001) Limitless lands and limited knowledge: coping with uncertainty and ignorance in northern Australia. In: Handmer, J.W., Norton, T.W. & Dovers, S.R. (eds.) *Ecology, Uncertainty and Policy: Managing Ecosystems for Sustainability.* Pearson Education Limited, Harlow, pp. 83–115.

Woinarski, J.C.Z. & Fisher, A. (2003) Conservation and the maintenance of biodiversity in the rangelands. *Rangeland Journal* 25, 157–171.

Woinarski, J.C.Z., Fisher, A. & Milne, D. (1999) Distribution patterns of vertebrates in relation to an extensive rainfall gradient and variation in soil texture in the tropical savannas of the Northern Territory, Australia. *Journal of Tropical Ecology* 15, 381–398.

Woinarski, J.C.Z. & Gambold, N. (1992) Gradient analysis of a tropical herpetofauna: distribution patterns of terrestrial reptiles and amphibians in Stage III of Kakadu National Park, Australia. *Wildlife Research* 19, 105–127.

Woinarski, J.C.Z., Hempel, C., Cowie, I. et al. (2006a) Distributional patterns of plant species endemic to the Northern Territory, Australia. *Australian Journal of Botany* 54, 627–640.

Woinarski, J.C.Z., McCosker, J.C., Gordon, G. et al. (2006b) Monitoring change in the vertebrate fauna of central Queensland, Australia, over a period of broad-scale vegetation clearance, 1975–2002. *Wildlife Research* 33, 263–274.

Woinarski, J.C.Z., Milne, D.J. & Wanganeen, G. (2001) Changes in mammal populations in relatively intact landscapes of Kakadu National Park, Northern Territory, Australia. *Austral Ecology* 26, 360–370.

Woinarski, J.C.Z., Williams, R.J., Price, O. & Rankmore, B. (2005) Landscapes without boundaries: wildlife and their environments in northern Australia. *Wildlife Research* 32, 377–388.

Woodroffe, C.D. (1993) Late Quaternary evolution of coastal and lowland riverine plains of Southeast Asia and northern Australia: an overview. *Sedimentary Geology* 83, 163–175.

Yibarbuk, D., Whitehead, P. J., Russell-Smith, J. et al. (2001) Fire ecology and Aboriginal land management on central Arnhemland, northern Australia: a tradition of ecosystem management. *Journal of Biogeography* 28, 325–343.

Young, G.R., Bellis, G.A., Brown, G.R. & Smith, E.S.C. (2001) The crazy ant *Anoplolepis gracilipes* (Smith) (Hymenoptera: Formicidae) in east Arnhem Land, Australia. *Australian Entomologist* 28, 97–104.

Zvereva, E.L. & Kozlov, M.V. (2006) Consequences of simultaneous elevation of carbon dioxide and temperature for plant-herbivore interactions: a metaanalysis. *Global Change Biology* 12, 27–41.

9

Livestock Grazing and Wildlife Conservation in the American West: Historical, Policy and Conservation Biology Perspectives

Thomas L. Fleischner

Environmental Studies Program, Prescott College, AZ, USA

Introduction

Grazing by domesticated livestock, primarily cattle, is the most ubiquitous land use in the western United States. Approximately 70% of the 11 westernmost states in the United States (those including and west of the Rocky Mountains) is grazed by livestock, at least part of the year (CAST 1974; Longhurst et al. 1982; Crumpacker 1984), including approximately 90% of federal land in these states (Armour et al. 1991). Livestock grazing occurs in more than 75% of the ecoregions delineated by the World Wildlife Fund (Ricketts et al. 1999) in the American West. It represents a primary ecological influence in more than half of these ecoregions.

The term *rangelands* is applied to most of the diverse ecological communities in this region – coniferous forests, broadleaf riparian forests, deserts, sandstone canyons and grasslands – if livestock are prevalent. There is no such thing as 'rangelands' in an ecological sense. While the term connotes open grasslands, it is in fact a catchword that implies a predominant form of land use – grazing by domesticated livestock – rather than any type of ecological community.

Wild Rangelands: Conserving Wildlife While Maintaining Livestock in Semi-Arid Ecosystems,
1st edition. Edited by J.T. du Toit, R. Kock, and J.C. Deutsch.
© 2010 Blackwell Publishing

While the term *grazing* sometimes is applied strictly to eating grasses, livestock in the arid and semi-arid American West feed on a wide variety of plant life forms – including forbs, shrubs and small trees. In this review, 'grazing' will refer to herbivory in this broader sense.

History and policy of grazing in the American West

Cattle first arrived in what is now the United States in 1540 when the Spanish explorer Coronado came north from Mexico into present-day Arizona, New Mexico, Colorado and as far east as Kansas. He was soon followed by missionaries who extolled the virtues of pastoralism; by 1700, many of these missions were major livestock centres (Stewart 1936; Stoddart & Smith 1943; Brand 1961). The trappings of American ranch culture – brands, seasonal roundups, rodeos and cattlemen's associations – were also imported from Mexico (Brand 1961). The region west of the Rocky Mountains incrementally became American territory in the nineteenth century and livestock interests have been embedded in the political, social and economic fabric of the region ever since. By the 1880s, immense herds of livestock roamed the West, provoking rampant economic speculation in the eastern United States and Great Britain on 'the beef bonanza' (Brisban 1881). By 1880, Utah was home to almost a hundred thousand cattle; New Mexico, a third of a million; Texas, over four million (Stewart 1936). But get-rich-quick schemes abruptly ran up against two hard climatic realities of the American West: aridity and unpredictability. The ecological consequences of aridity became apparent from the dramatic decline of forage plants. Climatic unpredictability traumatized the fledgling livestock industry when severe winter blizzards alternated with hot, dry summers in the second half of the 1880s. As much as 85% of herds perished in some regions; bones littered the ground from the prairies of the north to the deserts of the south (Bahre & Shelton 1996; Fleischner 2002; Figure 9.1).

These ecological events and human tragedies set the stage for eventual reform of livestock grazing practices in the western United States. Faint stirrings about the need for regulation began to be heard in the ranching community in the early twentieth century, but many ranchers remained steadfastly against any governmental role in their industry. As a result, livestock grazing was the last major form of land use in the American West to be regulated by the government. Long after federal oversight of mining and

(a)

(b)

Figure 9.1   State of Arizona rangelands in 1903: (a) Bones from dead cattle and horses, Robles Ranch; livestock bones were sold for fertilizer; (b) Dead cattle, near Avicaca (Both photographs by D.A. Griffiths, Arizona Experiment Station botanist. Courtesy of the National Archives, Washington, D.C.; 15.1.a – March 21, 1903 – Photograph No. R83-FB-2145; 15.1.b – April 10, 1903 – Photograph No. R83-FB-1760).

timber harvest were established and national parks and forests were created, range management was essentially non-existent.

Finally, in 1934, amidst continuing controversy, the pivotal Taylor Grazing Act was passed, which asserted the federal government's responsibility to manage livestock grazing (Foss 1960; Stout 1970). The need for range reform became increasingly visible, as evidence of the Dust Bowl began to literally dominate the atmosphere of the country. Passage of the bill was encouraged by what one senator called 'the most tragic, the most impressive lobbyists that have ever come to this capital' – some of the worst dust clouds in history, that had blown over a thousand miles from the overgrazed prairies of the Dust Bowl (Foss 1960). Livestock management practices were radically overhauled in the wake of the Taylor Grazing Act. Formal allotments were established, but because ranges had been seriously overstocked, not all ranchers received the new federal permits. While some ranchers were granted long-term access to the new grazing allotments, others were excluded from their former grazing lands. A new federal agency was established, which soon was named the U.S. Grazing Service and eventually transformed into the Bureau of Land Management (BLM) (Foss 1960; Muhn & Stuart 1988; Klyza 1996).

From the outset, this new agency promoted the ideal of 'home rule on the range', granting extraordinary regulatory authority to ranchers – in effect, allowing ranchers to determine the rules that would govern them. Ranchers representing newly established grazing district advisory boards were formed into a National Advisory Board Council. A member of that council later recollected that ranchers wrote the entire Federal Range Code at the council's first meeting, with government officials polite enough to offer to leave so as not to interfere (Foss 1960). This indicates the uniquely privileged role that livestock interests have played in American politics and policy.

Over a quarter of senators represent western states where livestock grazing is prevalent and tend to be unified in its defense. Quite a few of them have been ranchers themselves. These politicians often are appointed to Congressional committees that oversee livestock grazing policy. Congresspersons from other parts of the nation, where the BLM has no jurisdiction, have nothing to gain politically by contradicting the interests – reliably pro-ranching – of the western delegations. Consequently, a small group of Western politicians have historically exerted disproportionate influence over federal rangeland policy. Political scientist Foss (1960) referred to this as a *special private government*. BLM's most dependable supporters in Congress are those Western representatives who tend to have the strongest agenda to be pushed on it (Klyza 1996).

Similarly, the control of regulatory policy by the very same group it is supposed to regulate, as occurred with district and national advisory boards, is an example of a captured policy pattern. This represents one feature of *interest-group liberalism* – 'a system of self-government in which economic interests, organized in groups, are delegated authority over policy-making in their policy realm' (Klyza 1996). No other interest group has succeeded at this as thoroughly as ranchers. This notion of ranchers' political advantage is echoed by legal historians (Scott 1967; Donahue 1999). According to Scott (1967): 'the American cattle industry is unique in American history . . . . Because the industry developed and was strong before the law making and enforcement agencies . . . were developed it made its own law. . . . The ranchers sought the benefit of the legislatures and received it. . . . Finally, if there was no other way to preserve the needs of the cattle industry, the participants ignored or disobeyed . . . laws.' Any reflection on contemporary livestock grazing policy must be seen in the light of this long history of political privilege. Moreover, literature, film, and more recently, television have embedded a romantic view of cowboys and ranching into American popular culture as the genre of 'Westerns' developed in the twentieth century.

Land use that had been the last to be regulated by government was also the last to be monitored by citizen groups. Bernard DeVoto, writing in the 1930s, was one of the few early voices to speak critically of livestock grazing and its associated culture in the American West (DeVoto 1936; Stegner 1988). The 1960s and 1970s were a time of great political tumult in the United States of America and out of this unrest a revitalized environmental movement burst forth – its genesis often dated to the first Earth Day in 1970. This movement coincided with a general upsurge in outdoor and wilderness recreation. The more arid portions of the American West, which had generally been ignored by the public, began to be noted for their scenic and recreational values. As more people paid more attention to the canyon and desert country of the intermountain West (the region between the Sierra–Cascades ranges and the Rocky Mountains), they began to notice the omnipresence of livestock. Incrementally, critics began to voice concern about the management of these lands. Writings that questioned the status quo of range management began to appear in the popular media in the 1970s and 1980s (for example, Miller 1972; Ferguson & Ferguson 1983; Fradkin 1979). In what would have been unthinkable a few years earlier, the sportsmen's magazine *Outdoor Life* editorialized in 1985: 'fish and wildlife's biggest enemy is the excessive livestock grazing being done on more than 200 million acres . . . .' (Williamson 1985).

At the same time, more scientists began to pay attention to range management issues. Academic programs in range management within public universities had begun to develop in the 1920s–40s. The first comprehensive textbook (Stoddart & Smith 1943) had appeared in 1943 (its authors stressed that it dealt with range management, not range conservation, because conservation 'implies disuse . . . and disuse is waste') and a professional organization, the Society for Range Management, was founded in 1948. But for several decades most scientific attention came from people affiliated with the livestock industry or agricultural colleges, where livestock production was assumed to be of paramount importance. In the 1980s and 1990s, however, as ecological science became more sophisticated and conservation biology developed a broad focus on biodiversity, more widespread scientific attention began to be paid to 'rangelands'.

Wildlife conservation on rangelands: ecology and conservation biology play a larger role

Effects on ecological integrity

Wildlife and fisheries scientists were among the first to declare disturbing ecological effects of livestock grazing. A federal government symposium in 1977 concluded that livestock grazing was 'the single most important factor limiting wildlife production in the West' (Smith 1977). In 1979, an interagency committee in Oregon and Washington, composed of state and federal biologists, concluded that livestock grazing was the most important factor degrading fish and wildlife habitat in the 11 western states (Oregon-Washington Interagency Wildlife Committee 1979). In the 1990s, three professional scientific societies came out with position statements that enumerated ecological concerns with grazing practices: the American Fisheries Society (Armour et al. 1991), Society for Conservation Biology (Fleischner et al. 1994) and The Wildlife Society (1996). These and other scientists were concerned with what they perceived as the role of livestock in disrupting ecological integrity – an ecosystem's composition, structure, function (Angermeier & Karr 1994; Trombulak et al. 2004).

Livestock grazing affects different species and communities in distinct ways. Failure to recognize these natural differences – and a resulting oversimplified perspective on vegetation change – can lead observers to faulty conclusions concerning the effects of grazing, as 'good' or 'bad'. Not only do different species of grazers cause different impacts, but plant species, even related ones, can respond dissimilarly to grazing. Some species benefit from grazing-related disturbance, while it is deleterious to others. Often, species that benefit are habitat generalists (Ohmart 1996). For example, populations of the American robin (*Turdus migratorius*), one of the most widespread bird species in North America, increased in heavily grazed riparian habitat, while other species that require dense vegetation declined (Schulz & Leininger 1991). Species that prefer open habitats with lower vegetative density can benefit from grazing (Taylor 1986; Saab et al. 1995). In southern Arizona grassland, birds typical of more xeric habitats were more prevalent on livestock-impacted sites than on adjacent livestock exclosures (Bock et al. 1984). However, in ponderosa pine (*Pinus ponderosa*) forest and savanna in northern Arizona, cattle grazing reduced nesting success of a ground-nesting sparrow (dark-eyed junco, *Junco hyemalis*) by 75%. Livestock grazing created a less favorable microclimate, exposed nests to predators and, in some cases, damaged them directly through trampling (Walsberg 2005). Presumably, a canopy-nesting species in this same forest may have been unaffected by the same livestock grazing activity. In general, understorey bird species are especially impacted by livestock grazing (Krueper et al. 2003).

It has been suggested that livestock can be utilized as a wildlife management tool (Bokdam & Wallis de Vries 1992; Hobbs & Huenneke 1992). Severson (1990) clarified that such applications may be very limited. Because two species in the same community often respond differently to livestock grazing (Hobbs & Huenneke 1992), determination of its success or failure as a management tool depends on which species is used as a criterion. Thus, assessing the possible utility of livestock as a management tool must be context and species specific.

In a study of historic livestock grazing impacts at Chaco Culture National Historic Park in northern New Mexico – one of the longest continuously grazed regions of North America – vegetation at different sites responded differently to long-term (50+ years) protection from grazing, depending on specific site characteristics (Floyd et al. 2003; Floyd et al. In Press). Grasses were favoured in alluvial canyon bottoms, shrubs in upland sites and dense biological soil crusts on certain substrates. This variation reflected the inherent ecological potentials of the different sites – based on edaphic and topographic conditions, as well as residual plant propagules, degree of disturbance and details of land management history. Here, as elsewhere, simplistic conclusions of unilaterally positive or negative effects of grazing could be misleading.

Diverse taxa – including all vertebrate classes, vascular plants and cyanobacteria – have been observed to undergo negative effects from livestock grazing, including decreases in population size of individual species and reduction of species richness (reviewed in Fleischner 1994). Livestock grazing can influence plant communities through removal and structural alteration of vegetation (Krueper 1993; Saab et al. 1995; Dobkin et al. 1998; Krueper et al. 2003); trampling and compaction of soils and consequent effects on water availability and alteration of foraging guilds and disruption of successional patterns and nutrient cycling (Fleischner 1994). These influences on ecosystem composition, structure and function affect animals through direct and indirect effects on food resources, alteration of nesting habitat (including microclimate) and greater exposure to predation (Ammon & Stacey 1997; Walsberg 2005).

Distinct community types also respond differently to livestock grazing. In a review of the effects of grazing on neotropical migrant landbirds, Bock et al. (1993b) found an increasingly negative effect on bird abundances in grassland, riparian woodland and intermountain shrub-steppe community types. Almost equal numbers of grassland bird species had positive and negative responses to livestock grazing, while six times as many shrub-steppe species had negative responses as positive. Grasslands, the more resilient habitat, are much rarer than shrub-steppe in the West (Kuchler 1985).

Functioning and structure of both terrestrial and aquatic communities can also be dramatically altered by livestock grazing. Livestock grazing is considered one of three primary factors (along with fire suppression and logging) involved in changing the structure of ponderosa pine (*P. ponderosa*) forests, one of the most widespread forest types in the West, from open, park-like stands with dense grass cover to communities characterized by dense pine reproduction and lack of grasses (Rummell 1951; Cooper 1960; Covington & Moore 1994). Stand structure and soil dynamics of mixed conifer (*P. ponderosa*, *Pseudotsuga menziesii*) forests were similarly altered by long-term livestock grazing (Belsky & Blumenthal 1997).

The introduction of non-native grazing mammals to ecosystems usually involves dramatic alteration of soil and geomorphic characteristics (reviewed in Fleischner 1994; Trimble & Mendel 1995; also see Belsky & Blumenthal 1997). Trampling by livestock compacts the soil, decreasing its capacity for water infiltration (Gifford & Hawkins 1978); consequently, heavily grazed habitats have less capacity to hold water, thereby exacerbating the greatest limiting factor in arid and semi-arid ecosystems.

Livestock grazing has been a major contributing factor to stream channel entrenchment ('arroyo-cutting') in the West (Bryan 1925; Cooperrider & Hendricks 1937; Leopold 1946, 1951; Ohmart & Anderson 1982; Hereford & Webb 1992).

The influences of livestock grazing on species composition and on physical habitat characteristics interact to create community-scale alterations. At Capitol Reef National Park in central Utah, paleoecologists determined that the most dramatic vegetation change during the past 5400 years occurred in the past two centuries. They suggested that livestock grazing was the precipitating factor for this historic habitat change (Cole et al. 1997). Further south, climatologists and plant ecologists working along the United States–Mexico border attributed increasing soil surface temperatures and albedo to livestock grazing-related land degradation. They noted a positive feedback loop: grazing-related degradation leading to increases of local temperatures and potential evapotranspiration levels, which in turn reinforces the degradation (Balling et al. 1998).

Changes to forest stand structure and soil dynamics, described above, also altered fire patterns in these ecosystems (Belsky & Blumenthal 1997). By selectively foraging on herbaceous understories, livestock have opened up habitat for young trees, allowing greater recruitment – ultimately leading to much higher tree densities. This replacement of fine fuels with dense stands of small trees has dramatically increased fuel loads and thus, fire intensities. In the mountains along the United States–Mexico borderlands, the frequency of surface fires decreased dramatically between 1870 and 1900 – a change initially caused by livestock grazing and subsequently by a combination of grazing and fire suppression. On the Mexican side of the border, where livestock grazing was less intense and fire suppression more sporadic, surface fires continued well into the twentieth century (Swetnam et al. 2001). Livestock grazing has contributed less directly to increases in fire frequency in riparian habitats. Fire was historically rare in native riparian communities of the Southwest (Bahre 1985; Swetnam 1990), but riparian fire has become more common where the invasive alien tree, tamarisk (*Tamarix*), has become established (Busch & Smith 1993). The spread of tamarisk has been aided by livestock grazing (discussed below), which in turn has increased riparian fire frequency.

The influence of livestock grazing on two ecological features of the region – riparian habitat and biological soil crusts – have special importance for wildlife conservation and will be looked at in detail below.

Riparian habitats – essential to wildlife

Riparian ecosystems are the most critical wildlife habitats in the rangelands of the American West (Thomas et al. 1979). Riparian communities are among the most productive habitats in the American West. This vegetation type covers only 0.1% of the Western landscape (Ohmart 1996), yet provides habitat for more species of birds than all other habitats combined (Knopf et al. 1988). Approximately three-quarters of the vertebrate species in Arizona and New Mexico depend on riparian habitats for at least a portion of their life cycles (Johnson et al. 1977; Johnson 1989). One regional analysis concluded that more than three-quarters of the bird species of Southwestern deserts were dependent in some manner on water-related habitat; over half were completely dependent (Johnson et al. 1977; Chaney et al. 1990; Rich 2002). Riparian areas have been found to harbour more than 10 times the number of migrant birds as adjacent uplands (Stevens et al. 1977). Over 60% of the bird species identified as Neotropical migrants by the Partners in Flight program used western riparian areas during the breeding season or as migratory stopovers (Krueper 1993). In the Interior Columbia River Basin of the Pacific Northwest, over 60% of 132 species of neotropical migrants used riparian habitats – far more than any other habitat type (Saab & Rich 1997). Even xeroriparian habitats – normally dry stream corridors that intermittently carry floodwaters through low deserts – support 5–10 times the bird densities and species diversity of surrounding desert uplands (Johnson & Haight 1985). The critical importance of healthy riparian habitats for wildlife conservation in the region cannot be overemphasized.

But livestock has become concentrated in riparian habitats (Ames 1977; Kennedy 1977; Thomas et al. 1979; Roath & Krueger 1982; Van Vuren 1982; Gillen et al. 1984) for the same reasons as wildlife – the greater availability of water, shade and food than in adjacent dry country. Heavy-bodied herbivores have many impacts on riparian zones, however. In 1990, the U.S. Environmental Protection Agency concluded that riparian conditions throughout the West were the worst in history (Chaney et al. 1990). Several reviews have summarized the effects of livestock grazing on riparian habitats and their wildlife (Platts 1979, 1981; Kauffman & Krueger 1984; Fleischner 1994; Ohmart 1996; Belsky et al. 1999). In a comparison of five southern Arizona streams, Rucks (1984) determined that livestock grazing was the major factor degrading broadleaf riparian forest to scrub. Ohmart (1996) proposed three conceptual stages in riparian habitat degradation that occur over the span of approximately

200 years. Because habitat change continues over a longer time frame than that of human lives, degradation is often imperceptible to casual observers.

Young shoots of riparian trees such as cottonwood (*Populus* spp.) and willow (*Salix* spp.) are foraged by livestock, greatly reducing regeneration and converting riparian forests into even-aged stands of older trees (Szaro 1989). When livestock was removed in the early 1990s from the Fremont River corridor flowing through Capitol Reef National Park, Utah, cottonwood seedlings quickly recolonized sandbars from which they had largely been absent for decades. Livestock foraging in lush riparian vegetation compete with native herbivorous species, often eating the most nutritive elements of the vegetation. Grazers also remove organic debris, which eliminates cover for ground-dwelling species, such as snakes and lizards (Jones 1981, 1988; Szaro et al. 1985). Livestock grazing has been a principal factor contributing to the decline of native fishes in the West (Miller 1961; Armour et al. 1991). Cattle activities, especially damaging to fish habitat, are the removal of vegetative cover and the trampling of overhanging streambanks (Behnke & Zarn 1976; Platts 1981). Reduced vegetative cover leads to increased water temperatures and trampling-induced loss of streamside pools, which provide cover and increases predation risk.

Livestock grazing is the most widespread and pervasive threat to riparian habitats in the arid West. Other threats include water diversion and pumping, the introduction of alien species, recreation and timber harvest (Chaney et al. 1990; Fleischner 1994; Ohmart 1994; Dobkin et al. 1998). Disturbance by domesticated herbivores contributes to increases in non-native plant species (Parker et al. 2006). Livestock spread alien plants by dispersing seeds, opening up habitat for new species and reducing competition from the native species by eating them. Invasions of alien grass species in North America have been most severe in the more arid parts of the West, where invasion by many species (e.g. *Bromus tectorum*, *Bromus rubens*, *Bromus mollis*, *Bromus diandrus*, *Taeniatherum asperum* and *Avena* spp.) was associated with livestock grazing (Gould 1951; Mack 1981; D'Antonio & Vitousek 1992; Belsky & Gelbard 2000). Long-term livestock grazing in riparian areas has promoted establishment of tamarisk (*Tamarix*; Everitt 1980; Ohmart & Anderson 1982; Minckley & Brown 1994). Tamarisk stands have lower species richness of native avifauna than the native cottonwood-willow forests that they often replace (Ohmart & Anderson 1982; Strong & Bock 1990; Rosenberg et al. 1991).

One of the simplest and most effective means of conserving native wildlife in the American West is excluding livestock from riparian corridors. Ohmart

(1996), in a comprehensive review of wildlife conservation in riparian zones, concluded that 'the best way to manage riparian habitats is not to graze them.' The Smithsonian Migratory Bird Center (n.d.) stated that livestock grazing 'remains the single most destructive force that can be practically and significantly reduced' to benefit neotropical migrant birds. A recent synthesis of research in seven riparian ecosystems in five western states concluded that 'reducing cattle grazing is likely to produce the greatest benefits for bird species dependent on western deciduous riparian habitats' (Tewksbury et al. 2002).

It is encouraging to note that riparian habitats begin to recover relatively quickly when grazing disturbance is removed. Studies along several Western streams have reported increases in native wildlife and fish populations in a decade or less after livestock have been fenced out (Winegar 1977; Dahlem 1979; Duff 1979; Keller et al. 1979; Keller & Burham 1982; Krueper et al. 2003). Aquatic habitat often heals more quickly than adjacent riparian vegetation (Knopf & Cannon 1982; Szaro & Pase 1983).

A century ago, the San Pedro River of southern Arizona was unincised and marshy along much of its length (Hendrickson & Minckley 1985), but by the 1970s, it had been severely degraded ecologically. Less than 20 years after removing livestock from the San Pedro, it has one of the highest bird diversities of any area of its size in North America (CEC 1999). The removal of cattle led to rapid, substantial recovery of both vegetation and bird populations, including several species of special conservation concern; the speed and magnitude of recovery was surprising – and encouraging (Krueper et al. 2003; Figure 9.2). However, numerous studies have concluded that such vibrant restoration of riparian ecosystem health requires complete removal of livestock (Ames 1977; Dahlem 1979; Davis 1982; Chaney et al. 1990; Kovalchik & Elmore 1992).

Biological soil crusts

Biological soil crusts provide critical ecosystem functions (Belnap & Lange 2003), including fixing carbon in sparsely vegetated areas. Such carbon contributions help keep interspaces between vascular plants fertile and support other microbial populations (Beymer & Klopatek 1991). The availability of nitrogen is an important factor limiting primary production in arid habitats throughout the world. In the Great Basin Desert of the western United States, nitrogen is second only to moisture in importance (James & Jurinak 1978). In desert shrub and grassland communities that support few nitrogen-fixing

(a)

(b)

(c)

(d)

Figure 9.2   **San Pedro River, Cochise Co., Arizona, before and after exclusion of livestock in late 1987. Left and right photo in each pair taken at the same location: (a) June 1987; (b) June 1991 (less than 4 years recovery); (c) 1984; (d) 1997 (10 years recovery) (U.S. Dept. of Interior, Bureau of Land Management files; see Krueper et al. (2003) for more information).**

plants, biological soil crusts can be the dominant source of nitrogen (Rychert et al. 1978; Harper & Marble 1988; Evans & Ehleringer 1993; Evans & Belnap 1999). Nitrogen inputs are highly dependent on temperature, moisture and species composition of the crusts (Belnap & Lange 2003); therefore, both prevailing climate and the legacy of disturbances influence fixation rates (Belnap 1995, 1996). Additionally, crusts stabilize soils (Belnap & Gillette 1997, 1998; Warren 2003), retain moisture and provide seed germination sites. Soil crusts are effective in capturing aeolian dust deposits, contributing to a 2- to 13-fold increase in nutrients in southeastern Utah (Reynolds et al. 2001). The presence of soil crusts generally increases the amount and depth of rainfall infiltration (Loope & Gifford 1972; Brotherson & Rushforth 1983; Harper & Marble 1988; Johansen 1993). Thus, biological soil crusts play critical roles regarding the two most important limiting factors in arid landscapes: water and nitrogen.

These crusts, however, are easily damaged by livestock, which are heavier and often more abundant and concentrated than most native mammals. Contrary to popular misconception, bison (*Bison bison*) were rare or absent in much of the area west of the Rocky Mountains during most of the Holocene (McDonald 1981). Bison were present in the northern Rockies region and to the east, Northeast and Northwest of the Great Basin (Hall 1981; Mack & Thompson 1982; Van Vuren & Bray 1985; Van Vuren 1987; Zeveloff 1988; Van Vuren & Dietz 1993), but absent altogether from Arizona (Cockrum 1960; Hoffmeister 1986), western New Mexico (Bailey 1971) and most of California (Jameson & Peeters 1988). Soil crusts became established where large, heavy grazing mammals were absent; the Great Plains east of the Rockies hosted enormous herds of bison – and little biological soil crust was left (Mack & Thompson 1982). Thus, the turf-forming grasslands of the Great Plains can support large herbivores more sustainably than can the bunchgrass communities west of the Rocky Mountains.

Not surprisingly, then, livestock grazing has been correlated with the loss of biological soil crust cover and species richness (Johansen et al. 1981; Anderson et al. 1982; Jeffries & Klopatek 1987; Belnap & Eldridge 2003). Crusts can be severely damaged even while they (Belnap 1993) and the more conspicuous vascular plant communities (Kleiner & Harper 1972; Cole 1990) appear healthy. Nitrogenase activity has been reduced by 80–100% in the crust under a single human footprint (Belnap 1994; Belnap et al. 1994) and nitrogen content in the leaves of dominant plant species was lower in trampled than untrampled areas (Harper & Pendleton 1993). If a single foot can grind nitrogen fixation to a halt, the impact of herds of much larger animals for more than a century is easily imagined. In the Chaco Canyon area of New Mexico, the cover of nitrogen-fixing crusts was significantly higher in areas that had been protected from livestock grazing for half a century than in those still being grazed (Floyd et al. 2003).

Other management considerations

The focus on livestock production on such a vast geographic scale has a number of less direct impacts on wildlife conservation. Fencing, which is a fundamental livestock management tool, can functionally fragment habitats by impeding the movement of native species, such as pronghorn (*Antilocapra americana*). Water diversions and the introduction of alien plants, such as crested wheatgrass (*Agropyron cristatum*; see Menke & Bradford 1992 for an example) are sometimes undertaken as '*range improvements*'. The livestock industry has historically played a large role in the elimination of native predators, including wolves (*Canis lupus*), coyotes (*Canis latrans*) and mountain lions (*Puma concolor*). Some of the most energetic opposition to predator reintroductions, such as those of the Mexican grey wolf (*Canis lupus baileyi*) and Rocky Mountain grey wolf (*Canis lupus occidentalis*), continues to come from livestock interests.

Within land management agencies (especially the BLM and the U.S. Forest Service), livestock production has often been prioritized over wildlife conservation. This has been reflected in budgets and staffing, directed ultimately by Congressional dictates, which are prone to the political influences described above. For example, during 4 years in the late 1980s, the BLM directed only 3% of its total appropriation toward wildlife habitat management, compared to 34% for management of consumptive uses (range, energy and minerals and timber). During the same period the Forest Service allocated 4% of its appropriation to wildlife and fish habitat management and 26% to timber management (USGAO 1991). Moreover, field personnel of land management agencies (especially BLM) often lack faith that their conservation work will be supported by agency management if it is opposed by ranchers (USGAO 1988). Political and agency orientation towards livestock production can lead to contradictory management directions. For example, a U.S. Forest Service analysis of sensitive vertebrate species identified livestock grazing as one of five factors jeopardizing the northern goshawk (*Accipiter gentilis*) in the Southwest (Finch 1992), yet the goshawk management recommendations (Reynolds et al. 1992) released by the same office in the same year did not even mention livestock grazing. Such predilections by agencies reflect similar biases within the range management discipline; a respected 500-page textbook (Holecheck et al. 1989) devotes a single paragraph to nongame wildlife.

Livestock grazing in the American West amounts to a massive experiment without a control (Bock et al. 1993a; Noss 1994). The majority of the region has been devoted to this single land use, with only rare sites left ungrazed to allow comparisons. Not only are such livestock exclosures often too small to illuminate larger landscape effects, but virtually all of them had also been grazed by livestock in the past (thus, more accurately referred to as '*no longer grazed*' rather than '*ungrazed*' sites). Because initial impacts tend to be the most severe, these formerly grazed exclosures probably underestimate the impacts of livestock grazing (Fleischner 1994). Bock et al. (1993a) called for a system of federal livestock exclosures, whereby 20% of all federal leases would be left ungrazed for comparative study. At present, relatively few large-scale livestock exclosures exist in the southwestern United States – prominent among them being the Audubon Research Ranch (Bock et al. 1984), Buenos Aires National Wildlife Refuge, Canyonlands National Park, Chaco Culture National Historic Park (Floyd et al. 2003), Grand Canyon National Park, Mesa Verde National Park (Smith 2003) and Organ Pipe Cactus National Monument.

In essence, livestock grazing poses two fundamental threats to the health of native wildlife populations in the American West: degradation of riparian habitat and destruction of biological soil crusts. Riparian habitat has wildlife conservation importance that is far beyond its limited geographic area – in many cases, it truly has hemispheric significance. Fortunately, experience over the past quarter century suggests that the restoration of this habitat – which accomplishes crucial wildlife conservation – is possible. The removal of livestock from riparian zones is relatively simple logistically, if not always politically. Damage to soil crusts, however, is less easily healed and has much longer term (and thus less obvious) ecological effects. Nevertheless, with rest from livestock grazing pressures, there are indications that crusts can begin to restore themselves (Floyd et al. 2003).

Current concerns, conflicts and potential

There is general consensus that livestock grazing has exerted enormous ecological influence on landscapes of western North America in the past. But there is significantly less agreement about contemporary grazing practices and about what should happen in the immediate future. Some ecologists and environmental activists insist that livestock grazing should be terminated on public lands, while others defend ranching as the key to the maintenance of open space and traditional cultural values of the region. Environmental groups continue to clamour for major reform of livestock grazing policy (e.g. Williams 2006). Not long ago Bruce Babbitt, former U.S. Secretary of Interior and member of an Arizona ranching family, said 'I am now convinced that livestock do not belong in arid deserts. If it gets less than 10 inches of rainfall, cattle do not belong there. I am here to say the presumption that grazing is the dominant use of our public lands is the artifact of a distant past and must be replaced.'

Many ranch operations have ceased, when children of ranch families have had less interest than their parents and grandparents in maintaining livestock businesses. These social changes occur in the context of shifting local and regional economies in the West (Power 1996). Even with below-market value grazing fees on government allotments, it can be difficult to make a living on ranching alone. Simultaneously, property values in many rural and exurban areas have escalated rapidly in recent years. Thus, selling ranch property for real estate subdivision can be an alluring financial temptation.

Ranchers often contend that any increase in regulatory restriction or grazing fees will force them to sell off their land for real estate development. As a result, 'cows versus condos' has become a primary rallying cry for creating coalitions of landowners and environmentalists concerned about the maintenance of open space (Sheridan 2001). In the face of an onslaught of real estate speculation and exurban development throughout the rural West, it seems self-evident that the habitat fragmentation that accompanies subdivision of ranches works against the conservation of biodiversity (Jensen 2001). Conflicting views abound, however. Some feel that, in certain cases, subdivisions create less of an ecological impact than livestock grazing and other agriculture (Wuerthner 1994). Others insist the reverse – that keeping rural areas inhabited by ranchers and other agriculturalists is essential to protecting wildlife (Brown & McDonald 1995; Knight et al. 1995; Maestas et al. 2003). Both views, however, represent an oversimplified dichotomy (Siegel 1996): subdivision of land *and* livestock grazing can be detrimental to biodiversity. Moreover, there is enormous variation on how both land uses are undertaken, so it can be misleading to think in terms of such a simple choice as 'cows vs. condos'.

A recent study (Bock et al. 2006) in southeastern Arizona teased apart the effects of livestock foraging from that of real estate development/home construction by looking at populations of native rodents, the numerically dominant vertebrates in the region. Contrary to assumptions of the 'cows vs. condos' framing, rodent species richness was completely unaffected by proximity to exurban development. However, rodent species richness and abundance were negatively affected by livestock grazing – whether from typical cattle ranching or from horses on 'ranchettes', where the density of livestock (including horses) is actually often higher than on a ranch. Vegetation and soil change from livestock (cattle and horse) activity, rather than development *per se*, degraded the native faunal diversity. If we are monitoring mammals larger than rodents, we might expect different results. Nevertheless, this indicates that popular sentiments such as 'cows vs. condos' can obscure ecological complexities.

When reflecting upon such conflicts and controversies, it is useful to consider the psychological and social contexts of stakeholders. Social psychologists, observing cultural conflicts about rangeland issues, interviewed ranchers about a rangeland/endangered species conflict and noted that 'ranchers describe themselves as responsible stewards attuned to and part of nature, while they describe nonranchers as ignorant, irresponsible, insincere, and separate from nature' (Opotow & Brook 2003). While these researchers did not interview 'nonranchers', it is easy to imagine the same biases occurring in reverse. Thomashow (1995) defined a person's ecological identity as 'all the different ways people construe themselves in relationship to the earth as manifested in personality, values, actions, and sense of self'. Each person's ecological identity, he suggests, has three primary roots: childhood memories of special places, perceptions of disturbed places and contemplation of wild places. In a rangeland conflict, a rancher and an environmental activist both consider their lives deeply informed by their experience of, and passion for, nature. Yet their ecological identities might clash dramatically: the rancher's special childhood memory might involve working with family and livestock on a glorious summer day, while the activist's might stem from a remote national park, conspicuously removed from the presence of domesticated animals. Without understanding such fundamentally different human orientations to 'rangelands' – each valid in its own way – we are unlikely to progress towards the understanding and compassion necessary for the resolution of the conflict.

Social (group) identity also plays a key role in environmental conflicts. Social identities, such as 'rancher' and 'environmentalist', can allow individuals to stereotype and vilify those with different views through 'moral exclusion' – the sense that others can be excluded from the normal scope of fairness and civility (Opotow 1990). This moral exclusion can befall human political antagonists or other species considered unworthy of concern. One way some local groups have overcome these sorts of antagonisms is through the creation of a unifying 'overarching identity', such as the common purpose of protecting open space.

Conclusion

Three primary changes are needed to resolve ecological damage and social discord stemming from the legacy of livestock grazing in the American

West – two involving ecological restoration and one social. Two ecological restoration efforts are necessary for long-term conservation of wildlife in the arid and semi-arid regions of the American West. First, riparian areas – the most essential wildlife habitat in the region – must be restored. Abundant evidence suggests that the removal of livestock is necessary for restoration, but that ecosystem health can revive relatively quickly when this occurs. Second, to ensure long-term vitality of the region, biological soil crusts must be allowed to recolonize uplands. Because this involves much larger tracts of land and shows progress much more slowly, this will take greater political will to accomplish.

Finally, to resolve social conflicts on this issue, all parties involved need to strive for greater compassion (literally, 'feeling with') for people holding divergent views. Just as ranchers, environmental activists, land managers and scientists must all acknowledge some validity in each others' perspectives, the fundamental differences must be recognized, rather than pretending that all parties can achieve all their policy goals. Ultimately, difficult decisions must be made. These must be guided by the most accurate information possible, and communication between interest groups must be as honest and clear as possible. I conclude this review as I did an earlier one (Fleischner 1994): the future of livestock grazing in western North America is 'ultimately . . . a question of human values, not of science'. Clear and compassionate communication is needed more than ever.

References

Ames, C.R. (1977) Wildlife conflicts in riparian management: grazing. In: Johnson, R.R. & Jones, D.A. (technical coordinators) *Importance, Preservation and Management of Riparian Habitat: A Symposium*, General Technical Report RM-43. United States Forest Service, Rocky Mountain Forest and Range Experiment Station, Fort Collins, pp. 49–51.

Ammon, E.M. & Stacey, P.B. (1997) Avian nest success in relation to past grazing regimes in a montane riparian system. *Condor* 99, 7–13.

Anderson, D.C., Harper, K.T. & Holmgren, R.C. (1982) Factors influencing development of cryptogamic soil crusts in Utah deserts. *Journal of Range Management* 35, 180–185.

Angermeier, P.L. & Karr, J.R. (1994) Biological integrity versus biological diversity as policy directives. *BioScience* 44, 690–697.

Armour, C.L., Duff, D.A. & Elmore, W. (1991) The effects of livestock grazing on riparian and stream ecosystems. *Fisheries* 16, 7–11.

Bahre, C.J. (1985) Wildfire in southeastern Arizona between 1859 and 1890. *Desert Plants* 7, 190–194.

Bahre, C.J. & Shelton, M.L. (1996) Rangeland destruction: cattle and drought in southeastern Arizona at the turn of the century. *Journal of the Southwest* 38, 1–22.

Bailey, V. (1971) *Mammals of the Southwestern United States*. Dover Publications, New York. [Republication of Bailey, V. (1931). *Mammals of New Mexico*. North American Fauna, No. 53. Bureau of Biological Survey, Washington, D.C.]

Balling, R.C., Jr., Klopatek, J.M., Hildebrandt, M.L., Moritz, C.K. & Watts, C.J. (1998) Impacts of land degradation on historical temperature records from the Sonoran Desert. *Climatic Change* 40, 669–681.

Behnke, R.J. & Zarn, M. (1976) *Biology and Management of Threatened and Endangered Western Trouts*, General Technical Report RM-28. Rocky Mountain Forest and Range Experiment Station, United States Forest Service, Fort Collins.

Belnap, J. (1993) Recovery rates of cryptobiotic crusts: inoculant use and assessment methods. *Great Basin Naturalist* 53, 89–95.

Belnap, J. (1994) Potential role of cryptobiotic soil crusts in semiarid rangelands. In: Monsen, S.B. & Kitchen, S. (eds.) *Proceedings of a Symposium on Ecology, Management and Restoration of Intermountain Annual Rangelands*, General Technical Report INT-313. United States Forest Service, Intermountain Research Station, Ogden, pp. 179–185.

Belnap, J. (1995) Surface disturbances: their role in accelerating desertification. *Environmental Monitoring and Assessment* 37, 39–57.

Belnap, J. (1996) Soil surface disturbances in cold deserts: effects on nitrogenase activity in cyanobacterial-lichen soil crusts. *Biology and Fertility of Soils* 23, 362–367.

Belnap, J. & Eldridge, D.J. (2003) Disturbance and recovery of biological soil crusts. In: Belnap, J. & Lange, O.L. (eds.) *Biological Soil Crusts: Structure, Function and Management*. Springer-Verlag, Berlin, pp. 363–383.

Belnap, J. & Gillette, D.A. (1997) Disturbance of biological soil crusts: impacts on potential wind erodibility of sandy desert soils in southeastern Utah. *Land Degradation and Development* 8, 355–362.

Belnap, J. & Gillette, D.A. (1998) Vulnerability of desert biological soil crusts to wind erosion: the influences of crust development, soil texture and disturbance. *Journal of Arid Environments* 39, 133–142.

Belnap, J., Harper, K.T. & Warren, S.D. (1994) Surface disturbance of cryptobiotic soil crusts: nitrogenase activity, chlorophyll content and chlorophyll degradation. *Arid Soil Research and Rehabilitation* 8, 1–8.

Belnap, J. & Lange, O.L. (2003) *Biological Soil Crusts: Structure, Function and Management*. Springer-Verlag, Berlin.

Belsky, A.J. & Blumenthal, D.M. (1997) Effects of livestock grazing on stand dynamics and soils in upland forests of the interior West. *Conservation Biology* 11, 315–327.

Belsky, A.J. & Gelbard, J. (2000) *Livestock Grazing and Weed Invasions in the Arid West.* Oregon Natural Desert Association, Portland.

Belsky, A.J., Matzke, A. & Uselman, S. (1999) Survey of livestock influences on stream and riparian ecosystems in the western United States. *Journal of Soil and Water Conservation* 54, 419–431.

Beymer, R.J. & Klopatek, J.M. (1991) Potential contribution of carbon by microphytic crusts in pinyon-juniper woodlands. *Arid Soil Research and Rehabilitation* 5, 187–198.

Bock, C.E., Bock, J.H., Kenney, W.R. & Hawthorne V.M. (1984) Responses of birds, rodents and vegetation to livestock exclosure in a semidesert grassland site. *Journal of Range Management* 37, 239–242.

Bock, C.E., Bock, J.H. & Smith, H.M. (1993a) Proposal for a system of federal livestock exclosures on public rangelands in the western United States. *Conservation Biology* 7, 731–733.

Bock, C.E., Saab, V.A., Rich, T.D. & Dobkin, D.S. (1993b) Effects of livestock grazing on Neotropical migratory landbirds in western North America. In: Finch, D.M. & Stangel, P.W. (eds.) *Status and Management of Neotropical Migratory Birds*, General Technical Report RM-229. United States Forest Service, Rocky Mountain Forest and Range Experiment Station, Fort Collins, pp. 296–309.

Bock, C.E., Jones, Z.F. & Bock, J.H. (2006) Rodent communities in an exurbanizing southwestern landscape (U.S.A.). *Conservation Biology* 20, 1242–1250.

Bokdam, J. & Wallis de Vries, M.F. (1992) Forage quality as a limiting factor for cattle grazing in isolated Dutch nature reserves. *Conservation Biology* 6, 399–408.

Brand, D.D. (1961) The early history of the range cattle industry in northern Mexico. *Agricultural History* 35, 132–139.

Brisban, J.S. (1881) *The Beef Bonanza, or, How to Get Rich on the Plains.* J.B. Lippincott & Co., Philadelphia.

Brotherson, J.D. & Rushforth, S.R. (1983) Influence of cryptogamic crusts on moisture relationships of soils in Navajo National Monument, Arizona. *Great Basin Naturalist* 43, 73–78.

Brown, J.H. & McDonald, W. (1995) Livestock grazing and conservation on Southwestern rangelands. *Conservation Biology* 9, 1644–1647.

Bryan, K. (1925) Date of channel trenching (arroyo cutting) in the arid Southwest. *Science* 62, 338–344.

Busch, D.E. & Smith, S.D. (1993) Effects of fire on water and salinity relations of riparian woody taxa. *Oecologia* 94, 186–194.

Chaney, E., Elmore, W. & Platts, W.S. (1990) *Livestock Grazing on Western Riparian Areas.* United States Environmental Protection Agency Region 8, Denver.

Cockrum, E.L. (1960) *The Recent Mammals of Arizona: Their Taxonomy and Distribution.* University of Arizona Press, Tucson, Arizona.

Cole, D.N. (1990) Trampling disturbance and recovery of cryptogamic soil crusts in Grand Canyon National Park. *Great Basin Naturalist* 50, 321–325.

Cole, K.L., Henderson, N. & Shafer, D.S. (1997) Holocene vegetation and historic grazing impacts at Capitol Reef National Park reconstructed using packrat middens. *Great Basin Naturalist* 57, 315–326.

Commission for Environmental Cooperation [CEC] (1999) *Ribbon of Life: An Agenda for Preserving Transboundary Migratory Bird Habitat on the Upper San Pedro River.* Commission for Environmental Cooperation, Montreal.

Cooper, C.F. (1960) Changes in vegetation, structure and growth of ponderosa pine forests since white settlement. *Ecological Monographs* 30, 129–164.

Cooperrider, C.K. & Hendricks, B.A. (1937) *Soil Erosion and Streamflow on Range and Forest Lands of the Upper Rio Grande Watershed in Relation to Land Resources and Human Welfare,* Technical Bulletin 567, United States Department of Agriculture, Washington, D.C.

Council for Agricultural Science and Technology [CAST] (1974) Livestock grazing on federal lands in the 11 western states. *Journal of Range Management* 27, 174–181.

Covington, W.W. & Moore, M.M. (1994) Southwestern ponderosa pine forest structure: changes since European settlement. *Journal of Forestry* 92, 39–47.

Crumpacker, D.W. (1984) Regional riparian research and a multi-university approach to the special problem of livestock grazing in the Rocky Mountains and Great Plains. In: Warner, R.E. & Hendrix, K.M. (eds.) *California Riparian Systems: Ecology, Conservation and Productive Management.* University of California Press, Berkeley, pp. 413–422.

Dahlem, E.A. (1979) The Mahogany Creek watershed – with and without grazing. In: Cope, O.B. (ed.) *Proceedings of the Forum-Grazing and Riparian/Stream Ecosystems.* Trout Unlimited, Denver, pp. 31–34.

D'Antonio, C.M. & Vitousek, P.M. (1992) Biological invasions by exotic grasses, the grass/fire cycle and global change. *Annual Review of Ecology and Systematics* 23, 63–87.

Davis, J.W. (1982) Livestock vs. riparian habitat management – there are solutions. In: Nelson, L., Peek, J.M. & Dalke, P.D. (eds.) *Proceedings of the Wildlife-Livestock Relationships Symposium.* University of Idaho, Moscow, pp. 175–184.

DeVoto, B. (1936) *Forays and Rebuttals.* Little, Brown and Co., Boston.

Dobkin, D.S., Rich, A.C. & Pyle, W.H. (1998) Habitat and avifaunal recovery from livestock grazing in a riparian meadow system of the northwestern Great Basin. *Conservation Biology* 12, 209–221.

Donahue, D.L. (1999) *The Western Range Revisited: Removing Livestock from Public Lands to Conserve Native Biodiversity. Volume 5.* Legal History of North America Series. University of Oklahoma Press, Norman.

Duff, D.A. (1979) Riparian habitat recovery on Big Creek, Rich County, Utah. In: Cope, O.B. (ed.) *Proceedings of the Forum-Grazing and Riparian/Stream Ecosystems.* Trout Unlimited, Denver, pp. 91–92.

Evans, R.D. & Belnap, J. (1999) Long-term consequences of disturbance on nitrogen dynamics in an arid ecosystem. *Ecology* 80, 150–160.

Evans, R.D. & Ehleringer, J.R. (1993) A break in the nitrogen cycle in aridlands? Evidence from N^{15} of soils. *Oecologia* 94, 314–317.

Everitt, B.L. (1980) Ecology of saltcedar – a plea for research. *Environmental Geology* 3, 77–84.

Ferguson, D. & Ferguson N. (1983) *Sacred Cows at the Public Trough.* Maverick Publications, Bend.

Finch, D.M. (1992) *Threatened, Endangered and Vulnerable Species of Terrestrial Vertebrates in the Rocky Mountain Region,* General Technical Report RM-215. United States Forest Service, Rocky Mountain Forest and Range Experiment Station, Fort Collins.

Fleischner, T.L. (1994) Ecological costs of livestock grazing in western North America. *Conservation Biology* 8, 629–644.

Fleischner, T.L. (2002) Land held hostage: a history of grazing and politics. In: Wuerthner, G. & Matteson, M. (eds.) *Welfare Ranching: The Subsidized Destruction of the American West.* Island Press, Washington, D.C., pp. 33–38.

Fleischner, T.L., Brown, D.E., Cooperrider, A.Y., Kessler, W.B. & Painter E.L. (1994) Society for Conservation Biology Position Statement: livestock grazing on public lands of the United States of America. *Society for Conservation Biology Newsletter* 1(4), 2–3.

Floyd, M.L., Fleischner, T.L., Hanna, D. & Whitefield, P. (2003) Effects of historic livestock grazing on vegetation at Chaco Culture National Historic Park, New Mexico. *Conservation Biology* 17, 1703–1711.

Floyd, M.L., Hanna, D., Fleischner, T.L. & Shattuck, B. (In Press). Revisiting trends in vegetation recovery following protection from grazing, Chaco Culture National Historic Park, New Mexico. In: van Riper, C., III & Sherbrooke, E.K. (eds.) *The Colorado Plateau IV: Integrating Research into Resources Management on the Colorado Plateau.* University of Arizona Press, Tucson.

Foss, P.O. (1960) *Politics and Grass: The Administration of Grazing on the Public Domain.* Greenwood Press, New York.

Fradkin, P. (1979) The eating of the West. *Audubon* January, 94–121.

Gifford, G.F. & Hawkins, R.H. (1978) Hydrologic impact of grazing on infiltration: a critical review. *Water Resources Research* 14, 305–313.

Gillen, R.L., Krueger, W.C. & Miller, R.F. (1984) Cattle distribution on mountain rangeland in northeastern Oregon. *Journal of Range Management* 37, 549–553.

Gould, F.W. (1951) *Grasses of the Southwestern United States.* University of Arizona Press, Tucson.

Hall, E.R. (1981) *The Mammals of North America, Volume 2*, 2nd edn. John Wiley and Sons, New York.

Harper, K.T. & Marble, J.R. (1988) A role for nonvascular plants in management of arid and semiarid rangeland. In: Tueller, P.T. (ed.) *Vegetation Science Applications for Rangeland Analysis and Management*. Kluwer Academic Publishers, Dordrecht, pp. 135–169.

Harper, K.T. & Pendleton, R.L. (1993) Cyanobacteria and cyanolichens: can they enhance availability of essential minerals for higher plants? *Great Basin Naturalist* 53, 59–72.

Hendrickson, D.A. & Minckley, W.L. (1985) Ciénegas – vanishing climax communities of the American Southwest. *Desert Plants* 6(3), 130–176.

Hereford, R. & Webb, R.H. (1992) Historic variation of warm-season rainfall, southern Colorado Plateau, southwestern USA. *Climatic Change* 22, 239–256.

Hobbs, R.J. & Huenneke, L.F. (1992) Disturbance, diversity and invasion: implications for conservation. *Conservation Biology* 6, 324–337.

Hoffmeister, D.F. (1986) *Mammals of Arizona*. University of Arizona Press, Tucson.

Holecheck, J.L., Pieper, R.D. & Herbel, C.H. (1989) *Range Management: Principles and Practices*. Prentice-Hall, Englewood Cliffs.

James, D.W. & Jurinak, J.J. (1978) Nitrogen fertilization of dominant plants in the northeastern Great Basin Desert. In: West, N.E. & Skujins, J. (eds.) *Nitrogen in Desert Ecosystems*. Dowden, Hutchinson and Ross, Inc., Stroudsburg, pp. 219–231.

Jameson, E.W., Jr. & Peeters, H.J. (1988) *California Mammals*. University of California Press, Berkeley.

Jeffries, D.L. & Klopatek, J.M. (1987) Effects of grazing on the vegetation of the blackbrush association. *Journal of Range Management* 40, 390–392.

Jensen, M.N. (2001) Can cows and conservation mix? *BioScience* 51, 85–90.

Johansen, J.R. (1993) Cryptogamic crusts of semiarid and arid lands of North America. *Journal of Phycology* 29, 140–147.

Johansen, J.R., Rushforth, S.R. & Brotherson, J.D. (1981) Subaerial algae of Navajo National Monument, Arizona. *Great Basin Naturalist* 41, 433–439.

Johnson, A.S. (1989) The thin green line: riparian corridors and endangered species in Arizona and New Mexico. In: Mackintosh, G. (ed.) *In Defense of Wildlife: Preserving Communities and Corridors*. Defenders of Wildlife, Washington, D.C., pp. 35–46.

Johnson, R.R. & Haight, L.T. (1985) Avian use of xeroriparian ecosystems in the North American warm deserts. In: Johnson, R.R., Ziebell, C.D., Patton, D.R., Ffolliott, P.F. & Hamre, F.H. (technical coordinators.) *Riparian Ecosystems and their Management: Reconciling Conflicting Uses*, General Technical Report RM-120. United States Forest Service, Rocky Mountain Forest and Range Experiment Station, Fort Collins, pp. 156–160.

Johnson, R.R., Haight, L.T. & Simpson, J.M. (1977) Endangered species vs. endangered habitats: a concept. In: Johnson, R.R. & Jones, D.A. (technical coordinators.) *Importance, Preservation and Management of Riparian Habitat: A Symposium*, General Technical Report RM-43. United States Forest Service, Rocky Mountain Forest and Range Experiment Station, Fort Collins, pp. 68–79.

Jones, K.B. (1981) Effects of grazing on lizard abundance and diversity in western Arizona. *Southwestern Naturalist* 26, 107–115.

Jones, K.B. (1988) Comparison of herpetofaunas of a natural and altered riparian ecosystem. In: Szaro, R.C., Severson, K.E. & Patton, D.R. (technical coordinators.) *Management of Amphibians, Reptiles and Small Mammals in North America*, General Technical Report RM-166. United States Forest Service, Rocky Mountain Forest and Range Experiment Station, Fort Collins, pp. 222–227.

Kauffman, J.B. & Krueger, W.C. (1984) Livestock impacts on riparian ecosystems and streamside management implications: a review. *Journal of Range Management* 37, 430–437.

Keller, C., Anderson, L. & Tappel, P. (1979) Fish habitat changes in Summit Creek, Idaho, after fencing the riparian area. In: Cope, O.B. (ed.) *Proceedings of the Forum – Grazing and Riparian/Stream Ecosystems*. Trout Unlimited, Denver, pp. 46–52.

Keller, C.R. & Burnham, K.P. (1982) Riparian fencing, grazing and trout habitat preference on Summit Creek, Idaho. *North American Journal of Fisheries Management* 2, 53–59.

Kennedy, C.E. (1977) Wildlife conflicts in riparian management: water. In: Johnson, R.R. & Jones, D.A. (technical coordinators.) *Importance, Preservation and Management of Riparian Habitat: A Symposium*, General Technical Report RM-43. United States Forest Service, Rocky Mountain Forest and Range Experiment Station, Fort Collins, pp. 52–58.

Kleiner, E.F. & Harper, K.T. (1972) Environment and community organization in grasslands of Canyonlands National Park. *Ecology* 53, 229–309.

Klyza, C.M. (1996) *Who Controls Public Lands?: Mining, Forestry, and Grazing Policies, 1870–1990*. University of North Carolina Press, Chapel Hill.

Knight, R.L., Wallace, G.N. & Riebsame, W.E. (1995) Ranching the view: subdivisions versus agriculture. *Conservation Biology* 9, 459–461.

Knopf, F.L. & Cannon, R.W. (1982) Structural resilience of a willow riparian community to changes in grazing practices. In: Nelson, L., Peek, J.M. & Dalke, P.D. (eds.) *Proceedings of the Wildlife-Livestock Relationships Symposium*. University of Idaho, Moscow, pp. 198–207.

Knopf, F.L., Johnson, R.R., Rich, T., Samson, F.B. & Szaro, R.C. (1988) Conservation of riparian ecosystems in the United States. *Wilson Bulletin* 100, 272–284.

Kovalchik, B.L. & Elmore, W. (1992) Effects of cattle grazing systems on willow-dominated plant associations in central Oregon. In: Clary, W.P., McArthur, E.E.,

Bedunah, D. & Wambolt, C.L. (compilers.) *Proceedings of the Symposium on Ecology and Management of Riparian Shrub Communities*, General Technical Report INT-289. United States Forest Service, Intermountain Research Station, Ogden, pp. 111–119.

Krueper, D. (1993) Effects of land use practices on western riparian systems. In: Finch, D.M. & Stangel, P.W. (eds.) *Status and Management of Neotropical Migratory Birds*, General Technical Report RM-229. United States Forest Service, Rocky Mountain Forest and Range Experiment Station, Fort Collins, pp. 321–330.

Krueper, D., Bart, J. & Rich, T.D. (2003) Response of vegetation and breeding birds to the removal of cattle on the San Pedro River, Arizona (U.S.A.). *Conservation Biology* 17, 607–615.

Kuchler, A.W. (1985) *Potential Natural Vegetation*. National Atlas of the United States of America (map). United States Geological Survey, United States Department of Interior, Reston.

Leopold, A. (1946) Erosion as a menace to the social and economic future of the Southwest. *Journal of Forestry* 44, 627–633.

Leopold, L.B. (1951) Vegetation of southwestern watersheds in the nineteenth century. *Geographical Review* 41, 295–316.

Longhurst, W.M., Hafenfeld, R.E. & Connolly, G.E. (1982) Deer-livestock relationships in the western states. In: Nelson, L., Peek, J.M. & Dalke, P.D. (eds.) *Proceedings of the Wildlife-Livestock Relationships Symposium*. University of Idaho, Moscow, pp. 409–420.

Loope, W.L. & Gifford, G.F. (1972) Influence of a soil microfloral crust on select properties of soils under piñyon-juniper in southeastern Utah. *Journal of Soil Water and Conservation* 27, 164–167.

Mack, R.N. (1981) Invasion of Bromus tectorum L. into western North America: an ecological chronicle. *Agro-Ecosystems* 7, 145–165.

Mack, R.N. & Thompson, J.N. (1982) Evolution in steppe with few large, hooved mammals. *American Naturalist* 119, 757–773.

Maestas, J.D., Knight, R.L. & Gilgert, W.C. (2003) Biodiversity across a rural land-use gradient. *Conservation Biology* 17, 1425–1434.

McDonald, J.N. (1981) *North American Bison: Their Classification and Evolution*. University of California Press, Berkeley.

Menke, J. & Bradford, G.E. (1992) Rangelands. *Agriculture, Ecosystems and Environment* 42, 141–163.

Miller, J.N. (1972) The nibbling away of the West. *Reader's Digest*, 101, 107–111.

Miller, R.R. (1961) Man and the changing fish fauna of the American Southwest. *Papers of the Michigan Academy of Science, Arts and Letters* 46, 365–404.

Minckley, W.L. & Brown, D.E. (1994) Wetlands. In: Brown, D.E. (ed.) *Biotic Communities: Southwestern United States and Mexico*. University of Utah Press, Salt Lake City, pp. 223–287.

Muhn, J. & Stuart, H.R. (1988) *Opportunity and Challenge: The Story of BLM.* Bureau of Land Management, United States Department of Interior, Washington, D.C.

Noss, R.F. (1994) Cows and conservation biology. *Conservation Biology* 8, 613–616.

Ohmart, R.D. (1994) The effects of human-induced changes on the avifauna of western riparian habitats. In: Jehl, J.R., Jr. & Johnson, N.K. (eds.) *A Century of Avifaunal Change in Western North America.* Studies in Avian Biology, No. 15. Cooper Ornithological Society, Lawrence, pp. 273–285.

Ohmart, R.D. (1996) Historical and present impacts of livestock grazing on fish and wildlife resources in western riparian habitats. In: Kraussman, P. (ed.) *Rangeland Wildlife.* Society for Range Management, Denver, pp. 246–279.

Ohmart, R.D. & Anderson, B.W. (1982) North American desert riparian ecosystems. In: Bender, G.L. (ed.) *Reference Handbook on the Deserts of North America.* Greenwood Press, Westport, pp. 433–479.

Opotow, S. (1990) Moral exclusion and injustice: an introduction. *Journal of Social Issues* 46, 1–20.

Opotow, S. & Brook, A. (2003) Identity and exclusion in rangeland conflict. In: Clayton, S. & Opotow, S. (eds.) *Identity and the Natural Environment: The Psychological Significance of Nature.* The MIT Press, Cambridge, pp. 249–272.

Oregon-Washington Interagency Wildlife Committee (1979) *Managing Riparian Ecosystems for Fish and Wildlife in Eastern Oregon and Eastern Washington.* Oregon-Washington Interagency Wildlife Committee, Washington, D.C. Available from Washington State Library, Olympia.

Parker, J.D., Burkepile, D.E. & Hay, M.E. (2006) Opposing effects of native and exotic herbivores on plant invasions. *Science* 311, 1459–1461.

Platts, W.S. (1979) Livestock grazing and riparian/stream ecosystems – an overview. In: Cope, O.B. (ed.) *Proceedings of the Forum: Grazing and Riparian/Stream Ecosystems.* Trout Unlimited, Denver, pp. 39–45.

Platts, W.S. (1981) *Influence of Forest and Rangeland Management on Anadromous Fish Habitat in Western North America: Effects of Livestock Grazing, No. 7,* General Technical Report PNW-124. United States Forest Service, Pacific Northwest Forest and Range Experiment Station, Portland.

Power, T.M. (1996) *Lost Landscapes and Failed Economies: The Search for a Value of Place.* Island Press, Washington, D.C.

Reynolds, R., Belnap, J., Reheis, M., Lamothe, P. & Luiszer, F. (2001) Eolian dust in Colorado Plateau soils: nutrient inputs and recent change in source. *Proceedings of the National Academy of Sciences* 98, 7123–7127.

Reynolds, R.T., Graham, R.T., Reiser, M.H. et al. (1992) *Management Recommendations for the Northern Goshawk in the Southwestern United States,* General Technical Report RM-217. United States Forest Service, Rocky Mountain Forest and Range Experiment Station, Fort Collins.

Rich, T.D. (2002) Using breeding land birds in the assessment of western riparian systems. *Wildlife Society Bulletin* 30, 1128–1139.

Ricketts, T.H., Dinerstein, E., Olson, D.M. et al. (1999) *Terrestrial Ecoregions of North America: A Conservation Assessment*. World Wildlife Fund/Island Press, Washington, D.C.

Roath, L.R. & Krueger, W.C. (1982) Cattle grazing influence on a mountain riparian zone. *Journal of Range Management* 35, 100–103.

Rosenberg, K.V., Ohmart, R.D., Hunter, W.C. & Anderson, B.W. (1991) *Birds of the Lower Colorado River Valley*. University of Arizona Press, Tucson.

Rucks, M.G. (1984) Composition and trend of riparian vegetation on five perennial streams in southeastern Arizona. In: Warner, R.E. & Hendrix, K.M. (eds.) *California Riparian Systems: Ecology, Conservation and Productive Management*. University of California Press, Berkeley, pp. 97–107.

Rummell, R.S. (1951) Some effects of livestock grazing on ponderosa pine forest and range in central Washington. *Ecology* 32, 594–607.

Rychert, R.C., Skujins, J., Sorensen, D. & Porcella, D. (1978) Nitrogen fixation by lichens and free-living microorganisms in deserts. In: West, N.E. & Skujins, J. (eds.) *Nitrogen in desert ecosystems*. Dowden, Hutchinson and Ross, Inc., Stroudsburg, pp. 20–30.

Saab, V.A., Bock, C.E., Rich, T.D. & Dobkin, D.S. (1995) Livestock grazing effects in western North America. In: Martin, T.E. & Finch, D.M. (eds.) *Ecology and Management of Neotropical Migratory Birds*. Oxford University Press, New York, pp. 311–353.

Saab, V.A. & Rich, T.D. (1997) *Large-Scale Conservation Assessment for Neotropical Migratory Landbirds in the Interior Columbia Basin*, General Technical Report PNW-GTR-399. Pacific Northwest Forest and Range Experiment Station, United States Forest Service, Portland.

Schulz, T.T. & Leininger, W.C. (1991) Nongame wildlife communities in grazed and ungrazed montane riparian sites. *Great Basin Naturalist* 51, 286–292.

Scott, V.W. (1967) The range cattle industry: its effect on western land law. *Montana Law Review* 28, 155–183.

Severson, K.E. (1990) Summary: livestock grazing as a wildlife habitat management tool. In: Severson, K.E. (technical coordinator.) *Can Livestock be Used as a Tool to Enhance Wildlife Habitat?*, General Technical Report RM-194. United States Forest Service, Rocky Mountain Forest and Range Experiment Station, Fort Collins, pp. 3–6.

Sheridan, T.E. (2001) Cows, condos and the contested commons: the political ecology of ranching on the Arizona-Sonora borderlands. *Human Organization* 60, 141–152.

Siegel, J.J. (1996) "Subdivisions versus agriculture": from false assumptions come false alternatives. *Conservation Biology* 10, 1473–1474.

Smith, D.A. (2003) "Only man is vile" In: Floyd, M.L. (ed.) *Ancient Piñon-Juniper Woodlands: A Natural History of Mesa Verde Country*. University Press of Colorado, Boulder, pp. 321–336.

Smith, R.J. (1977) Conclusions. In: Townsend, J.E. & Smith, R.J. (eds.) *Proceedings of a Seminar on Improving Fish and Wildlife Benefits in Range Management*, FWS/OBS-77/1. Fish and Wildlife Service, Biological Services Program, U.S. Department of Interior, Washington, D.C., pp. 117–118.

Smithsonian Migratory Bird Center. *Western Rivers: Magnets for Migrants*, Fact Sheet No. 5. Smithsonian Migratory Bird Center, National Zoo, Washington, D.C. http://nationalzoo.si.edu/ConservationAndScience/MigratoryBirds/Fact_Sheets/

Stegner, W. (1988) *The Uneasy Chair: A Biography of Bernard DeVoto*. Peregrine Smith Books, Salt Lake City.

Stevens, L.E., Brown, B.T., Simpson, J.M. & Johnson, R.R. (1977) The importance of riparian habitat to migrating birds. In: Johnson, R.R. & Jones, D.A. (technical coordinators.) *Importance, Preservation and Management of Riparian Habitat: A Symposium*, General Technical Report RM-43. United States Forest Service, Rocky Mountain Forest and Range Experiment Station, Fort Collins, pp. 156–164.

Stewart, G. (1936) History of range use. In: The Western Range, 74th Congress, 2d Session, Senate Document No. 199.

Stoddart, L.A. & Smith, A.D. (1943) *Range Management*. McGraw-Hill, New York.

Stout, J.A., Jr. (1970) Cattlemen, conservationists and the Taylor Grazing Act. *New Mexico Historical Review* 45, 311–332.

Strong, T.R. & Bock, C.E. (1990) Bird species distribution patterns in riparian habitats in southeastern Arizona. *Condor* 92, 866–885.

Swetnam, T.W. (1990) Fire history and climate in the southwestern United States. In: Krammes, J.S. (ed.) *Effects of Fire Management on Southwestern Natural Resources*, General Technical Report RM-191. Rocky Mountain Forest and Range Experiment Station, United States Department of Agriculture/Forest Service, Fort Collins, pp. 6–17.

Swetnam, T.W., Baisan, C.H. & Kaib, J.M. (2001) Forest fire histories of the sky islands of La Frontera. In: Bahre, C.J. & Webster, G. (eds.) *Changing Plant Life of La Frontera: Observations on Vegetation Change in the United States-Mexico Borderlands*. University of New Mexico Press, Albuquerque, pp. 95–119.

Szaro, R.C. (1989) Riparian forest and scrubland community types of Arizona and New Mexico. *Desert Plants* 9(3-4), 69–138.

Szaro, R.C., Belfit, S.C., Aitkin, J.K. & Rinne, J.N. (1985) Impact of grazing on a riparian garter snake. In: Johnson, R.R., Ziebell, C.D., Patton, D.R., Ffolliott, P.F. & Hamre, F.H. (technical coordinators.) *Riparian Ecosystems and their Management: Reconciling Conflicting Uses*, General Technical Report RM-120.

United States Forest Service, Rocky Mountain Forest and Range Experiment Station, Fort Collins, pp. 359–363.

Szaro, R.C. & Pase, C.P. (1983) Short-term changes in a cottonwood-ash-willow association on a grazed and an ungrazed portion of Little Ash Creek in Central Arizona. *Journal of Range Management* 36, 382–384.

Taylor, D.M. (1986) Effects of cattle grazing on passerine birds nesting in riparian habitat. *Journal of Range Management* 39, 254–258.

Tewksbury, J.J., Black, A.E., Nur, N., Saab, V.A., Logan, B.D. & Dobkin, D.S. (2002) Effects of anthropogenic fragmentation and livestock grazing on western riparian bird communities. *Studies in Avian Biology* 25, 158–202.

Thomas, J.W., Maser, C. & Rodiek, J.E. (1979) Riparian zones in managed rangelands – their importance to wildlife. In: Cope, O.B. (ed.) *Proceedings of the Forum: Grazing and Riparian/Stream Ecosystems*. Trout Unlimited, Denver, pp. 21–31.

Thomashow, M. (1995) *Ecological Identity: Becoming a Reflective Environmentalist*. The MIT Press, Cambridge.

Trimble, S.W. & Mendel, A.C. (1995) The cow as a geomorphic agent – a critical review. *Geomorphology* 13, 233–253.

Trombulak, S.C., Omland, K.S., Robinson, J.A. et al. (2004) Principles of conservation literacy: recommended guidelines for conservation literacy from the Education Committee of the Society for Conservation Biology. *Conservation Biology* 18, 1180–1190.

United States General Accounting Office [USGAO] (1988) *Public Rangelands: Some Riparian Areas Restored but Widespread Improvement will be Slow*, GAO/RCED-88-105. United States General Accounting Office, Washington, D.C.

United States General Accounting Office [USGAO] (1991) *Public Land Management: Attention to Wildlife is Limited*, GAO/RCED-91-64. United States General Accounting Office, Washington, D.C.

Van Vuren, D. (1982) Comparative ecology of bison and cattle in the Henry Mountains, Utah. In: Nelson, L., Peek, J.M. & Dalke, P.D. (eds.) *Proceedings of the Wildlife-Livestock Relationships Symposium*. Forest, Wildlife and Range Experiment Station, University of Idaho, Moscow, pp. 449–457.

Van Vuren, D. (1987) Bison west of the Rocky Mountains: an alternative explanation. *Northwest Science* 61, 65–69.

Van Vuren, D. & Bray, M.P. (1985) The recent geographic distribution of *Bison bison* in Oregon. *Murrelet* 66, 56–58.

Van Vuren, D. & Dietz, F.C. (1993) Evidence of *Bison bison* in the Great Basin. *Great Basin Naturalist* 53, 318–319.

Walsberg, G.E. (2005) Cattle grazing in a national forest greatly reduces nesting success in a ground-nesting sparrow. *Condor* 107, 714–716.

Warren, S.D. (2003) Biological soil crusts and hydrology in North American deserts. In: Belnap, J. & Lange, O. (eds.) *Biological Soil Crusts: Structure, Function and Management*, Ecological Studies 150. Springer-Verlag, Berlin, pp. 327–337.

The Wildlife Society (1996) The Wildlife Society position statement on livestock grazing on federal rangelands in the western United States. *Wildlifer*, 274, 10–13.

Williams, T. (2006) Sacred cows. *Audubon*, 42–48. http://audubonmagazine.org/content/content0603.html

Williamson, L. (1985) Where the grass is greenest. *Outdoor Life*, 30–31.

Winegar, H.H. (1977) Camp Creek channel fencing – plant, wildlife, soil and water responses. *Rangeman's Journal* 4, 10–12.

Wuerthner, G. (1994) Subdivisions versus agriculture. *Conservation Biology* 8, 905–908.

Zeveloff, S.L. (1988) *Mammals of the Intermountain West*. University of Utah Press, Salt Lake City.

10

Guanaco Management in Patagonian Rangelands: A Conservation Opportunity on the Brink of Collapse

Ricardo Baldi[1,2], Andrés Novaro[1,2], Martín Funes[2], Susan Walker[1,2], Pablo Ferrando[3], Mauricio Failla[4] and Pablo Carmanchahi[1]

[1]Consejo Nacional de Investigaciones Científicas y Técnicas (CONICET), Argentina
[2]Patagonian and Andean Steppe Program, Wildlife Conservation Society, Argentina
[3]Secretaría de Ambiente y Desarrollo Sustentable de la Nación, Argentina
[4]Dirección de Fauna Silvestre de Río Negro, Argentina

The conservation problem

Background information

The guanaco is one of four South American camelid species. Guanaco and the vicuña (*Vicugna vicugna*) are wild, whereas the llama (*Lama glama*) and alpaca (*Vicugna pacos*) are domestic. With the distribution of vicuñas limited to high altitudes of the Andes, guanacos are the only large native herbivores widely distributed across South America. Recent DNA studies have demonstrated the

Wild Rangelands: Conserving Wildlife While Maintaining Livestock in Semi-Arid Ecosystems, 1st edition. Edited by J.T. du Toit, R. Kock, and J.C. Deutsch.

existence of only two guanaco subspecies, *Lama guanicoe cacsilensis*, found from 8 to 20°S and *Lama guanicoe guanicoe*, found from 22 to 55°S, occupying a range of arid lands from sea level to 4000 m above sea level (Marin et al. 2006). The northern subspecies is critically endangered, with fewer than 5000 animals in Peru and northern Chile (Wheeler 2006); the southern population may number 600,000. It is estimated that 96% of the total guanaco population is in Argentina (Franklin 1982; Torres 1985), mainly in the country's Patagonia region (Figure 10.1).

The guanaco was the only ungulate to inhabit the Patagonian steppe from the end of the Pleistocene (10,000–12,000 years ago) until the introduction

Figure 10.1 **Original and current geographic range of the guanaco (taken from Franklin et al. 1997).**

of domestic livestock (Franklin 1982). It is estimated that the original guanaco population numbered 30–50 million (Raedeke 1979), but during European colonization this population declined severely (Franklin & Fritz 1991). Accounts of nineteenth century naturalists and travellers suggest that guanacos were abundant. Darwin (1845) wrote that guanacos were 'very common over the whole of the temperate parts of the continent'. He saw herds of up to 500 on the shores of the Santa Cruz River in southern Patagonia. During Musters' (1871) voyage across Patagonia, he reported seeing herds of 3000–4000 and Prichard (1902) wrote, 'literally thousands appeared on the summits of the surrounding ridges'.

Although guanacos are still numerous and widely distributed (Redford & Eisenberg 1992; Franklin et al. 1997), they have continued their precipitous decline since the 1800s. Over hunting, range degradation due to sheep overstocking and interspecific competition have contributed to the guanaco's demise (Raedeke 1979; Franklin 1982; Cunazza et al. 1995; Baldi et al. 2001, 2004). Today, the guanaco occupies only 40% of its original range (Puig 1995; Franklin et al. 1997, Figure 10.1), distributed in small, relatively isolated populations. Although the species is not threatened with demographic extinction at a continental scale, it is ecologically extinct in most of its remaining range (Novaro et al. 2000). Some southern populations are at risk of local or regional extirpation (Cunazza et al. 1995), while some predictions indicate that the northern subspecies will become extinct in Peru within 30 years if current hunting rates continue (www.conopa.org).

Processes affecting guanaco conservation

The 700,000 km^2 of Patagonian grasslands and shrublands still harbour most of the world's guanaco population. In the late nineteenth century, European civilization regarded this region as wilderness, which the government in Buenos Aires was determined to conquer either by war against the nomadic Tehuelche native peoples or by settlement of the European colonists. It is in this context that British immigrants introduced the first sheep to Patagonia in the 1880s. Some estimates suggest that approximately 10 million guanacos remained at that time (Cabrera & Yepes 1940; Torres 1985), along with the two other large native herbivores, Darwin's rheas (one of South America's flightless, ostrich-like birds) and maras (a large, endemic deer-like rodent).

Within 50 years of sheep's introduction to Patagonia, their numbers peaked at 22 million.

The massive introduction of the domestic sheep affected the numbers and distribution of guanacos. Although sheep of the merino breed, the most common in Patagonia, weigh 40–60 kg compared to adult guanacos' 80–120 kg, the body sizes of both species place them in Jarman's (1974) 'intermediate selectivity' foraging-style category. The diets of both guanacos and sheep include high proportions of mono and dicotyledonous plants – roughly speaking, grasses and shrubs. Baldi et al. (2004) concluded that the potential for competition between guanacos and sheep in nine Patagonian sites was high, since the diets of the species overlap markedly, especially in summer when food is scarce. This interspecific competition also manifests itself spatially; guanaco abundance has been found to be negatively related to sheep density. At some sites, guanacos moved into paddocks from which ranchers had removed their sheep (Baldi et al. 2001). Across sites, sheep are up to an order of magnitude more abundant than guanacos, limiting the latter's food resources. Sheep density is positively related to the abundance of forage plants important to both herbivores, while guanaco density, presumably due to competition, is negatively related to the abundance of the same plants (Baldi et al. 2001).

The introduction of sheep to Patagonia has also dramatically impacted the landscape. Overgrazing has led to the desertification of approximately 30% of the steppe (del Valle 1998). Ranchers keep sheep densities high as long as the local productivity of forage species is sufficiently high. Under heavy grazing, large vegetation patches dominated by palatable shrubs and grasses were replaced by smaller patches with significantly lower species richness, dominated by unpalatable shrubs and dwarf shrubs (Bisigato & Bertiller 1997, but see Chapter 4). Sheep husbandry itself has declined in recent decades because expanses of the degraded land can no longer support the number of sheep they once did (Baldi et al. 2004).

The threat to guanacos is compounded by wild and domestic ruminants' shared susceptibility to disease (Thedford & Johnson 1989; Gulland 1995; Nettles 2001). It is well known that disease transmission between domestic and wild animals can have dramatic impacts on wildlife, as when rinderpest killed 90% of the African buffalo (*Syncerus caffer*) population and five million cattle in 10 years (Bengis 2002a,b). Ironically, wild species are usually blamed as the source of infection for domestic animals. Indeed, local opinion in Patagonia holds guanacos responsible for transmitting disease to sheep. Studies in mainland Patagonia have shown, however, that free-ranging guanacos are relatively disease-free but are themselves susceptible to the common diseases of domestic livestock (Karesh et al. 1998; Beldoménico et al. 2003; M. Uhart et al. unpublished observations, 2004). Castillo (2006) came to a similar conclusion based on parasite loads in free-ranging Peruvian guanacos.

In arid Patagonia, where over 95% of the land is privately owned, wire fences are an important source of guanaco mortality. Four-foot high, five-wire fences prevent sheep from moving between ranches or paddocks. While adult guanacos can jump over the fences, chulengos often get entangled in the upper wire and die. In Northeast Chubut, fences cause up to 15% of guanaco mortalities during the first year of life (Baldi, unpublished observations, 2005).

Guanacos, as well as other native species, were excluded from the model of development used in Patagonia. Nearly all Patagonian private ranches were dedicated to sheep husbandry throughout the twentieth century. Traditional European farming activities exported to Patagonia did not consider the use of native species complementary to domestic livestock production. Instead, guanacos were viewed as an obstacle to sheep ranching and killed, either illegally or in accordance with government authorization (Baldi et al. 1997). Since the 1960s, commercial demand has risen for chulengo pelts and guanaco fibre, particularly for export to Europe. In Argentina, more than 220,000 chulengos were killed legally for their pelts from 1976 to 1979 and the country exported more than 440,000 guanaco pelts from 1972 to 1979 (Ojeda & Mares 1982). In 1979, the value of these exports was $3.6 million. Between 1984 and 1994, the government of Chubut, a Patagonian province with significant guanaco populations, issued permits to kill 118,000 guanacos. The government based its quota on abundance estimates from landowners who claimed to have many guanacos on their properties. In a 10-year span, ranchers in Northeast Chubut were allowed to kill 38,000 guanacos, reducing the population to approximately 12,400 animals by 1995 (Baldi et al. 1997). Entire herds have been legally eradicated without any knowledge of population parameters.

Until recently, most guanaco hunting for commercial purposes or to clear range for sheep has taken place on horseback or from the few roads in Patagonia. In recent years, exploration and extraction of oil, gas and other minerals have caused the region's road network to expand (Radovani 2004). Consequently, large populations of guanaco and other wildlife have become accessible to humans and been decimated by poachers, mostly town dwellers with trucks who hunt for sport and/or meat. In the oil fields of the Auca

Mahuida area in northern Patagonia, guanaco numbers have declined by 92% on average since 1986 as road density has increased from 0.14 to 1.84 km/km^2 of habitat (Radovani 2004).

The advent of international regulations has slowed, but not stopped, guanaco hunting for commercial purposes. In 1992, guanacos were listed in CITES Appendix II and fibre and pelt exports banned until a sustainable management plan was implemented. Little changed in practice, however; landowners kept pressing for hunting permits, while poaching continued apace due to lack of human and material resources for law enforcement. Guanacos are hunted on almost all private sheep ranches. Whether to free up forage for sheep, to feed shepherd dogs, to sell meat illegally in poor Patagonian neighbourhoods or to sell chulengo pelts illegally to tourists, guanacos are still killed by the thousands throughout Patagonia every year.

Consequences for population spatial structure and functionality

With ranching monopolizing Patagonia's most productive areas – and the associated competition with sheep, hunting and habitat modification – guanacos have been relegated to marginal lands with poor vegetation (Baldi et al. 2001, 2004). Guanaco densities are generally low in Patagonia; most density estimates range below 2 animals/km^2. High-density populations occur on the few protected areas or on land that is abandoned or unsuitable for ranching. Separating these infrequent dense populations are hundreds of kilometres (Figure 10.2) where guanacos are either absent or found at very low densities (Baldi et al. 1997, 2001; Novaro et al. 2000).

Such spatial fragmentation threatens guanaco persistence. Human activities and habitat loss often prevent movement between populations. This loss of connectivity produces small, closed and isolated populations under increased risk of collapse from loss of genetic variation or environmental or demographic stochasticity. For example, one of Patagonia's highest density guanaco populations, comprising approximately 40 guanacos/km^2, is found in a small, 1800-hectare reserve under effective protection; just across a fence is a property where sheep ranching and guanaco hunting take place, with a guanaco density of 12 animals/km^2 (Baldi et al. 2001). After an unusually dry year in which precipitation decreased to only 20% of the regional average, over 80% of the reserve's guanacos died of starvation (R. Baldi, unpublished observations, 2004). Thus, population density, reserve size, habitat condition

Figure 10.2 **Distribution of high-density (>8 animals/km^2) and large (<8 animals/km^2 but >1000 individuals) populations of guanacos and populations under pressure of exploitation in Patagonia.**

and contrasting management conditions are all interacting factors affecting viability of guanaco populations.

At the landscape level, one consequence of low density and spatial fragmentation is the loss of ecological functionality. After the great Pleistocene extinctions, guanacos greatly impacted plant distribution, abundance and composition and served as the major prey species for pumas. Guanaco densities are now so low at most sites that they no longer fill these ecological roles. A study in Northwest Patagonia (Novaro et al. 2000) found the availability of native prey species, including the guanaco, to the carnivore guild to be so low as to be under-represented in the diets of pumas and fox. In contrast, introduced prey species formed major components of the native carnivores' diets (Figure 10.3a). The authors concluded that the guanaco and other native prey species were ecologically extinct at the sites, since they had lost their functionality as prey. Moreover, Novaro and Walker (2005) found a population density threshold of approximately 8 guanacos/km^2; above this level, the role of guanacos in pumas' diets was significantly higher than below the threshold, where guanacos were just an occasional prey (Figure 10.3b). We believe identifying relatively easy-to-measure functionality thresholds such as these (see Chapter 2) is crucial to defining conservation targets within goals to restore healthy populations with better chances for long-term persistence.

Current pressures and a new challenge: is mixed management of guanacos and sheep possible?

Although still seen as a pest by many sheep ranchers, the high value of guanaco fibre on the international market has influenced some landowners whose properties harbour high densities of the animal. Attempts to capture, shear and release wild guanacos started in the late 1990s and have grown rapidly since 2003, particularly in the Patagonian province of Río Negro (Figure 10.2). The animals are herded mainly on horseback (sometimes on motorcycle), into a trap, shorn and released. The trap consists of two arms up to 2000 m long, funnelling towards an entrance 300–500 m wide to either a single round corral or a series of successive holding corrals, separated by wooden or mesh gates (Montes et al. 2006).

Since 2003, approximately 11,000 guanacos have been captured on seven ranches in Río Negro (Figure 10.2, Table 10.1); this figure represents 32% of the guanaco population on average per ranch (range 3–52%, see Table 10.2

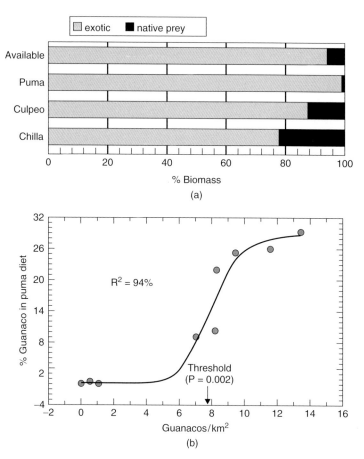

Figure 10.3 (a) **Ecological extinction of native wildlife as prey for native carnivores (taken from Novaro et al. 2000) and (b) Functionality of guanaco populations as prey for pumas as a function of guanaco population density (taken from Novaro & Walker 2005).**

for density estimates). Of the guanacos captured, over 9000 were shorn and released, producing 4900 kg of fibre (Table 10.1). Approximately 15% (n = 1604) of those captured were chulengos, 18% of which were moved elsewhere to start guanaco farms. Contrary to managed wild populations, guanacos on farms are kept in captivity and therefore lose their role in ecosystem functioning. In Chubut, the number of guanaco farms has grown from one to

Table 10.1 Current use of guanaco populations for fibre production.

Ranch ID	1	2	3	4	5	6	7
Period	Feb 03–Feb 06	Jan 03–Jan 06	Jan 04–Jan 06	Nov 03–Oct 06	Oct 05–May 06	Oct 03–Oct 04	Jan 05–Nov 05
No. of round-up events	12	11	10	10	6	5	4
Guanacos captured	2640	2241	2314	543	2733	174	284
Guanacos sheared	2407	1664	2033	451	2172	146	268
Chulengos captured	181	432	345	131	471	28	16
Chulengos sent to farms	56	64	22	131	12	1	0
Average % mortality per round-up (range)	0.78 (0–3.6)	0.61 (0–2.8)	1.96 (0–9.1)	0.85 (0–7.7)	1.13 (0–2.3)	5.00 (0–20)	0.93 (0–2.2)
Total fibre obtained (kg)	599	727	1013	589	1615	142	213
Average fibre per animal (kg)	0.23	0.43	0.47	0.54	0.47	No data	0.51

Table 10.2 **Livestock (sheep) and wild (guanaco) resources across ranches with guanaco shearing.**

Ranch ID	1	2	3	4	5	6	7
Total area (hectares)	51,000	40,000	31,922	39,845	30,200	5603	23,700
Sheep density (per km^2)	8	25	0	2	0	9	21
Guanaco density (per km^2)	12	20	49	7	28	62	35
% area with sheep	90	100	0	30	0	64	92
% area with guanaco management	10–20	13–25	8–16	8–13	10–20	27–36	8

nine since 2002 (S. Rivera, personal communication, 2006), while 12 farms are legally operating in Río Negro. Captive guanaco management has a mounting effect on wild populations, as chulengos are taken from the wild by the hundreds every year to augment farm stocks.

Except for two ranches of wild, managed guanacos where sheep have been removed, most initiatives for live guanaco shearing and release are combined with traditional sheep ranching (Table 10.2). On an average, owners dedicate 18% of the area of these properties (n = 7) to guanaco management, though generally the size of the guanaco-management area varies annually. In ranches that combine the two activities (n = 5), an average of 75% of the ranch area is dedicated to sheep (Table 10.2). Overall, 31,000 hectares, or 15% of Río Negro's 222,000 hectares of ranchland is dedicated to wild-guanaco management.

The demand for guanaco shearing is increasing, raising new questions about its impacts on affected populations. The Santa Cruz Province is likely to begin live shearing, and governments in Río Negro and Chubut are under pressure to allow new shearing initiatives (B. Alegre, personal communication, 2006). To minimize the negative effects of shearing, the governments of Río Negro, Chubut, Neuquén and Mendoza are beginning to regulate the activity and its handling procedures. Government officials are usually present to supervise animal welfare, ear-tag for subsequent identification, take blood samples for stress indicators, assess mortality during the procedures and supervise the amount of fibre obtained and sold, as the species is under CITES Appendix II. Although mortality is reportedly low during shearing (Table 10.1), there is little data on the possible after-effects of capture, live shearing and release. All 10 male guanacos radio-tracked after an experimental shearing in Payunia Reserve in 2005 survived the first year (A. Novaro et al., unpublished observations, 2007). However, 4 of 10 male guanacos radio-tracked after a Río Negro shearing died in the first month, presumably from capture myopathy (A. Rey et al., personal communication, 2006).

The conservation approach

A conservation strategy for the Patagonian rangelands

As in other rangeland ecosystems, most wildlife of the Patagonian rangelands occurs outside protected areas, sharing habitat with livestock and humans. We estimate that even if current protected areas in Patagonia and the southern Andean steppe were effective, only 10% of the guanaco population would be protected (S. Walker et al., unpublished observations, 2005). We therefore advocate managing for conservation throughout Patagonia as a means to restore ecosystem functionality and connectivity among wild guanaco populations. To this end, it is imperative that private ranches adopt wildlife-friendly practices; that students, the government and non-governmental organizations (NGOs) receive scientific training and conservation tools and that people and resources are devoted to conservation increase.

We believe that conservation-oriented management of guanacos can favour the recovery of wild populations, relieve pressure on native habitat and even facilitate limited range recovery. While consolidating existing protected areas and creating new ones is important, new conservation models account for the fact that most protected areas are not protected effectively and that range-lands where most guanacos live will likely remain under-represented in the protected-area system. The ideal conservation model for the Patagonian range-lands would allow human use of ecosystem services but prioritize the long-term persistence of wildlife and the reversal of habitat degradation. Sheep ranching has thus far failed to beget this vision. The industry has collapsed; today, sheep in the region number less than a third of what they did 50 years ago because of high mortality and low reproduction rates (Soriano & Movia 1986). Most dra-matically, overstocking has produced irreversible habitat loss and degradation across large areas of Patagonia (Beeskow et al. 1995; del Valle 1998).

There is potential for the profitable and sustainable use of guanacos, but to achieve it, ranchers must reduce their sheep stocks. Existing guanaco density estimates for shearing ranch operations are preliminary and lack a standardized methodology, but point to a higher density of guanacos than sheep at such ranches (Table 10.2). This contrasts with the usual situation in Patagonia, where sheep densities are up to an order of magnitude higher than those of guanacos (Baldi et al. 2001). Moreover, sheep densities at ranches with guanaco shearing (Table 10.2) are considerably lower than at non-shearing, traditional ranches (Baldi et al. 2001, 2004; Novaro et al. 2000). This is an expected finding, as shearing is only legal and profitable on ranches with abundant guanacos, which as we discussed above, are not compatible with high sheep densities (Baldi et al. 2001). In some cases, sheep density is low because of limited water resources or inappropriate range conditions; in others, ranchers decided to focus primarily on guanacos. These differences account for the contrast among proportions of ranch area devoted to raising sheep, as in Table 10.2.

Political context of guanaco management in Argentina

Argentina is politically organized into provinces with their own constitutions and governments. Under the country's federal system, however, these must operate in compliance with the national government and its constitution. Most of the country's natural resource management decisions and implementation take place at the provincial level of government.

In 2006, Argentine national and provincial authorities, scientists and NGOs crafted a National Management Plan for Guanacos (Baldi et al. 2006). The overall goal of the plan was to ensure the 'conservation of guanacos', specifically the viability of wild, ecologically functional guanaco populations and the persistence of the species throughout its geographic range (Baldi et al. 2006). Clear conservation goals, as defined in the Plan, are essential to a sound management strategy that is likely to be properly implemented. The Patagonian provinces harbouring most of the country's guanacos support the plan.

Each provincial government has a wildlife department charged with making and enforcing wildlife management decisions. Often, these departments lack the staff, training and capacity needed to enforce laws, especially to control poaching. This shortage is particularly felt in expansive Patagonia, where large private ranches dominate the landscape.

Argentina's political agenda often ignores conservation issues and even sees them as obstacles to development. Moreover, national policy frequently runs counter to conservation-oriented management, favouring the intensification of established activities – such as sheep ranching, mining and oil exploration – that jeopardize natural resources.

Combining sheep ranching with guanaco management: a preliminary analysis

Wild guanaco management for fibre production on sheep ranches may contribute to the long-term conservation of Patagonian rangelands, but determining whether mixed guanaco–sheep production is feasible requires assessing the economic and ecological implications of guanaco use.

Two different approaches towards the economic analysis of guanaco management through live shearing have been attempted.

Following trial live-shearing operations in Santa Cruz, Río Negro and Neuquén in 2000–2001 (Montes et al. 2006), the total cost of each shearing was calculated and the cost per guanaco shorn estimated. These shearings had a low trapping rate due to poor organization, lack of experience and the ranches' relatively low density of guanacos (5.5–15 individuals/km^2, Montes et al. 2006). Results ranged widely for the cost per guanaco shorn, from $36 for 21 guanacos to an astronomic $1476 per guanaco, when only two were shorn after several round-up attempts (M. Funes & P. Carmanchahi, unpublished observations, 2006). The price for the more valuable fibre from the back, flanks and external and upper parts of the legs was about $150 per kg. Given that on an average almost 330 g of fibre per guanaco was harvested (Montes et al. 2006), we estimate that each guanaco produced a potential benefit of approximately $50. Comparing this figure with the cost per guanaco captured, only the most successful operation showed a slight positive cost–benefit outcome.

A different and more comprehensive analysis stems from 27 live shearings on six ranches in Río Negro during 2003–6 (a larger scale than the 2000–1 cases) (P. Ferrando et al., in preparation). This analysis is based on a model that has become the norm at ranches where live shearing is practised. The Río Negro data were analysed for three consecutive periods: (i) September 2003–February 2004, (ii) September 2004–February 2005 and (iii) September 2005–February 2006. Costs were categorized as fixed or variable, but only for those costs considered incremental, because most ranches involved in guanaco management still produce livestock as their main activity.

The model was developed using the following parameters:

Fixed costs: corral facilities, calculated as the average expense for all materials employed in each ranch, with a depreciation of 6 years, as well as the cost of assembling and disassembling the corrals. These costs were analysed under two different scenarios, a large-scale and a medium-scale operation. The large-scale had a fixed cost of $22,500 and the medium scale a cost of $10,000.

Variable costs: personnel for round-up and shearing. This included 22 horsemen over 4 days, representing $2700, plus nine shearers, adding a variable cost depending on the number of guanacos sheared.

Income variables: income was a function of the average fibre obtained per individual, estimated at 400 g/guanaco and the average market price of $135/kg, during the last 3 years (range $70–230). All values are presented in US dollars.

With this information, we estimated the equilibrium as the point at which profits absorb the cost of the activity, or the point at which equilibrium exists between costs and benefits, without any net profit. Equilibrium for the operation with the highest fixed cost was 123 guanacos (Figure 10.4), whereas equilibrium for the operation with the lowest fixed cost was 84 guanacos.

Using this model, net income for each ranch can be roughly estimated based on the number of guanacos captured and shorn during the 3-year period, using average values for income and cost parameters for purposes of simplification. We can then look at the net income per hectare, based on the fraction of each ranch committed to guanaco management (Table 10.2). We conducted this analysis only for the ranch with the highest fixed cost.

The ranches with the lowest net income per hectare earned $0.3–0.5/hectare, an intermediate group of ranches earned $2.3–7.7/hectare and one ranch obtained the highest value, almost $20/hectare, although only in 1 year.

Although insufficient information exists to perform a similar analysis of sheep husbandry on the same ranches, preliminary analysis can be attempted. Sheep wool in Río Negro currently averages $2.7/kg for a ewe producing 4.5 kg of wool on average. The recommended stocking rate for this zone of Río Negro, based on the land's carrying capacity, is about one sheep every 4 hectares. This means the gross income related to sheep breeding is about $3 per hectare, close to the low end of the net income estimated for guanaco

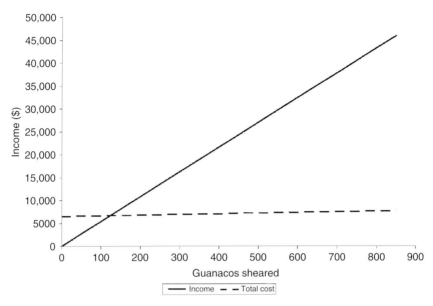

Figure 10.4   **Preliminary analysis of the cost–benefit ratio of shearing guanacos.**

management. Therefore, guanaco management on Patagonian rangelands has the potential to contribute economic benefits per unit of land as high as, and potentially higher than, sheep wool production. Moreover, the economic magnitude of guanaco management is potentially high, as the 4900 kg of fibre exported produced a total income of over $650,000 within 3 years (average market price $135/kg).

Three additional qualifications apply to this analysis: (i) the profit in sheep ranching derives not just from wool, but also from meat, and in traditional ranching operations at least 15% of male lambs are sold for slaughter (SSD 2000); (ii) complete cost–benefit analysis of sheep ranching usually takes into account additional costs necessary for the operation of the ranch, which are not usually considered in guanaco-management analysis, being based only on incremental costs; and (iii) the current average price for guanaco fibre (~$135 per kg), a key determinant of the relative profitability of guanaco and sheep production systems, may not be sustainable in the long term as more guanaco fibre enters the market.

Future prospects

The challenge of managing functional populations

Because profits from guanaco live shearing depend on the number of animals captured, shorn and released, efforts are limited to sites where guanaco densities are relatively high and habitat conditions are appropriate for successful round-ups. As we showed above (see Figure 10.2), high-density – and probably ecologically functional – populations are scarce and isolated by barriers (usually human-made). So far, data on shearing indicate that over 10,000 guanacos have been directly affected by herding, handling and release in Argentine Patagonia. However, this use takes place in high-density or large populations (see Figure 10.2). On the basis of density and population distribution data (Figure 10.2, Table 10.2), we estimate that the number of guanacos in populations currently exploited for shearing is approximately 71,000 – almost 15% of remaining guanacos and 35% of high-density or large populations. Given the increased occurrence of shearing, these figures indicate that many of the best-conserved guanaco populations may soon be subject to capture and shearing.

The use of wild guanacos for complementary income may be desirable to reduce overstocking with sheep, to increase the value of natural resources and to promote connectivity by maintaining high-density, functional guanaco populations between protected areas. For this strategy to be successful, however, it is necessary to assess the effects of capture and shearing on these populations. So far, there is no evidence on the ecological sustainability of live-shearing guanacos. Sustainable management will therefore require careful evaluation of the activity's effects both on individuals and populations.

The vicuña also yields a valuable product from the wild without the need to harvest individuals (Vilá & Lichtenstein 2006). Vicuña capture, shearing and release was the traditional system used by the Incas and other pre-Columbian cultures in the high Andes. After excessive hunting in the eighteenth and nineteenth centuries, the species was saved from extinction; populations are currently stabilizing in part of their range (Laker et al. 2006). For over a decade, vicuña live capture, shearing and release has taken place in Peru (Sahley et al. 2007) and Chile. It has occurred in Argentina since 2003 (Arzamendia & Vilá 2006). Vicuña mortality during capturing and handling has decreased substantially after the introduction of animal welfare practices (Gimpel & Bonacic 2006). However, systematic surveys of post-release mortality are rare. The first assessments of the effects of shearing by marking, releasing and comparing captured/shorn vicuñas with captured/unshorn vicuñas took place in Chile. Gimpel and Bonacic (2006) found overall mortality of captured vicuñas to be 2.5%, but mortality of shorn animals was 12.9% after accounting for capture effects. Most (75%) of the deaths occurred within 49 days of release.

Slight changes in demographics, social structure or health from capture and shearing could impact guanaco populations. We still do not know guanaco mortality rates during the first months after shearing. Is there a differential effect on juveniles and adults, or between the sexes? Does the activity affect female fecundity? Hundreds of guanacos are herded together for shearing. Although they are supposedly managed according to animal care regulations, how do these stressful conditions affect individuals during and after release? There is evidence that at least three guanacos died because of myopathy within hours or days after being released from fences where they had become entangled (M. Uhart & R. Baldi, unpublished observations, 2004). What are the post-release impacts on social structure that have important consequences for population dynamics (Dobson & Poole 1998)?

The guanaco breeding system is a resource–defence polygyny system where a dominant male controls a harem or family group of several females and their calves. Lower rank males aggressively excluded from reproduction by harem males wander in bachelor groups or alone, occasionally challenging harem males for control of the breeding groups (Franklin 1983). Family groups form relatively stable, territorial associations occupying $4-9\ km^2$ home ranges in eastern Patagonia (Burgi 2005), whereas ranges are one to two times larger in Northwest Patagonia (A. Novaro et al., unpublished observations, 2005). The number of individuals per group influences not only home range size but also the way the individuals invest their time budgets in feeding and watching behaviour, likely an antipredatory strategy (Marino & Baldi 2008). If guanacos require a minimum group size for success, like many other species in which the Allee effect has been recorded (Courchamp et al. 1999), disturbance by shearing may cause inverse density dependence and put local populations at risk. What are the chances of reuniting mothers and calves after shearing? When chulengos are very young, mother–calf separation accounts for up to 23% of chulengo mortality (R. Baldi, unpublished observations, 2004). Except for the two radio-telemetry studies mentioned above, no other tracking of sheared guanacos has taken place and no government or private funding is available in Argentina to carry out such a study.

Studies of the effects of vicuña shearing, while still limited, may have relevance for guanacos. A study in Chile found that change in the number of adults in vicuña family groups was pronounced and persisted up to 3 months after capture and release (Gimpel & Bonacic 2006). Moreover, up to 40% of the calves were separated from their original groups after management. Recent estimates showed that two populations increased in numbers during a 3-year period after shearing, while the calf–female ratio was similar for shorn and unshorn females (Sahley et al. 2007). However, the authors stressed that capture and shearing in that case were conducted in the spring, with effects expected to be less severe than in winter. But shearing schedules are usually based on logistics, not biological factors (Sahley et al. 2007). This is especially true for guanacos, for which capture and shearing can occur in spring before the birth season, or in late summer when chulengos are still with their mothers and females must face winter temperatures after being shorn.

We believe careful evaluation of the potential consequences of live shearing on guanaco populations can lead to a successful adaptive management plan. To this end, current shearing initiatives must be conducted as experiments, in which intensive studies, including short- and long-term animal tracking, inform our knowledge of the activity's impacts and allow us to modify management techniques. For results to be conclusive, these studies must scrutinize such variables as timing, frequency of shearing, number of guanacos and types of social groups sheared. Provinces should not approve new guanaco shearing unless incorporated into a programme of intensive evaluation. To apply the precautionary principle, and until more information is available, ongoing shearing operations should be limited. If all sheared guanacos die or do not reproduce after shearing, the demographic consequences would be the same as after a harvest. The only guanaco population used for harvesting is in the Chilean Tierra del Fuego (see Figure 10.2). Begun in 2003, this culling aims to reduce the population from 39,000 to 30,000 by 2010, based on a logistic growth model. Culling takes place at five sites that are home to approximately 50% of the region's guanaco population, killing 20% of local guanacos (range 8–39%) (Morales Pavez 2004; Soto Volcart 2005). Because the puma, the guanaco's main predator, is absent from the sites mentioned above but occurs at most Patagonian shearing sites, we propose that no more than 15% of the guanaco population at each site be captured for shearing per year. We also believe that obtaining accurate estimates of local population densities and structure before and after shearing should be a priority. Until more is known, population declines in two consecutive years should lead to immediate discontinuation of shearing. Furthermore, resources must be found for researching the effects of shearing on guanacos and for establishing shearing management guidelines, including defining the conditions under which shearing should begin and end.

There is also an urgent need for economic assessments of the options available to sheep ranchers whose properties harbour abundant guanacos. If mixed management of sheep and wild guanacos is possible, as preliminary figures suggest, it is crucial to determine the threshold number of sympatric sheep and guanacos to produce the highest economic benefit and lowest impact on the land and on other biodiversity components. It is likely these numbers will vary spatially and temporally according to primary productivity and local conditions. At the same time, government subsidies must shift from traditional sheep ranching to alternative management based on sound science.

A conservation opportunity

We believe there is still time to improve the demographic viability and ecological functionality of guanaco populations and thus restore significant parts of the Patagonian rangelands. What is needed are the human capacity to design and implement a sound management plan and participation from private citizens and governments. However, if momentum is not translated into effective action, the last opportunity to conserve guanacos will be lost. Habitat degradation and fragmentation, competition with exotic herbivores and poaching drove the catastrophic guanaco population decline of the nineteenth and twentieth centuries. Setting the stage for these factors were weak institutions for natural resources management and law enforcement, lack of interest in native species, land tenure patterns, Patagonian land-use practices and a shortage of protected areas. Until today, traditional sheep ranching has proved unsustainable, at least under current management schemes and as a single and central activity for Patagonian ranches.

Immediate action is necessary to achieve conservation-oriented management of wild guanacos in Patagonia. Action must be focused towards specific conservation goals for wild populations and their habitat. Government agencies like the Institute of Technology for Agriculture and Livestock Production (INTA) promote 'guanaco farms' in Patagonia. These farms receive calves from wild populations. Although claiming to favour conservation, these farms cannot foster wild, functional guanaco populations. It is doubtful that such farms can foster economic sustainability either; in Argentina's Northern Andes, INTA vicuña captive breeding programmes have been unsuccessful because they did not benefit small local farmers or promote positive attitudes towards vicuña conservation on the part of local people (Lichtenstein 2006). The captive vicuñas did not fare well either, experiencing high mortality rates and low breeding success. This loss for farmers and conservation is mirrored today in Chile and Peru, where similar captive vicuña programmes are ongoing (Lichtenstein 2006). Successful conservation of South American camelids will clearly hinge on properly planned and implemented management of wild populations (Baldi et al. 2006; Vilá & Lichtenstein 2006).

We envision a network of functional guanaco populations throughout Patagonia, which must be protected at specified reserve sites. These guanacos will be distributed across a connecting matrix where functional populations can coexist with livestock at lower densities, while guanacos can be managed as an alternative way of production for human development. Productive activities such as sustainable live shearing, culling or tourism are still a possibility, depending on local conditions. Either we use the available expertise and social and economic interests to ensure the long-term persistence of wild guanaco populations, to stop habitat degradation and to benefit the local economy by diversifying productive initiatives, or it will be too late. Because we only have one chance; we have no time left for indifference.

References

Arzamendia, Y. & Vilá, B. (2006) Estudios etoecológicos de vicuñas en le marco de un plan de manejo sustentable: Cieneguillas, Jujuy. In: Vilá, B. (ed.) *Investigación, Conservación y Manejo de Vicuñas*. Proyecto MACS, Buenos Aires, pp. 69–83.

Baldi, R., Albon, S.D. & Elston, D.A. (2001) Guanacos and sheep: evidence for continuing competition in arid Patagonia. *Oecologia* 129, 561–570.

Baldi, R., Campagna, C. & Saba, S. (1997) Abundancia y distribución del guanaco (*Lama guanicoe*) en el NE del Chubut, Patagonia Argentina. *Mastozoología Neotropical* 4(1), 5–15.

Baldi, R., de Lamo, D., Failla, M. et al. (2006) *Plan Nacional de Manejo del Guanaco (Lama guanicoe)*. República Argentina. Secretaría de Ambiente y Desarrollo Sustentable de la Nación, Buenos Aires.

Baldi, R., Pelliza-Sbriller, A., Elston, D. & Albon, S.D. (2004) High potential for competition between guanacos and sheep in Patagonia. *Journal of Wildlife Management* 68(4), 924–938.

Beeskow, A.M., Elissalde, N.O. & Rostagno, C.M. (1995) Ecosystem changes associated with grazing intensity on the Punta Ninfas rangelands of Patagonia, Argentina. *Journal of Range Management* 48, 517–522.

Beldoménico, P.M., Uhart, M., Bono, M.F., Marull, C., Baldi, R. & Peralta, J.L. (2003) Internal parasites of free-ranging guanacos from Patagonia. *Veterinary Parasitology* 118, 71–77.

Bengis, R.G. (2002a) Infectious diseases of wildlife: detection, diagnosis and management (part one). *Revue Scientifique et Technique de l'OIE* 21(1), 1–210.

Bengis, R.G. (2002b) Infectious diseases of wildlife: detection, diagnosis and management (part two). *Revue Scientifique et Technique de l'OIE* 21(2), 211–404.

Bisigato, A.J. & Bertiller, M.B. (1997) Grazing effects on patchy dryland vegetation in northern Patagonia. *Journal of Arid Environments* 36, 639–653.

Burgi, M.V. (2005) *Radio de Acción y Uso de Hábitat en Hembras de Guanaco (Lama guanicoe) en el NE de Chubut*. Tesis de Lincenciatura en Ciencias Biológicas. Facultad de Ciencias Naturales, Universidad Nacional de la Patagonia, Argentina.

Cabrera, A.L. & Yepes, J. (1940) *Mamíferos Sudamericanos: Vida, Costumbres y Descripción*. Ediar, Compañía Argentina de Editores, Buenos Aires.

Castillo, V.H. (2006) Contribución al estudio del parasitismo gastrointestinal en guanacos (Lama guanicoe cacsilensis) silvestres del Perú. Resúmenes y Trabajos del IV Congreso Mundial sobre Camélidos. Eje 3, 34–39.

Courchamp, F.T., Clutton-Brock, T.H. & Grenfell, B.T. (1999) Inverse density dependence and Allee effect. *Trends in Ecology and Evolution* 14, 405–410.

Cunazza, C., Puig, S. & Villalba, L. (1995) Situación actual del guanaco y su ambiente. In: Puig, S. (ed.) *Técnicas para el Manejo del Guanaco*. International Union for Conservation of Nature and Natural Resources, Gland, pp. 27–53.

Darwin, C. (1845) *Journal of Researches into the Natural History and Geology of the Countries Visited During the Voyage of HMS Beagle Round the World*. Ward Lock and Co. Ltd., London.

Dobson, A. & Poole, J. (1998) Conspecific aggregation and conservation biology. In: Caro, T. (ed.) *Behavioral Ecology and Conservation Biology*. Oxford University Press, Oxford, pp. 193–208.

Franklin, W.L. (1982) Biology, ecology and relationship to man of the South American camelids. In: Mares, M.A. & Genoways, H.H. (eds.) *Mammalian Biology in South America*. Volume 6, *Special Publication Series*. University of Pittsburgh Press, Pittsburgh, pp. 457–489.

Franklin, W.L. (1983) Contrasting socioecologies of South America's wild camelids: the vicuña and the guanaco. In: Eisenberg, J.F. & Kleiman, D.G. (eds.) *Advances in the Study of Mammalian Behavior*, Special Publication 7. The American Society of Mammalogists, Shippensburg, 573–629.

Franklin, W.L, Bas, F., Bonacic, C.F., Cunazza, C. & Soto, N. (1997) Striving to manage Patagonia guanacos for sustained use in the grazing agroecosystems of southern Chile. *Wildlife Society Bulletin* 25, 65–73.

Franklin, W.L. & Fritz, M.A. (1991) Sustained harvesting of the Patagonia guanaco: is it possible or too late? In: Robinson, J.G. & Redford, K.H. (eds.) *Neotropical Wildlife Use and Conservation*. University of Chicago Press, Chicago and London, pp. 317–336.

Gimpel, J. & Bonacic, C. (2006) Manejo sostenible de la vicuña bajo estándares de bienestar animal. In: Vilá, B. (ed.) *Investigación, Conservación y Manejo de Vicuñas*. Proyecto MACS, Buenos Aires, pp. 113–132.

Gulland, F.M.D. (1995) The impact of infectious diseases on wild animal populations: a review. In: Grenfell, B.T. & Dobson, A.P. (eds.) *Ecology of Infectious Diseases in Natural Populations*. Cambridge University Press, Cambridge, pp. 20–51.

Jarman, P.J. (1974) The social organisation of antelope in relation to their ecology. *Behaviour* 48, 215–267.

Karesh, W.B., Uhart, M.M., Dierenfeld, E.S. et al. (1998) Health evaluation of free-ranging guanaco (*Lama guanicoe*). *Journal of Zoo and Wildlife Medicine* 29, 134–141.

Laker, J., Baldo, J., Arzamendia, Y. & Yacobaccio, H.D. (2006) La vicuña en los Andes. In: Vilá, B. (ed.) *Investigación, Conservación y Manejo de Vicuñas*. Proyecto MACS, Buenos Aires, pp. 37–50.

Lichtenstein, G. (2006) Manejo de vicuñas en cautiverio: el modelo de criaderos del CEA INTA Abrapampa (Argentina). In: Vilá, B. (ed.) *Investigación, Conservación y Manejo de Vicuñas*. Proyecto MACS, Buenos Aires, pp. 133–146.

Marin, J.C., Sportorno, A. & Wheeler, J.C. (2006) Sistemática molecular y filogeografía de camélidos sudamericanos: implicancias para su conservación y manejo. In: Vilá, B. (ed.) *Investigación, Conservación y Manejo de Vicuñas*. Proyecto MACS, Buenos Aires, pp. 85–100.

Marino, A. & Baldi, R. (2008) Vigilance patterns of territorial guanacos (*Lama guanicoe*): the role of reproductive interests and predation risk. *Ethology* 114, 413–423.

Montes, M.C., Carmanchahi, P.D., Rey, A. & Funes, M.C. (2006) Live shearing free-ranging guanacos (*Lama guanicoe*) in Patagonia for sustainable use. *Journal of Arid Environments* 64, 616–625.

Morales Pavez, R.A. (2004) *Revisión de la Dinámica Poblacional del Guanaco (Lama guanicoe) en el Sector Centro-Sur de la Isla de Tierra del Fuego, Chile*. Memoria de título, Facultad de Medicina Veterinaria, Universidad de Concepción, Chile.

Musters, G.C. (1871) *Vida Entre los Patagones: Un año de Excursiones por Tierras no Frecuentadas, Desde el Estrecho de Magallanes Hasta el río Negro*, 1997 edn. El Elefante Blanco, Buenos Aires.

Nettles, V.F. (2001) Wildlife-livestock disease interactions. *United States Animal Health Association Newsletter* 28(5), 11–12.

Novaro, A.J., Funes, M.C. & Walker, R.S. (2000) Ecological extinction of native prey of a carnivore assemblage in Argentine Patagonia. *Biological Conservation* 92, 25–33.

Novaro, A.J. & Walker, R.S. (2005) Human-induced changes in the effect of top carnivores on biodiversity in Patagonia. In: Ray, J.C., Berger, J., Redford, K.H. & Steneck, R. (eds.) *Large Carnivores and the Conservation of Biodiversity*. Island Press, Washington, D.C., pp. 267–287.

Ojeda, R.A. & Mares, M.A. (1982) Conservation of South American mammals: Argentina as a paradigm. In: Mares, M.A. & Genoways, H.H. (eds.) *Mammalian Biology in South America, Volume 6*. Special Publication Series. University of Pittsburgh Press, Pittsburgh, pp. 505–521.

Prichard, H. (1902) Field notes upon some of the larger mammals of Patagonia made between September 1900 and June 1901. *Proceedings of the Zoological Society of London* 1, 272–277.

Puig, S. (ed.) (1995) Uso de los recursos ambientales por el guanaco. *Técnicas Para el Manejo del Guanaco*. International Union for Conservation of Nature and Natural Resources, Gland, pp. 119–134.

Radovani, N. (2004) *Parámetros Poblacionales del Guanaco (Lama guanicoe) en el Area Protegida Auca Mahuida: Efectos de las Picadas Petroleras y la Topografía*. Tesis de Licenciatura en Ciencias Biológicas, Facultad de Ciencias Exactas y Naturales, Universidad de Buenos Aires, Argentina.

Raedeke, K.J. (1979) *Population Dynamics and Socioecology of the Guanaco (Lama guanicoe) of Magallanes, Chile*. PhD Dissertation, University of Washington, United States of America.

Redford, K.H. & Eisenberg, J.F. (1992) *Mammals of the Neotropics, Volume 2: The Southern Cone: Chile, Argentina, Uruguay, Paraguay*. University of Chicago Press, Chicago and London.

Sahley, C.T., Torres Vargas, J. & Sanchez Valdivia, J. (2007) Biological sustainability of live shearing of vicuña in Peru. *Conservation Biology* 21, 98–105.

Sistema de Soporte de Decisiones para la Producción Ganadera Sustentable en la Provincia de Río Negro (2000) Coordinator of simulation models: H. Méndez Casariego. Instituto Nacional de Tecnologia Agropecuaria Estación Experimental Agropecuaria Bariloche. http://www.inta.gov.ar/bariloche/ssd/rn

Soriano, A. & Movia, C.P. (1986) Erosión y desertización en la Patagonia. *Interciencia* 11, 77–83.

Soto Volcart, N. (2005) Evaluacón del primer año de caza, monitoreo de la población 2004 y determinación de cuota de extracción año 2005. Report to Servicio Agrícola y Ganadero, Chile.

Thedford, T.R. & Johnson, L.W. (1989) Infectious diseases of new-world camelids. *Veterinary Clinics of North America: Food Animal Practice* 5, 145–157.

Torres, H. (1985) *Distribución y Conservación del Guanaco, Informe Especial N° 2*. International Union for Conservation of Nature and Natural Resources Publication Services Unit, Cambridge.

del Valle, H.F. (1998) Patagonian soils: a regional synthesis. *Ecología Austral* 8, 103–123.

Vilá, B. & Lichtenstein, G. (2006) Manejo de vicuñas en la Argentina: experiencias en las provincias de Salta y Jujuy. In: Bolkovic, M.L. & Ramadori, D. (eds.) *Manejo de Fauna Silvestre en la Argentina, Programas de Uso Sustentable*. Dirección de Fauna Silvestre, Secretaría de Ambiente y Desarrollo Sustentable, Buenos Aires, pp. 121–135.

Wheeler, J.C. (2006) Guanaco 1: working to save Peru's endangered guanacos. *Camelid Quarterly* 5, 39–42.

11

Multiple Use of Trans-Himalayan Rangelands: Reconciling Human Livelihoods with Wildlife Conservation

Charudutt Mishra[1,2], *Sumanta Bagchi*[1,3],
Tsewang Namgail[1,4] *and Yash Veer Bhatnagar*[1,2]

[1]Nature Conservation Foundation, Karnataka, India
[2]Snow Leopard Trust, WA, USA
[3]Department of Biology, Syracuse University, NY, USA
[4]Department of Environmental Sciences, Wageningen University, Wageningen, The Netherlands

Introduction

The rain-shadow of the Himalayan Mountains in South and Central Asia constitutes the Trans-Himalayas, a vast rangelands system (2.6 million km^2; Figure 11.1) comprising the Tibetan plateau and its marginal mountains. This high-altitude arid landscape north of the main Himalayan range is contiguous with the Eurasian steppes and has a history of pastoralism dating back at least 3 millennia (Handa 1994; Schaller 1998). Historically, nomadic pastoralism in the region presumably involved low-intensity grazing, much like the Central Asian steppes (Blench & Sommer 1999). The Trans-Himalayas is amongst the least productive of graminoid-dominated ecosystems on earth in terms of above-ground graminoid biomass (Mishra 2001). Yet, the Trans-Himalayan rangelands harbour a surprisingly rich assemblage of wild mountain ungulate

Wild Rangelands: Conserving Wildlife While Maintaining Livestock in Semi-Arid Ecosystems,
1st edition. Edited by J.T. du Toit, R. Kock, and J.C. Deutsch.

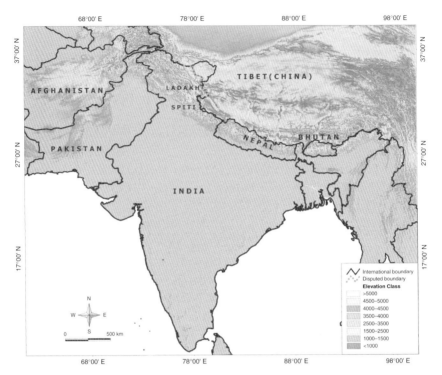

Figure 11.1 **The Trans-Himalayan landscape including the Tibetan plateau and its marginal mountains. The areas north of the Himalayan chain along India's northern boundary (above elevations of 3000 m), extending into Tibet, form the Trans-Himalayan rangelands.**

species and are home to endangered carnivores such as the snow leopard (*Uncia uncia*).

Within India, the Trans-Himalayas is spread over approximately 186,000 km² (Rodgers et al. 2000). In accordance with the country's preservationist policies (national and regional), wherein conservation is essentially viewed as a policing responsibility of the state (Mishra 2000), 8.2% of this region has been designated as wildlife protected areas (Rodgers et al. 2000). These protected areas are, however, by themselves inadequate in conserving Trans-Himalayan wildlife. Compared to the country's other terrestrial ecosystems, where most extant wildlife populations survive inside protected areas, the Himalayan and Trans-Himalayan landscapes are unique in that wildlife populations are not restricted to protected areas here, but occur across the landscape. At the same time, livestock grazing and associated use of rangelands is pervasive across the Trans-Himalayan landscape, including protected areas (Mishra 2001). Today, as local production systems become integrated with national and even global markets, the region is witnessing rapid changes in land use, grazing practices and herd composition (Mishra 2000; Mishra et al. 2003b; Namgail et al. 2007a). Because of these changes, the wildlife in the region faces a variety of threats, and given the unique topographic, climatic, ecological, political and cultural contexts of the Trans-Himalayas, current conservation approaches seem inadequate to address the region's important needs.

In this chapter, we first introduce the Trans-Himalayan rangelands in terms of their floristics and production, and provide an overview of their wildlife value. Then we briefly describe the livestock production systems and outline ongoing socio-economic changes and their consequences for conservation. We review the current understanding of rangeland dynamics and the response of rangeland vegetation to grazing, and outline the impacts of pastoralism on wildlife conservation. We assess the sustainability of the Trans-Himalayan rangelands and their wildlife in the context of the changing climate, socio-economy and land use. Finally, we highlight the need for an alternate approach to conserving wildlife in the Trans-Himalayas, and outline our recent efforts in this direction.

Rangeland vegetation

The Tibetan plateau attained its present average elevation (3500–5500 m) by the Miocene (8 million years ago) and through the Pleistocene and became progressively arid with open steppe vegetation (Harrison et al. 1992). Today, the two important vegetation formations in the region include open or desert steppe dominated by grasses and sedges (e.g. *Stipa, Leymus, Festuca, Carex*) at altitudes of up to 4600 m and dwarf shrub steppes between 4000 and 5000 m dominated by shrubs such as *Caragana, Artemisia, Lonicera* and *Eurotia*. Mesic sites such as river valleys and areas along springs and glaciers are often covered by sedge meadows (*Carex, Kobresia*). Vegetation occurs up to 5200 m, but becomes sparse above 4800 m and is limited to forbs such as *Saussurea* and cushionoid plants such as *Thylacospermum*. The important plant families include Gramineae, Cyperaceae, Brassicaceae, Fabaceae, Ranunculaceae and Leguminoceae.

Plant species richness and composition vary considerably over space and along gradients of altitude, soil moisture and soil texture (Klimes 2003; Rawat & Adhikari 2005). Kachroo et al. (1977) report 611 vascular plants from the Ladakh region ($\sim$90,000 km^2), from 190 genera and 51 families. Rawat and Adhikari (2005) report 232 vascular plants from a 300 km^2 basin in eastern Ladakh belonging to 101 genera and 38 families. Eighty one species including 13 Gramineae and 6 Cyperaceae are reported from a 35 km^2 area around a single village in Spiti region (Mishra 2001). In general, hemicryptophytes (perennial grasses and sedges) and chamaephytes (dwarf forbs and matted shrubs) are the dominant life forms in these rangelands (77% of all plants, Rawat & Adhikari 2005). The soils are generally alkaline (pH 7–9) and the texture varies from sandy to sandy-clay, with 4–6% organic carbon content and 0.1–0.5% total nitrogen (Rawat & Adhikari 2005).

Plant growth is restricted to a short season (May–August) because of low temperatures during the rest of the year and available soil moisture is an important limiting factor for plant growth during the growing season (Mishra 2001). Inter-annual variation in primary production is high (Bai et al. 2004) and seems dependent on precipitation. For instance, aboveground net primary production (ANPP) in Spiti's rangelands was c. 21 g/m^2 ($\pm$48% CV) in 2005, whereas it was 34 g/m^2 ($\pm$43% CV) during 2006 when there was higher precipitation (S. Bagchi, unpublished observations).

Wildlife of the Trans-Himalayan rangelands

The Trans-Himalayan rangelands support 20 species (7 families) of wild herbivores, 13 species (4 families) of wild carnivores (Table 11.1), and over 275 species (41 families) of birds (Pfister 2004). The region has many high-altitude wetlands, which serve as breeding habitats for migratory waterfowl such as Bar-headed Goose *Anser indicus*, Brown-headed Gull *Larus brunnicephalus*, and Black-necked Crane *Grus nigricollis*. Prominent resident birds include the Tibetan Snowcock *Tetraogallus thibetanus*, Tibetan Partridge *Perdix hodgsoniae*, Tibetan Sandgrouse *Syrrhaptes thibetanus*, Golden Eagle *Aquila chrysaetos*, and Lammergeier *Gypaetus barbatus*.

The Trans-Himalayas is one of the few places on earth that continues to support a relatively intact assemblage of Pleistocene large herbivores (Table 11.1) alongside a suite of domestic ungulates (Schaller 1977). Wild ungulates form the most significant group of wildlife in the region that shares

Table 11.1   **List of mammal species recorded in Trans-Himalayan rangelands and their IUCN Red List categories.**

Order/Family	Genus	Species	Common name	IUCN Status
Artiodactyla				
Bovidae				
Subfamily	*Capra*	*C. ibex siberica*	Asiatic ibex	Low risk
Caprinae	*Ovis*	*O. ammon hodgsoni*	Tibetan argali	Near threatened
		O. vignei vignei	Ladakh urial	Endangered
	Pseudois	*P. nayaur*	Bharal	Least concern
	Pantholops	*P. hodgsoni*	Tibetan antelope	Endangered
Subfamily: Antilopinae	*Procapra*	*P. picticaudata*	Tibetan gazelle	Least concern
Subfamily: Bovinae	*Bos*	*B. grunniens*	Wild yak	Vulnerable
Perissodactyla				
Equidae	*Equus*	*E. kiang*	Tibetan wild ass	Low risk
Rodentia				
Sciuridae	*Marmota*	*M. caudata caudata*	Long-tailed marmot	Low risk
		M. bobak himalayana	Himalayan marmot	Low risk
Muridae	*Alticola*	*A. roylei*	Royle's Mountain vole	Low risk
		A. argentatus	Silvery mountain vole	Low risk
		A. stoliczkanus	Stoliczka's mountain vole	Low risk
Lagomorpha				
Leporidae	*Lepus*	*L. oiostolus*	Woolly hare	Low risk
		L. capensis tibetanus	Cape hare	Low risk
Ochotonidae	*Ochotona*	*O. curzoniae*	Plateau pika	Low risk
		O. ladacensis	Ladakh pika	Low risk
		O. macrotis	Large-eared pika	Low risk
		O. nubrica	Nubra pika	Low risk
		O. roylei	Royle's pika	Low risk

(Cont'd)

Table 11.1   (*Continued*)

Order/Family	Genus	Species	Common name	IUCN Status
Carnivora				
Canidae	*Canis*	*C. lupus laniger*	Tibetan wolf	Least concern
	Cuon	*C. alpinus laniger*	Wild dog	Endangered
	Vulpes	*V. vulpes montana*	Red fox	Least concern
	Vulpes	*V. ferrilata*	Tibetan fox	Least concern
Felidae	*Uncia*	*U. Uncia*	Snow leopard	Endangered
	Lynx	*L. lynx isabellinus*	Eurasian lynx	Near threatened
	Otocolobus	*O. manul nigripectus*	Pallas's cat	Near threatened
Ursidae	*Ursus*	*U. arctos isabellinus*	Brown bear	Low risk
Mustelidae	*Lutra*	*L. lutra monticola*	Eurasian otter	Near threatened
	Martes	*M. foina intermedia*	Stone marten	Low risk
	Mustela	*M. altaica temon*	Mountain weasel	Low risk
		M. erminea whiteheadi	Stoat	Low risk
		M. siberica	Himalayan weasel	Low risk

forage resources with livestock and most of these species, including the wild yak *Bos grunniens*, represent an important genetic resource for potential livestock improvement. These mountain ungulates underwent adaptive radiation by evolving ecological and phenotypic diversity in the late Miocene, occupying the mountainous ecological niches created in the aftermath of the collision of the Eurasian and the Indian tectonic plates and the consequent rise of the Himalaya. Within the Indian Trans-Himalayas, there is a preponderance of large herbivores belonging to the tribe Caprini (Table 11.1), which appeared during the late Miocene (Ropiquet & Hassanin 2004).

Trans-Himalayan mountain ungulates are thought to have evolved in sympatry and diverged in their morphology and resource use patterns (Schaller 1977). For example, ibex *Capra ibex* have relatively muscular legs and stocky bodies that help them negotiate steep cliffs when escaping from predators, while the Tibetan argali *Ovis ammon* have longer legs that enable them to outrun predators. Mountain ungulates differ in the use of the terrain: ibex occupy very steep and broken areas, bharal *Pseudois nayaur* prefer rolling areas in the vicinity of cliffs and species such as the Tibetan argali, the Tibetan gazelle *Procapra picticaudata* and kiang *Equus kiang* occur on plateaus (Bhatnagar 1997; Bagchi et al. 2004; Namgail et al. 2004). Most species show limited local migration (up to 10 km) seasonally, often moving to relatively higher altitudes (by ~500 to 1000 m) in summer and to relatively snow-free patches in winter. The chiru *Pantholops hodgsonii* is the only species where, in some populations, females show long-distance latitudinal migration (several hundred kilometres), moving North in summer and returning to lower latitudes in autumn (Schaller 1998). Several species of smaller herbivores such as pikas *Ochotona* spp. and voles *Alticola* spp. also occur in these rangelands (Table 11.1) and seem to play a role in maintaining vegetation diversity at local scales through soil disturbance (Bagchi et al. 2006).

Graminoids form a significant proportion of the diet of most Trans-Himalayan wild and domestic large herbivores (Schaller 1998; Mishra et al. 2002, 2004; Bagchi et al. 2004), though species are known to expand their diet breadth to include forbs and shrubs particularly during winter, which is a period of lean resource availability (Mishra et al. 2004). The only exception, presumably, is the small-sized Tibetan gazelle *P. picticaudata*, feeding predominantly on forbs (Schaller 1998). Given the region's relatively low graminoid biomass, interspecific competition, rather than facilitation, is expected to be the dominant form of interaction amongst Trans-Himalayan grazer species (Mishra 2001). Preliminary evidence on the possible role of competition in structuring this wild grazer assemblage is seen in a morphological pattern within this guild, where a proportional regularity in body masses is evident, with each Trans-Himalayan wild grazer species, on average, being about twice as large as the nearest smaller one (Mishra et al. 2002). Such morphological patterns within species guilds are thought to be brought about by competition through character displacement, which is a co-evolutionary mechanism, and species-sorting, which is an ecological outcome (Dayan & Simberloff 1998; Prins & Olff 1998).

Pastoralism in the Trans-Himalaya

For several millennia, the Trans-Himalayan rangelands have been used by Sino-Tibetan speaking pastoral and agro-pastoral communities of the Mongoloid stock. Within India, the present human population is largely Buddhist. Although the human population density is low ($\sim$1 person/km^2), populations are increasing with the breakdown of traditional systems of polyandry and primogeniture, as well as the influx of Tibetan refugees in some parts (Ahmed 1996; Mishra 2000; Mishra et al. 2003b; Namgail et al. 2007a). Production systems in the Trans-Himalayas include sedentary agro-pastoralism up to altitudes of c. 4500 m, and nomadic pastoralism up to 5200 m. Livestock are owned by individual families, while the herding systems are variable, ranging from individual to co-operative (between a few families) to communal herding managed by village councils. Most of the grazing land is communally owned, though individual families may have usufruct grazing rights. Rotational grazing between pastures is practised and some pastures may be maintained exclusively for winter grazing (Ahmed 1996; Mishra et al. 2003b).

Livestock in the region includes goat, sheep, yak, cattle, horse, donkey and hybrids of yak and cattle, which provide various goods and services including wool, milk, butter, meat, dung for manure, transport and draught. In parts of the Indian Trans-Himalayas, there has been a history of cashmere or *pashmina* production, obtained from the underwool of the local *changra* goat, a trade that is intensifying rapidly today (Jina 1995; Rizvi 1999; Bhatnagar et al. 2006b).

The diversity of livestock species and associated herding practices followed by the Trans-Himalayan pastoralists reduce climatic risks (e.g. avalanches) to livestock and allow a more efficient exploitation of the rangelands (Mishra et al. 2003b). The wide range of body masses of livestock (mean adult body mass ranging from 34 kg for goats to 298 kg for yaks) and the combination of fore-gut and hind-gut fermenters allows the use of a range of forage in terms of plant species and quality (Mishra et al. 2002, 2004).

The current trends in livestock population or biomass densities in the Trans-Himalayan rangelands are variable. At localized scales, livestock biomass densities have declined, remained stable or increased, while at regional scales, there largely appears to be an increase in biomass density over the last few decades (Mishra 2000; Bhatnagar et al. 2006a; Namgail et al. 2007a). A relatively large scale ($\sim$10,000 km^2) overstocking of Trans-Himalayan rangelands is reported (Mishra et al. 2001), though such overstocking may be relatively more common in agro-pastoral systems where forage grown in crop-fields allows livestock populations to be supplement-fed in winter and by off-setting winter starvation mortality, and be maintained above the levels that the range-lands can support (Mishra 2001). Pastoralists tend to maximize herd sizes for several reasons, including maximizing short-term livestock production as well as maintaining herd stability, particularly in areas where they lose livestock to wild carnivores (Mishra et al. 2001).

Even in traditionally purely pastoral Trans-Himalayan areas such as eastern Ladakh, biomass densities of livestock have increased as a consequence of the increase in the number of herding families, increased demands for cashmere, loss of access to traditional pastures that lie across the border in China and influx of Tibetan refugees with their livestock herds (Bhatnagar et al. 2006b; Namgail et al. 2007a). There is also increased access to imported concentrated supplemental feed, as well as the forage that is now being grown locally as a consequence of ongoing sedentarization (Namgail et al. 2007a). This is presumably facilitating a further increase in livestock biomass density in these rangelands.

Grazing impacts and conflicts between pastoralism and wildlife conservation

Available information from the Spiti region of the Trans-Himalayas suggests a consumption of 44–47% of ANPP by livestock and native herbivores during the growth season itself (S. Bagchi 2006, unpublished observations). This level of consumption is comparable with the global average for grass-dominated ecosystems, despite the ANPP in the Trans-Himalayas being two standard deviations below the global average (Milchunas & Lauenroth 1993). These results suggest a relatively high grazing intensity in Trans-Himalayan rangelands and are consistent with observations of widespread overstocking (Mishra et al. 2001) and vegetation degradation in the rangelands (Mishra 2001). Of the total forage removal, the majority is consumed by livestock, given that livestock densities are often up to 10 times greater than wild ungulate densities in these rangelands (Mishra 1997). For instance, the relative extent of forage removal by the kiang – a large-bodied hindgut fermenter (mean adult body mass 275 kg) whose population in eastern Ladakh is believed to be very high – is estimated to be only 3–4%, compared to 96–97% consumption by local livestock (Bhatnagar et al. 2006b).

There is considerable diet overlap between livestock and Trans-Himalayan wild herbivores (Bagchi et al. 2004; Mishra et al. 2004) and a growing body of literature establishes the competitive effects of high-intensity livestock grazing on wild ungulates. Studies have documented both exploitative and interference competition between these groups (Bagchi et al. 2004; Mishra et al. 2004; Namgail et al. 2007c), resulting in population declines of wild ungulates as the livestock density increases (Mishra et al. 2004). Competition with livestock, together with collateral effects such as hunting in a few places, has led to local extinctions and drastic range reductions of Trans-Himalayan wild herbivores (Mishra et al. 2002). For example, the range of the Tibetan gazelle in the Ladakh region, over the last 100 years, has diminished from approximately 20,000 km^2 to less than 100 km^2 today (Bhatnagar et al. 2006a). Table 11.2 illustrates the high variation in species richness seen in Trans-Himalayan catchments, presumably brought about by both intrinsic characteristics of the habitat as well as anthropogenic factors that have led to species' declines. Table 11.2 also suggests that the smallest and largest Trans-Himalayan wild herbivores have been more vulnerable to local extinctions compared to medium-sized ones, which is consistent with extinction patterns reported from other ecosystems (Newmark 1995, 1996).

The high livestock density and associated declines in wild herbivore density in the Trans-Himalayan rangelands presumably has a cascading effect of intensifying the conflict between humans and endangered large carnivores such as the snow leopard and the wolf *Canis lupus* over livestock depredation. Retaliatory persecution in response to livestock losses is one of the most important threats to these carnivores (Mishra et al. 2003a). Given the high relative abundance of livestock when compared to wild ungulates in Trans-Himalayan rangelands, expectedly high levels of livestock depredation are reported, amounting to up to 12% of the livestock holding annually (Mishra 1997; Bagchi & Mishra 2006; Namgail et al. 2007b). Snow leopards seem to have a high dependence on livestock for food – in two Trans-Himalayan wildlife reserves, 40–60% of the snow leopard's diet was comprised of livestock (Bagchi & Mishra 2006). Effectively managing such conflicts in the face of high livestock density is a major challenge, but they will need to be addressed for these endangered large carnivores to be conserved.

In parts of the Trans-Himalayas, a decline in hunting has enabled some mountain ungulates such as the Ladakh urial *Ovis vignei* and the Tibetan argali to recover locally, though the species continue to be threatened by increasing livestock populations in their habitats (Chundawat & Qureshi 1999;

Table 11.2 Wild mountain ungulate species composition in 12 interspersed but isolated watershed catchments (each 200–400 km²) in India's western Trans-Himalayas. Species are arranged along a gradient of body mass from left to right. Although species' presence–absence is governed by inherent habitat characteristics, 6 of the 12 catchments examined have only one extant species, presumably representing community collapse in some areas due to grazing and collateral anthropogenic factors.

Catchment	Species of wild large herbivore							
	Tibetan gazelle (14 kg)	Chiru (32 kg)	Bharal (55 kg)	Urial (70 kg)	Ibex (76 kg)	Tibetan Argali (80 kg)	Kiang (275 kg)	Wild yak (413 kg)
Collapsed assemblages								
Hanle (lower)[a]			P				P	
Spiti (Pin Valley)					P			
Spiti (Tabo)			P					
Nubra (Kuber)					P			
Tso Mo Riri			P					
Kargil (Sangkoo)					P			
Diverse assemblages								
Hanle (Upper)	P		P				P	
Lower Indus (Sham)			P	P	P			
Upper Indus (Demchok)			P			P	P	
Hemis (Rumbak)			P	P	P	P		
Nubra (lower Shyok)				P	P			
Changchenmo		P	P				P	P

[a]P, presence of species.

Namgail et al. 2007c). A peculiar human–wildlife conflict has recently arisen in eastern Ladakh, where another such species, the kiang, is believed to be over-abundant, and is increasingly viewed as a forage competitor for local livestock (Bhatnagar et al. 2006b). A closer examination of this conflict has revealed that kiang densities here are not inflated and are comparable to those reported from Tibet (Schaller 1998). The misplaced perceptions of the over-abundance of kiang seem to have arisen, ironically, after its populations recovered over two decades following a drastic decline in the 1960s when a war was fought in the region between China and India (Bhatnagar et al. 2006b).

The future of Trans-Himalayan mountain ungulates remains uncertain. All species are potentially threatened by high-intensity livestock grazing, and of the eight extant species, only two have estimated populations of more than 10,000 within India, while four species number less than 500 individuals (Johnsingh et al. 2006). Most large mammal species are potentially threatened by livestock diseases. This is a little-understood but important conservation issue in the Trans-Himalayan rangelands, given that diseases such as *Peste des petits ruminants* and foot-and-mouth are increasingly being reported in the region's livestock (Bhatnagar et al. in press).

A well-documented consequence of overgrazing in water-limited ecosystems around the world is the catastrophic shifts in vegetation, rather than smooth successional changes, and the prevalence of stably degraded vegetation states (van de Koppel & Reitkerk 2000). The mechanism behind such threshold effects and discontinuous changes in vegetation is thought to be the interaction of grazing with a positive-feedback mechanism (habitat self-improvement by plants) that exists between vegetation and soils (van de Koppel et al. 1997). These feedback mechanisms lead to self-organized pattern formation in the vegetation (Klausmeier 1999; HilleRisLambers et al. 2001; Rietkerk et al. 2004); Trans-Himalayan rangelands commonly show such patterned vegetation (Mishra 2001). Vegetation in these rangelands can therefore be vulnerable to catastrophic shifts. Against this background, further research into vegetation dynamics in the Trans-Himalayas is critical, since the possibility of stably degraded vegetation states implies that even several years of protection following degradation due to overgrazing is unlikely to allow the vegetation to recover to original, more productive states. Such research would be critical to designing grazing policies that address the resilience of these rangelands, particularly during droughts.

A related aspect that also needs to be better understood is the response of these rangelands to climate change. Reports indicate that mean annual temperatures on the Tibetan plateau have been increasing at 0.16°C per decade since the 1950s and winter temperatures at 0.32°C per decade (Liu & Chen 2000), which can greatly influence local hydrological cycles. Average annual potential evapotranspiration (PET) has been declining since the 1960s (Shenbin et al. 2006), which may have a favourable impact on vegetation growth. At the same time, however, PET during the plant growth season seems to be increasing (Shenbin et al. 2006). The impact of the resultant shrinking glaciers and lakes on hydrological cycles remains unclear. Although simulations predict that climate change would increase this ecosystem's primary production, the expanse of shrub-steppe biome is likely to decline (Ni 2000). Coupled with such climate-induced changes are various socio-economic and herding-practice shifts (Mishra 2000; Mishra et al. 2003b; Namgail et al. 2007a) that collectively pose many challenges towards sustainable management of these rangelands for wildlife conservation as well as for pastoral livelihoods.

Towards better conservation management in Trans-Himalayan rangelands

Wildlife habitats and populations in most of India's terrestrial ecosystems today largely survive within wildlife protected areas that are often isolated and surrounded by rural and urban landscapes (Figure 11.2a). Wherever possible, these protected areas are further divided into core areas, where human presence and resource use are completely curtailed, and buffer areas, where, at least theoretically, some regulated human use is allowed. Following a similar approach, relatively large protected areas have also been declared on paper in the Trans-Himalayas, with an average size of 3035 km^2 compared to the national average of 267 km^2 (Rodgers et al. 2000). However, human use in the form of intensive livestock grazing and associated activities continues unabated in them and there are no truly inviolate core areas (Figure 11.2b), although they exist on paper in some cases. Furthermore, up to 30–40% of the land in some Trans-Himalayan parks is composed of areas that have hydrological and other kinds of importance but little biological value (such as permafrost areas and glaciers).

The Trans-Himalayas has traditionally been viewed as a low-productive ecosystem where wildlife inherently occurs in low densities and a region where Buddhist communities (again at very low densities) live in harmony with wildlife (see Mishra 2001 for a detailed discussion). Given that most

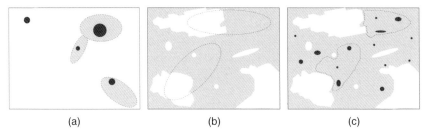

(a)   (b)   (c)

Figure 11.2 **A schematic representation of wildlife management in India. (a) The majority of India's terrestrial landscapes, where wildlife persists largely in insular protected areas, are further divided into core (no anthropogenic use; dark areas in the figure) and buffer zones (regulated anthropogenic use; grey in figure), surrounded by rural and urban landscapes (white). (b) In the Trans-Himalayas, where there is a near-complete absence of 'core' areas, but often-depleted wildlife populations persist across the entire landscape except permafrost areas (irregular white) and larger human settlements (white circles). (c) A more effective framework in the Trans-Himalayas would be to follow a landscape-level approach where each landscape unit is either a core or a buffer unit, with specific multiple-use objectives for each unit in the latter group. The protected area boundaries will need to be realigned to exclude white areas.**

of the landscape is grazed, ecological benchmarks that could show how the vegetation and wildlife would appear in the absence of livestock grazing are completely lacking (Mishra et al. 2002). Consequently, livestock grazing was not considered a serious conservation issue in this landscape until recent studies were undertaken, and this perhaps also explains the lack of serious efforts to establish core zones in Trans-Himalayan protected areas. Given people's traditional land use, lack of alternatives and a continued dependence on the landscape, establishing large core zones is also difficult. Thus, 'core' areas are missing in the Trans-Himalayas, while the 'buffer' areas, which form the bulk of the landscape today, have seen wildlife declines and species extinctions.

Nevertheless, the Trans-Himalayan (and higher Himalayan) landscape is unique when compared to most other terrestrial ecosystems in India, given that the bulk of the area comprises 'buffer' landscape units that continue to harbour wildlife populations, albeit in depleted states (Figure 11.2b). Keeping these realities in mind, we propose a slightly different conceptual framework for wildlife management and multiple use in the Trans-Himalayan landscape, which may have relevance for other rangeland and mountain systems as well.

Wildlife management in the Trans-Himalayas needs to look beyond protected areas and follow a larger landscape level approach, wherein, the first step would be to undertake biologically and socially meaningful landscape zonation (Bhatnagar et al. in press). This would involve identification of important landscapes (generally large spatial scales; $>1000\,\mathrm{km}^2$) and setting the management objectives for each landscape unit within the larger landscape (smaller spatial scales; ~ 10–$100\,\mathrm{km}^2$), based on its relative importance for wildlife conservation and human use. Each landscape unit may be demarcated based on a combination of geological, ecological and administrative characteristics. Such an exercise needs to be undertaken within as well as outside the existing parks, and on both government and community-owned lands.

Within this matrix of landscape units, based on zonation, a set of 'core' units needs to be established, interspersed among a series of 'buffer' landscape units, with each of the latter group having a variable set of multiple-use objectives. The guiding principles underlying the management objectives for wildlife populations for this mosaic of landscape units can be as follows:

1. In core landscape units, management objectives should aim to maintain wildlife populations (N_c) at carrying capacity (K) over the long term, enable conditions where birth rates (b_c) exceed rates of mortality (m_c) and rates of emigration (e_c) are considerably higher than immigration rates (i_c) to enable spill-over effects, that is,

$$N_c \approx K,\ b_c > m_c,\quad \text{and}\quad e_c \gg i_c$$

2. For each buffer landscape unit, it is at least conceptually important to estimate the desirable wildlife population size (N_b) – which will be a function of the trade-off between conservation and rangeland use objectives – and ensure that populations are maintained around that level:

$$N_b = K - f(A),\quad \text{and}\quad b_b + i_b \geq m_b + e_b$$

where $f(A)$ is a function by which the wildlife population size is reduced below carrying capacity as a result of an acceptable level of human anthropogenic pressure for each landscape unit.

The size and number of core landscape units, wherever feasible, should be large and adequately interspersed within a matrix of buffer landscape units to enable the conservation of viable wildlife populations. At a minimum, the coupled landscape-level guiding principle for core and buffer units should be to aim for the total spill-over from core units to at least offset the net individuals lost from buffer units due to mortality and emigration, that is,

$$\sum N_c(e_c - i_c) \geq \sum N_b(b_b - m_b - e_b)$$

Illustratively, this means that as the livestock grazing intensity in a buffer or multiple-use landscape unit increases, one can expect a decline in the density of wild ungulates. A stated desirable wild ungulate density can help guide the management of livestock grazing and vice-versa within any given buffer unit. As the need for pastoral production in any landscape unit increases within sustainable limits, it will need to be counter-balanced by the need to establish a core unit in the proximity such that the inequality condition above continues to hold. This dynamic management approach will also facilitate grazing and wildlife management that is sensitive to inter-annual as well as long-term climatic variation. Our preliminary efforts at creating such core landscape units at experimental scales in the Trans-Himalayas have shown considerable wildlife recovery and spill-over effects within a span of 3–4 years in even small-sized (5–20 km^2) core areas (Mishra et al. 2003b, C. Mishra 1998–2007, unpublished observations). In a landscape where most of the area represents potential wildlife habitat and is simultaneously subjected to widespread anthropogenic resource use, this suggests that even small-sized core areas, which are more feasible, would be immensely valuable for long-term wildlife conservation as long as their numbers satisfy the inequality criterion specified above.

In such a management system, the need to have large-sized protected areas diminishes. At the same time, the legal implications of existing protected areas – in particular, for preventing land diversion to environmentally damaging large-scale developmental projects (such as big dams) that are a serious threat to biodiversity in the Himalayan landscape (e.g. Menon et al. 2003) – cannot be ignored. It may, therefore, be prudent to maintain the existing Trans-Himalayan protected areas but manage them within the larger landscape level, dynamic framework outlined here. The protected area boundaries, however, will nevertheless need to be realigned to exclude regions that have less biological value, but form considerable parts of existing Trans-Himalayan protected areas (Figure 11.2c).

In addition to adopting this broad framework for management of Trans-Himalayan rangelands, much needs to be done regarding human–wildlife conflicts. Flexible and community-based management of human–wildlife conflicts and providing incentives to local communities for conservation will be critical (e.g. Mishra et al. 2003b). There are other needs, such as putting in place adequate veterinary health care, scientifically well-informed regional, landscape-level and local grazing guidelines and policies, habitat restoration and species recovery programmes (e.g. Bhatnagar et al. 2007; Namgail et al. 2008).

The importance of constantly generating scientific knowledge on wildlife ecology and human society is particularly underscored in the Trans-Himalayas, where, until just a few years ago, it was believed that the region did not face any major conservation issues and that there was harmony between pastoral production and wildlife conservation (Mishra 2001). It is therefore critical that an adaptive framework for wildlife management be followed, which actively supports research and monitoring, and constantly incorporates ecological and social feedback into management planning at the landscape and landscape unit levels.

To address these issues, since 2004, we have been working with the concerned state and central governments to develop, along these lines, a new scheme and policy for the conservation of Himalayan and Trans-Himalayan rangelands and their wildlife, known as *Project Snow Leopard* (Anon. 2008). We are hopeful that Project Snow Leopard, whose broad objectives are outlined in Table 11.3, became operational in 2009. We would like to re-emphasize that most conservation efforts in landscapes used by people carry conservation costs to indigenous communities, and it is critical that future conservation initiatives be undertaken with their consent, support and participation. This is particularly important in the Trans-Himalayan landscape, where conventional large protected areas are difficult to establish and to manage effectively. Active participation of local communities is inherent to the framework for the management of Trans-Himalayan rangelands we have proposed here, which is essentially a combination of protectionist and sustainable use conservation paradigms (Mishra et al. 2003b). We believe that such a landscape-level, knowledge-based, adaptive, participatory, and dynamic conservation framework will be a major step in securing the conservation of wildlife and the sustainability of Trans-Himalayan rangelands.

Table 11.3 **Broad objectives of Project Snow Leopard, a new conservation policy for India's Himalayan and Trans-Himalayan landscapes being currently proposed.**

Objectives

1. Facilitate a landscape-level approach to wildlife conservation
2. Rationalize the existing protected area network and improve protected area management
3. Develop a framework for wildlife conservation outside protected areas and promote ecologically responsible development
4. Encourage focused conservation and recovery programmes for endangered species such as the snow leopard
5. Promote stronger measures for wildlife protection and law enforcement
6. Promote better understanding and management of human–wildlife conflicts
7. Restore degraded landscapes in the Himalayan and Trans-Himalayan biogeographic regions
8. Promote a knowledge-based approach to conservation and an adaptive framework for wildlife management
9. Reduce existing anthropogenic pressures on natural resources
10. Promote conservation education and awareness

Acknowledgements

We gratefully acknowledge the Whitley Fund for Nature and the Ford Foundation for providing core support to our academic and conservation programmes. We thank R. Raghunath for help in preparing the map.

References

Ahmed, M. (1996) *We are Warp and Weft: Nomadic Pastoralism and the Tradition of Weaving in Rupshu.* PhD Dissertation, Oxford University, United Kingdom.

Anon. (2008) *The Project Snow Leopard.* Ministry of Environment & Forests, Government of India, New Delhi.

Bagchi, S. & Mishra, C. (2006) Living with large carnivores: predation on livestock by the snow leopard (*Uncia uncia*). *Journal of Zoology* 268, 217–224.

Bagchi, S., Mishra, C. & Bhatnagar, Y.V. (2004) Conflicts between traditional pastoralism and conservation of Himalayan ibex (*Capra sibirica*) in the Trans-Himalayan mountains. *Animal Conservation* 7, 121–128.

Bagchi, S., Namgail, T. & Ritchie, M.E. (2006) Small mammalian herbivores as mediators of plant community dynamics in the high-altitude arid rangelands of Trans-Himalaya. *Biological Conservation* 127, 438–442.

Bai, Y., Han, X., Wu, J., Chen, Z. & Li, L. (2004) Ecosystem stability and compensatory effects in Inner Mongolia grassland. *Nature* 431, 181–184.

Bhatnagar, Y.V. (1997) *Ranging and Habitat Utilization by the Himalayan Ibex (*Capra ibex sibirica*) in Pin Valley National Park*. PhD Dissertation, Saurashtra University, India.

Bhatnagar, Y.V., Mishra, C. & McCarthy, T. (in press) Protected areas and beyond: wildlife conservation in the Trans-Himalaya. Proceedings of the Journal of the Bombay Natural History Society Centenary Seminar.

Bhatnagar, Y.V., Mishra, C. & Wangchuk, R. (2006a) Decline of the Tibetan gazelle in Ladakh. *Oryx* 40, 229–232.

Bhatnagar, Y.V., Wangchuk, R., Prins, H.H.T., van Wieren, S.E. & Mishra, C. (2006b) Perceived conflicts between pastoralism and conservation of the kiang *Equus kiang* in the Ladakh Trans-Himalaya, India. *Environmental Management* 38, 934–941.

Bhatnagar, Y.V., Seth, C.M., Takpa, J. et al. (2007) A strategy for conservation of Tibetan gazelle *Procapra picticaudata* in Ladakh. *Conservation and Society* 5, 262–276.

Blench, R. & Sommer, F. (1999) *Understanding Rangeland Biodiversity*, Working Paper 121. Overseas Development Institute, London.

Chundawat, R.S. & Qureshi, Q. (1999) *Planning Wildlife Conservation in Leh and Kargil Districts of Ladakh, Jammu and Kashmir*. Wildlife Institute of India, Dehradun.

Dayan, T. & Simberloff, D. (1998) Size patterns among competitors: ecological character displacement and character release in mammals, with special reference to island populations. *Mammal Review* 28, 99–124.

Handa, O.P. (1994) *Tabo Monastery and Buddhism in Trans-Himalaya: A Thousand Years of Existence of the Tabo Chos-Khor*. Indus Publishing Company, Shimla.

Harrison, T., Copeland, P., Kidd, W. & An, Y. (1992) Raising Tibet. *Science* 255, 1663–1670.

HilleRisLambers, R., Rietkerk, M., Prins, H.H.T., van den Bosch, F. & de Kroon, F. (2001) Vegetation pattern formation in semi-arid grazing systems. *Ecology* 82, 50–61.

Jina, P.S. (1995) *High Pasturelands of Ladakh Himalaya*. Indus Publishing Company, New Delhi.

Johnsingh, A.J.T., Mishra, C. & Bhatnagar, Y.V. (2006) Conservation status and research of mountain ungulates in India. Fourth World Conference on Mountain Ungulates, Munnar, Kerala, India, 6–8.

Kachroo, P., Sapru, B.L. & Dhar, U. (1977) *Flora of Ladakh: An Ecological and Taxonomic Appraisal*. Bishen Singh Mahendra Pal Singh, Dehradun.

Klausmeier, C.A. (1999) Regular and irregular patterns in semiarid vegetation. *Science* 284, 1826–1828.

Klimes, L. (2003) Life-forms and clonality of vascular plants along an altitudinal gradient in E Ladakh (NW Himalayas). *Basic and Applied Ecology* 4, 317–328.

van de Koppel, J. & Reitkerk, M. (2000) Herbivore regulation and irreversible vegetation change in semi-arid grazing systems. *Oikos* 90, 253–260.

van de Koppel, J., Rietkerk, M. & Weissing, F.J. (1997) Catastrophic vegetation shifts and soil degradation in terrestrial grazing systems. *Trends in Ecology & Evolution* 12, 352–356.

Liu, X. & Chen, B. (2000) Climatic warming in the Tibetan plateau during recent decades. *International Journal of Climatology* 20, 1729–1742.

Menon, M., Vagholikar, N., Kohli, K. & Fernandes, A. (2003) Large dams in the north-east – a bright future? *Ecologist Asia* 11, 3–8.

Milchunas, D.G. & Lauenroth, W.K. (1993) Quantitative effects of grazing on vegetation and soils over a global range of environments. *Ecological Monographs* 63, 327–366.

Mishra, C. (1997) Livestock depredation by large carnivores in the Indian trans-Himalaya: conflict perceptions and conservation prospects. *Environmental Conservation* 24, 338–343.

Mishra, C. (2000) Socioeconomic transition and wildlife conservation in the Indian trans-Himalaya. *Journal of the Bombay Natural History Society* 97, 25–32.

Mishra, C. (2001) *High Altitude Survival: Conflicts between Pastoralism and Wildlife in the Trans-Himalaya*. PhD Dissertation, Thesis, Wageningen University, The Netherlands

Mishra, C., Allen, P., McCarthy, T., Madhusudan, M.D., Bayarjargal, A. & Prins, H.H.T. (2003a) The role of incentive programs in conserving the snow leopard. *Conservation Biology* 17, 1512–1520.

Mishra, C., van Wieren, S.E. & Prins, H.H.T. (2003b) Diversity, risk mediation, and change in a Trans-Himalayan agropastoral system. *Human Ecology* 31, 595–609.

Mishra, C., Prins, H.H.T. & van Wieren, S.E. (2001) Overstocking in the Trans-Himalayan rangelands of India. *Environmental Conservation* 28, 279–283.

Mishra, C., van Wieren, S.E., Heitkonig, I.M.A. & Prins, H.H.T. (2002) A theoretical analysis of competitive exclusion in a Trans-Himalayan large-herbivore assemblage. *Animal Conservation* 5, 251–258.

Mishra, C., van Wieren, S.E., Ketner, P., Heitkonig, I.M.A. & Prins, H.H.T. (2004) Competition between livestock and bharal *Pseudois nayaur* in the Indian Trans-Himalaya. *Journal of Applied Ecology* 41, 344–354.

Namgail, T., Bagchi, S., Bhatnagar, Y.V. & Mishra, C. (2008) Distributional correlates of the Tibetan gazelle *Procapra picticaudata* in Ladakh, northern India: towards a recovery programme. *Oryx* 42, 107–112.

Namgail, T., Fox, J.L. & Bhatnagar, Y.V. (2004) Habitat segregation between sympatric Tibetan argali *Ovis ammon hodgsoni* and blue sheep *Pseudois nayaur* in the Indian Trans-Himalaya. *Journal of Zoology* 262, 57–63.

Namgail, T., Bhatnagar, Y.V., Mishra, C. & Bagchi, S. (2007a) Pastoral nomads of the Indian Changthang: production system, landuse and socioeconomic changes. *Human Ecology* 35, 497–504.

Namgail, T., Fox, J.L. & Bhatnagar, Y.V. (2007b) Carnivore-caused livestock mortality in Trans-Himalaya. *Environmental Management* 39, 490–496.

Namgail, T., Fox, J.L. & Bhatnagar, Y.V. (2007c) Habitat shift and time budget of the Tibetan argali: the influence of livestock grazing. *Ecological Research* 22, 25–31.

Newmark, W.D. (1995) Extinction of mammal populations in western North American national parks. *Conservation Biology* 9, 512–526.

Newmark, W.D. (1996) Insularization of Tanzanian parks and the local extinction of large mammals. *Conservation Biology* 10, 1549–1556.

Ni, J. (2000) A simulation of biomes on the Tibetan Plateau and their responses to global climate change. *Mountain Research and Development* 20, 80–89.

Pfister, O. (2004) *Birds and Mammals of Ladakh*. Oxford University Press, New Delhi.

Prins, H.H.T. & Olff, H. (1998) Species-richness of African grazer assemblages: towards a functional explanation. In: Newberry, D.M., Prins, H.H.T. & Brown, N.D. (eds.) *Dynamics of Tropical Communities*, British Ecological Society Symposium 37. Blackwell Science, Oxford, pp. 449–490.

Rawat, G.S. & Adhikari, B.S. (2005) Floristics and distribution of plant communities across moisture and topographic gradients in Tso Kar basin, Changthang plateau, eastern Ladakh. *Arctic, Antarctic, and Alpine Research* 37, 539–544.

Rietkerk, M., Dekker, S.C., de Ruiter, P.C. & van de Koppel, J. (2004) Self-organized patchiness and catastrophic shifts in ecosystems. *Science* 305, 1926–1929.

Rizvi, J. (1999) *Trans-Himalayan Caravans: Merchant Princes and Peasant Traders in Ladakh*. Oxford University Press, New Delhi.

Rodgers, W.A., Panwar, H.S. & Mathur, V.B. (2000) *Wildlife Protected Area Network in India: A Review (Executive Summary)*. Wildlife Institute of India, Dehradun.

Ropiquet, A. & Hassanin, A. (2004) Molecular phylogeny of caprines (Bovidae, Antilopinae): the question of their origin and diversification during the Miocene. *Journal of Zoological Systematics* 43, 49–63.

Schaller, G.B. (1977) *Mountain Monarchs: Wild Goat and Sheep of the Himalaya*. University of Chicago Press, Chicago.

Schaller, G.B. (1998) *Wildlife of the Tibetan Steppe*. University of Chicago Press, Chicago.

Shenbin, C., Yunfeng, L. & Thomas, A. (2006) Climatic change on the Tibetan Plateau: potential evapotranspiration trends from 1961–2000. *Climatic Change* 76, 291–319.

12

Herders and Hunters in a Transitional Economy: The Challenge of Wildlife and Rangeland Management in Post-socialist Mongolia

Katie M. Scharf[1], *María E. Fernández-Giménez*[2], *Batjav Batbuyan*[3] *and Sumiya Enkhbold*[4]

[1]Department of History, Yale University, CT, USA
[2]Department of Forest, Rangeland, and Watershed Stewardship, Colorado State University, CO, USA
[3]Center for Nomadic Pastoralism Studies, Mongolia
[4]Swiss Agency for Cooperation and Development, Mongolia

Introduction: the global significance of Mongolian pastoralism and biodiversity

Mongolia is one of the most pastoral nations on earth. Sandwiched between Siberia to the North and China to the South, its landscape is dominated by rangelands, which cover 83% of the country's 1.29 million km². Mongolia also has the lowest population density of any country in the world, with a higher proportion of its population living in rangeland areas (77% of 2.67 million people) than any other range-dominated country (Thornton et al. 2002; Kerven 2006). For centuries, semi-nomadic herders have practised extensive semi-nomadic livestock husbandry here, herding flocks of sheep,

Wild Rangelands: Conserving Wildlife While Maintaining Livestock in Semi-Arid Ecosystems,
1st edition. Edited by J.T. du Toit, R. Kock, and J.C. Deutsch.
© 2010 Blackwell Publishing

goats, cattle, horses, camels and yaks. Herding households move with their livestock from pasture to pasture in a pattern of seasonal moves adapted to the dry, continental climate. To appreciate the vulnerability of the steppe ecosystem and the comparative health of the Mongolian grasslands, one need only look across the border, where the steppe extends into Russia (Buryatia) or China (Inner Mongolia). There, sedentary populations raise 'improved' livestock breeds on cultivated fodder in dense settlements, causing erosion and desertification (Humphrey & Sneath 1999).

Mongolia's historically low population density and relative economic isolation have made it the last stronghold of many species of wildlife that are rare, endangered or declining elsewhere in their range. Here, it is possible to see the grey wolf (*Canis lupus*), red fox (*Vulpes vulpes*) and corsac fox (*Vulpes corsac*). Mounds of earth mark the burrows of marmots (*Marmota sibirica* and *Marmota baibacina*). Raptors, like the saker falcon (*Falco cherrug*), golden and steppe eagle (*Aquila chryseatos* and *Aquila nipalensis*) are a common sight. Mongolian gazelles (*Procapra gutterosa*) wander the steppe in one of the world's last great ungulate migrations. Huge salmon-like taimen (*Hucho taimen*) course through undammed rivers. The desert expanses harbour endangered wildlife such as wild Bactrian camels (*Camelus bactrianus*), Gobi bears (*Ursus arctos gobiensis*), Asiatic wild ass (*Equus hemionus*) and saiga antelope (*Saiga tatarica mongolica*). Taiga forests in northern Mongolia shelter red deer (*Cervus elaphus*), roe deer (*Capreolus pygargus*) and wild boar (*Sus scrofa*). Snow leopards (*Uncia uncia*), argali sheep (*Ovis ammon*) and ibex (*Capra sibirica*) roam the mountains in the western and southern parts of the country.

Today, the management of Mongolia's rangelands and wildlife has become a matter of global concern. Since the country began its transition from a centrally planned to a free-market economy in the early 1990s, threats to its pastoral economy and its biodiversity have multiplied. Herders have found it difficult, risky and in some cases even impossible to rely exclusively on livestock husbandry for their livelihoods. At the same time, the transition opened Mongolia's borders to Chinese markets for wild furs, game meat and medicinal products. Many Mongolians began to supplement their incomes by engaging in the wildlife trade, resulting in a catastrophic decline in rare and abundant wildlife populations. Marmots and argali sheep numbers dropped by an estimated 75%; red deer by 92% (Wingard & Zahler 2006). In addition to destroying biodiversity and disturbing the ecological balance of Mongolia's rangeland ecosystem, unsustainable hunting added to the insecurity of its pastoral economy, as some herders lose an important supplementary source of food, clothing and cash income.

From a management perspective, regulation of the pastoral economy and the wildlife trade pose similar challenges. Mongolia is unique among other countries with pastoral regimes in that its vast rangeland (including wildlife habitat) continues to be state owned (Fratkin & Mearns 2003). Local officials bear the primary responsibility for enforcing seasonal movement between pastures, limiting stocking rates, issuing hunting licenses and restricting the wildlife harvest to specific seasons, quotas and geographic areas, but they are typically understaffed, underpaid and poorly equipped. In the absence of clearly defined property rights and effective government regulation, access to both wildlife and forage is negotiated through an overlapping system of formal rights and informal practices. A herder's ability to hunt or trap is influenced by evolving norms relating to property, pasture and protected areas, and by a dynamic relationship between the herder, his neighbours, geography, the market and various agencies of the state. None of these institutions – formal or informal, alone or in combination – is capable of fully regulating hunting or pasture use. The consequences (in some areas) are a vicious cycle of declining mobility, increasing out-of-season and year-round grazing of seasonal pastures, escalating conflicts among herders and the widespread, unsustainable harvest of wildlife (Fernández-Giménez 2002). Before discussing strategies for breaking out of that cycle, we will first put this crisis in context by surveying the history of natural resource management in Mongolia.

Mongolian politics and natural resource management in historical perspective

For centuries, the challenge of rangeland management in Mongolia has been to govern a pastoral system of mobile, largely self-sufficient pastoralists within a political system based on fixed, territorial units (Lattimore 1940, 1951 reprint). While other states have exerted control over their nomadic subjects by sedentarizing them in villages and towns, Mongolia's ruling elites imposed spatial restrictions on herder mobility to facilitate taxation and conscription (Scott 1998; Fernández-Giménez 1999). For example, Chinggis Khan was the first to assign Mongolia's mobile tribal groups to fixed geopolitical units, dividing tribes and pasture lands into fiefdoms led by his political allies. Thereafter, pasture use was governed by customary law and formal regulation, which was imposed by secular princes or – after the reintroduction of the Tibetan Buddhist church in the late 1500s – by high-ranking lamas (Fernández-Giménez 1999). When Mongolia submitted to rule by China's Qing Dynasty in 1691, Qing colonial governors further sub-divided these fiefdoms into 100 military territorial units called *khoshuun*. Herders' access to pastures within the *khoshuun* was administered by des-ignated nobles or monasteries; migration between *khoshuun* was strictly forbidden.

The founding of the Mongolian People's Republic in 1924 undermined this system. Communist revolutionaries destroyed monasteries and – over fierce resistance – eventually established herding collectives (called *negdels*) in the 1950s (Bawden 1968; Fernández-Giménez 1999). The *negdel* system marked the height of state intervention in rangeland management. *Khoshuun* units were reorganized into 21 *aimags* (each one the equivalent of a state or province), which were subdivided into a total of 300 *soum* administrative units (each composed of 1000–2000 households), which were themselves partitioned into smaller subdivisions called *bags* (100–300 households). As *negdel* members, herders tended single-species flocks of state-owned livestock on pasture allocated by local administrators. Buoyed by finan-cial support from the Soviet Union, Mongolia's socialist government supplied *negdel* members with a salary, winter shelters, mechanical wells, emergency fodder, veterinary support and specialized animal breeding services, as well as transport to distant 'reserve' pastures during natural disasters (Mearns 1996; Fernández-Giménez 1999).

The breakup of the Soviet Union in the early 1990s precipitated the collapse of Mongolia's socialist government, and along with it, the *negdel* system. Mongolia's newly democratic government dismantled the *negdels* in 1992, discontinuing subsidies and direct supervision of the herding sector. Livestock, tractors and other *negdel* assets were privatized. At the same time, however, the new democratic regime preserved the legal basis for continued state intervention in rangeland management. For example, the Law on Land, approved by the Mongolian Parliament in 1994, retained state ownership of pastureland and gave local *soum* government officials formal powers to regulate the possession and use of campsites and key seasonal pastures. *Soum* governors were empowered to confer the so-called 'collective use' rights on groups of households (*khot ail*) or small administrative units (*bags*). The Land Law was amended in 2002 to allow individual households to obtain certificates of possession for critical winter and spring campsites and to authorize *soum* officials to impose fines for out-of-season grazing in winter and spring pastures.

In practice, implementation of these provisions has been uneven. In 1999, we observed that many *soum* governments had issued certificates of possession for campsites to household groups in the name of individual household heads. This practice appeared to confer collective use rights over campsites and adjacent pasture, but it threatened to marginalize poorer herding households whose names did not appear on certificates (Fernández-Giménez & Batbuyan 2004). In a survey of *aimag* land management that officials conducted in 7 of the 21 *aimags* in 2005, possession contracts over campsites used in winter and spring had been widely allocated, while possession of pasture remained unauthorized by law (Fernández-Giménez & Khishigbayar 2005). In an area we have studied in depth (Bayankhongor *aimag* in southwestern Mongolia), local officials have yet to exercise their legal authority to regulate management of seasonal pastures by enforcing seasonal movements and limiting stocking rates (Fernández-Giménez & Batbuyan 2004; Fernández-Giménez et al. 2006). In sum, Mongolia's rangelands are still formally owned by the central state, but weakly managed by local (*soum*) governments. Herders have shouldered the burden of production inputs and risks largely on their own, without ready access to transportation, markets and social services formerly provided by the collectives and with uncertain tenure over winter and spring campsites and other key pastoral resources, such as hayfields, wells and natural water sources.

The history of wildlife management in Mongolia follows a similar pattern. Generally speaking, the same territorial limitations used to govern mobile pastoralists have also been applied to wildlife management. Just as Chinggis Khan united tribes and pasture into fixed political units, he also set aside Mongolia's first hunting reserves, limiting their use to certain groups and imposing penalties to deter poachers. Bogd Khan Mountain (located just South of Mongolia's capital city, Ulaanbaatar) was placed under protection as a sacred and natural site in 1771, making it the world's first national park (Reading et al. 1999). By that time, Mongolian hunters were already participating in a trans-border wildlife trade. Russian fur traders set up shops in Ulaanbaatar in the late 1600s. As Mongolia came under Chinese (Qing Dynasty) rule in the seventeenth century, Qing emperors demanded annual tribute in the form of thousands of sable, lynx, red fox or squirrel pelts (Avirmed 1999). Chinese fur traders arrived in the 1720s to counter the Russian trading presence, opening shops near monasteries and *khushuu* offices. Wildlife products became an alternate form of currency, as herders exchanged furs and antlers for food and household supplies, including guns, bullets, gunpowder and traps (Sanjdorj 1980; Bergholz 1993). The volume of exported Siberian marmot furs peaked at about 2.5 million in 1911, the year of the Qing Dynasty's collapse (Wingard & Zahler 2006).

In the 1920s, communist revolutionaries began to assert greater state control over hunting and the wildlife trade, just as they did over livestock husbandry. Resolutions passed in 1924 and 1926 required licenses for hunting and prohibited certain techniques, such as the use of vehicles and military weapons to harvest Mongolian gazelles. In 1927, a state corporation obtained exclusive rights to purchase and export marmot and other furs (Avirmed 1999). When the Japanese occupation of Manchuria (1931) and Inner Mongolia (1937) interrupted Mongol–Chinese trade, Soviet Russia became the chief importer of Mongolian wildlife products (Bawden 1968). Between 100,000 and 150,000 gazelles (out of a population estimated in 1940 to be about 1 million) were hunted during World War II to provision Soviet troops; 2.49 million marmots were harvested in 1947 alone. After the war, the Mongolian government organized a network of hunting brigades with exclusive rights to harvest wildlife and guard against illegal hunters. Like the *negdels*, hunting brigades were administered by local governments and were outfitted with vehicles to conduct regular patrols. Twenty-five thousand people participated nationwide, hunting a variety of furbearing species for export to the Soviet Union and other socialist states (Scharf & Enkhbold 2002).

With the shift to a market economy in the early 1990s, the hunting brigades quickly disintegrated for lack of state support (Wingard & Zahler 2006). The Mongolian–Chinese border opened to private trade for the first time in decades, placing hunters in formerly remote areas in closer proximity to market demand for furs and game meat in increasingly affluent China. The combination of economic hardship and market access encouraged both urban and rural dwellers to enter the wildlife trade as an alternative source of income. The price of marmot skins surged in 1995, the same year that Mongolia's gun control laws were relaxed (Scharf & Enkhbold 2002). Markets in provincial centres swelled with guns, bullets, traps, furs, meat, antlers and organs used for medicinal purposes. In a nationwide study of Mongolia's wildlife trade conducted in 2005, Wingard and Zahler estimated that between 220,000 and 250,000 Mongolians were hunting wildlife for subsistence and the commercial market. The value of the (mostly illegal) trade in wildlife products was estimated at $100 million. This figure is slightly higher than the estimated value of exports (both legal and illegal) of cashmere, the most important 'cash crop' produced by Mongolia's pastoral economy. In short, hunters adapted quickly to the new market for wildlife products in the 1990s, just as herders adjusted to livestock privatizations. As the next section will show, pasture and wildlife management regimes have been unable to keep pace.

Managing hunting and herding in Mongolia today: common problems, common solutions?

Today, while Mongolia's pastureland remains the property of the state, no strong regulatory institutions – formal or informal – have emerged to take the place of the *negdels* and hunting brigades in governing the use of pasture and wildlife resources. In the herding context, customary and informal institutions, weakened during the collective era, have been unable to assume the functions of the *negdel*, such as mitigating trends of year-round grazing of pastures that formerly were used only seasonally, or mediating the growing number of conflicts among herders over pasture use. As a result, herding today is governed by a mix of lingering collective-era administrative structures, vested in *soum* and *bag* political units; contested notions of property rights; and geographical constraints like water scarcity and proximity of towns and markets (Fernández-Giménez 1999; Fernández-Giménez & Batbuyan 2004).

For example, according to traditional norms of pasture use, herders refrain from grazing winter and spring pastures out of season, rotating their livestock to distant pastures to conserve forage for use during winter and spring when plants are dormant and animals' physiological demands are the greatest (Fernández-Giménez 2000). This practice was expanded and enforced by *negdel* managers during the socialist era. The state developed and maintained tens of thousands of water points, which allowed for broad distribution of grazing pressure among otherwise inaccessible pasture. State-funded transport further increased access to remote pastures and market centres. A majority of those state-owned wells have since fallen into disrepair and transport services have been discontinued. Many herders have adapted to the new challenge of transporting and marketing their own livestock by moving less frequently, over shorter distances. Some have given up seasonal movement altogether, opting to stay close to the markets and services of settled areas. Grazing pressure is concentrated near towns and on pastures abutting natural water sources. Out-of-season grazing has become more common and incidents of trespass and conflict have increased.

These patterns of rangeland use have consequences for both the ecological health of the steppe and the well-being of herding households. Mongolia's rangelands evolved with large grazers and many plant communities of the steppe are quite resilient under grazing pressure. They quickly recover their productivity and species diversity when temporarily rested from grazing during favourable rainfall conditions. But repeated grazing of the same plants during the entire year with brief intervals for recovery weakens individual plants, reduces overall vegetation cover and productivity and can lead to undesirable changes in species composition and an overall loss of species diversity. Continuous grazing, especially under drought conditions, may push some sites beyond a threshold of rapid recovery once rains return. While there is general consensus that rangeland degradation is a serious and growing problem across Mongolia, estimates of the severity and spatial extent of rangeland degradation are varied and controversial, with little credible data to back them up.

These changes in rangeland ecology have had dire consequences for humans. Between 1999 and 2003, a series of drought years followed by several severe winters (known as *dzuuds*) led to the death of approximately 10 million head of livestock, or nearly 30% of Mongolia's national herd. In some areas, entire communities virtually disintegrated as desperate herders migrated to better pastures in the north and east of the country, or sought alternative employment in urban areas. At the national level, the total number of livestock in Mongolia has since rebounded to pre-*dzuud* levels, but individual households have been slow to recover. As observed in our June 2006 survey of households in two *soums* in Bayankhongor *aimag*, herders in *dzuud*-affected areas perceived that since 1999, their livelihoods had worsened while pasture conditions improved. Herders in areas unaffected by *dzuud* reported that their livelihoods had improved since 1999, but that pasture conditions had declined (Fernández Giménez et al. 2006).

As with rangelands, formal regulation of hunting and the wildlife trade has weakened in the post-socialist era. Recent quantitative surveys confirm that the market for wildlife has grown beyond government control. During the 2001 calendar year, we conducted a study of the wildlife trade in Mongolia's three eastern *aimags*: Dornod, Khentii, and Sukhbaatar (see Figure 12.1) (Scharf & Enkhbold 2002). As part of this study, we interviewed 675 herding households in survey areas in six *soums*. Additionally, monitors recorded daily observations of wildlife trade volumes in the open-air markets of each of

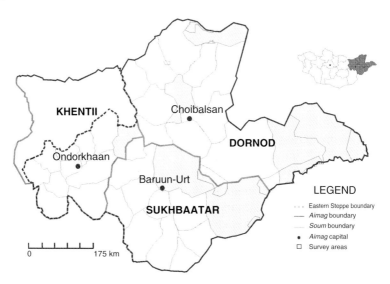

Figure 12.1 **Map of Mongolia's Eastern Steppe. Herding households were inter-viewed in the six survey areas shaded on the map.**

the eastern *aimag* capitals (Ondorkhaan, Choibalsan, Baruun-Urt). A total of 86,000 marmot skins, 177 gazelles, 226 wolves, 675 red foxes and almost 3000 corsac foxes were observed in the three *aimag* capital markets during the study period, which spanned the hunting season between August 2001 and March 2002 (see Table 12.1). The number of marmot skins observed in the three *aimags* alone was four times the number of marmot licenses sold in the same area. In their nationwide study, Wingard and Zahler (2006) estimated that annual wildlife trade volumes for the country as a whole exceeded 3 million marmot, 250,000 Mongolian gazelles, 200,000 corsac fox, 185,000 red fox, 170,000 red squirrel, 100,000 roe deer, 30,000 wild boar, 6000 red deer, 4500 Siberian ibex and 3000 Asiatic wild ass.

The consequence of this trade has been a catastrophic decline in Mongolia's wildlife populations. Rare species used in traditional Chinese medicine have been the hardest hit. Premium prices for red deer antler (*C. elaphus*) have caused a 92% decline in that population in Mongolia, from an estimated 130,000 deer in 1986 to only about 8000–10,000 remaining today. Similarly, the Mongolian population of saiga antelope (*Saiga tatarica mongolicus*), valued for the medicinal uses of its horn, is believed to have declined by 85%

Table 12.1 Quantity of wildlife observed in eastern *aimag* capital markets during hunting season, August 2001 – March 2002 (value in thousands of Mongolian tugriks [MNT]).

Species	Choibalsan			Baruun-Urt			Ondorkhaan			Total		
	Quantity	Mean price	Value	Quantity	Mean price	Value	Quantity	Mean price	Value	Quantity	Mean price	Value
Marmot	42,550	2478	105,458	2406	2006	49,559	19,233	1935	37,211	86,374	2223	192,021
Gazelle	89	2101	187				88	2238	197	177	2169	384
Grey wolf	150	9321	1398	6	8833	53	70	7609	533	226	8777	1984
Red fox	484	8172	3955	94	3736	351	97	5278	512	675	7139	4819
Corsac fox	1418	3947	5597	750	1946	1460	770	2270	1748	2938	2997	8804

from over 5000 animals in 1999 to less than 800 in 2004 (Millner-Gulland et al. 2001; Wingard & Zahler 2006), although population estimates have been questionable. Species valued for trophy hunting have also experienced steep declines. The argali (*O. ammon*) population has dropped by 75% in just 16 years, from an estimated 50,000 in 1975 to only 13,000–15,000 in 2001 (Amgalanbaatar et al. 2002). The sale of saker falcons (*F. cherrug*) for falconry has been a substantial source of revenue for the Mongolian government in recent years, generating about $2.2 million in 2003, equal to 8.8% of the total revenue derived from natural resources (Wingard & Zahler 2006). Saker falcon populations have decreased by 30% in just 5 years, from an estimated 3000 breeding pairs in 1999 to 2200 in 2004 (Wingard & Zahler 2006). Comparatively, 'abundant' species have also been depleted. The Siberian marmot (*M. sibirica*) population, for example, has declined by 75% in 12 years, from around 20 million in 1990 to about 5 million in 2002 (Batbold 2002).

What can be done to reverse the effects of hunting pressure on wildlife? And what can be done to make herding less precarious, from an economic and ecological standpoint? Can a coordinated solution address both of these problems? To answer that question, we must first take a closer look at the intersections between the wildlife trade and the pastoral economy, to understand the extent to which herders are engaged in hunting and the wildlife trade.

Herders as hunters: linkages between pastoralism and wildlife trade

To understand herders' participation in the current large-scale harvest of wildlife, we conducted a survey of 675 herding households in the three eastern *aimags* of Mongolia in 2001 (Scharf & Enkhbold 2002) (see Figure 12.1). Herding households account for almost one-third of Mongolian households (National Statistical Office of Mongolia 2006). Our exclusive focus on Mongolia's eastern provinces yielded certain regional deviations from the national norm. For example, we found that 47% of herding households in eastern Mongolia were engaged in hunting, although in some study sites the proportion was as high as 59%. These percentages are much higher than the national average of herding households who hunt, estimated by Wingard and Zahler (2006) at 11%. Some two-thirds of the 'herder–hunters' we interviewed hunted gazelle and a smaller proportion hunted wolves and foxes. Almost all harvested the Siberian marmot (Scharf & Enkhbold 2002). Despite these regional variations in hunter participation and species hunted, we expect that the motivations and constraints that shape eastern herders' hunting practices also influence herder–hunters elsewhere in Mongolia.

How do herders hunt, and why? We found that the time, labour and mobility constraints of herding livestock make herders opportunistic hunters. Technical and economic considerations further limit the type and number of animals that herders can hunt. For example, herders may take a few Mongolian gazelles in any season for personal consumption if a gazelle herd happens to range near enough to their campsite, but few herders are able to harvest gazelles on a commercial scale. A commercial gazelle harvest requires access to a vehicle, quality rifles and reliable buyers. Compared to lightweight furs of similar value, Mongolian gazelle carcasses are heavy, conspicuous and difficult to transport. Fines for poaching gazelle are steeper than those for other abundant species, preventing hunters from trading gazelles in open markets. Those who hunt illegally on a commercial scale must be organized and work in teams. They must sometimes travel long distances to locate and harvest hundreds to thousands of the migratory gazelles at one time; they must quickly smuggle the perishable carcasses to their buyers, usually through pre-arranged contacts at the Mongolian–Chinese border. As a result, the commercial harvest of Mongolian gazelles is conducted mostly by small groups of *soum* or *aimag* centre residents who can afford transport and weaponry.

By contrast, the commercial harvest of Siberian marmots – a non-migratory species – requires very little organization or investment, enabling many herders to participate (both legally and illegally). Because marmots do not migrate, planning a successful hunt becomes just a matter of ensuring that a herder's autumn campsite is in adequate proximity to areas of high marmot density. Marmots are hunted with guns or relatively inexpensive traps mostly in the fall, just before hibernation, when their fur is thickest. Marmot meat and oil are a prized supplement to the Mongolian diet. Marmot skins can be dried, hidden and sold months after harvest, whenever the herder makes his next trip to the *soum* centre. Comparatively low fines for illegal marmot hunting enable traders to collect the desired skins openly in *aimag* and *soum* centre markets, amassing thousands of skins from small-scale hunters.

Many herders in eastern Mongolia supplement their income by selling wildlife products, especially marmot pelts. In areas of wildlife habitat, the sale of hunted skins or meat contributes roughly one-fifth to one-half of

the income of surveyed herder–hunter households. Local officials have confirmed this perception. One interviewed *bag* leader in Chuluunkhoroot *soum*
(Dornod *aimag*) stated, 'If hunting is stopped, 70–80% of the total population in my *bag* will live below the poverty line'. In Erdenetsagaan *soum*
(Sukhbaatar *aimag*), a *bag* leader reported, 'In my *bag*, 21 households are
living – surviving – on what they catch hunting. Some of them go without
food for several days at a time'. In addition to providing a direct source of
cash income, hunting enables poorer households to conserve their breeding
livestock for reproduction instead of selling or consuming them, thus allowing
these households the opportunity to rebuild a viable herd for subsistence.
Wealthier households may hunt for similar reasons, to further accumulate
wealth in livestock by selling or consuming wildlife products rather than
domestic livestock.

   Herders' wealth (measured by the size of their livestock herds) is an
important determinant of hunting strategy and success (See Table 12.2). The
poorest herders have few livestock to tend and thus have the most time
(and the greatest need) to hunt, but they also own fewer traps and lack the
horses, motorcycles or other transportation necessary to access less-disturbed,
outlying habitat. Wealthy herders can afford to travel to more productive
hunting grounds and may have plenty of traps and quality firearms, but they
are unable to spare much time from tending their large herds. Herders of

Table 12.2 **Herders' estimates of marmot hunting practices (yield, days spent
hunting, equipment used) in 2001, by wealth class.**

Wealth class	Herd size	Percentage of all households	Percentage of hunting households	Days spent hunting	Transport, equipment used	Annual harvest
Wealthy	>501	9.5	9.3	50	Gun, motorcycle	33
Middle	151–200	30.5	32.5	87	Gun, trap, horse, motorcycle	73
Poor	51–150	30.5	32.3	150	Trap, horse, on foot	50
Very poor	<50	29.4	25.6	150	Trap, on foot	50

middle-wealth classes are best able to balance effort and capital inputs most effectively. We found that while the majority (78%) of very poor herders (owning fewer than 50 head of livestock) hunt, middle-income hunters comprise the most numerous group of hunters, and are typically responsible for the largest harvests (Scharf & Enkhbold 2002).

Technology, transport and labour inputs aside, the primary determinant of a hunter's success is his ability to secure exclusive access to areas of prime wildlife habitat. For the herder–hunter, utilization of wildlife habitat must be balanced with the spatial and seasonal demands of animal husbandry. In the case of marmots, for example, access to marmot habitat during the fall hunting season must be incorporated into the seasonal rotation among various pastures, water points, salt licks and other resources. Competition between hunters mirrors the classic dilemma of the pastoralist, who must assert exclusive rights over winter pastures while in absentia, grazing elsewhere during spring, summer and autumn. Similarly, the marmot hunter must prevent others from harvesting marmots in his desired hunting grounds, not only during the fall hunting season but also in the preceding months of breeding and fattening that guarantee a successful harvest. Both customary and formal campsite tenure extends informal rights to the surrounding pasture, with greater control and exclusivity of use near the camp and diminishing influence with increasing distance. Informal rights to marmot hunting grounds follow the same pattern. The herder must strike a balance: if he camps too close to marmot habitat, the presence of livestock and humans may prevent the marmots from feeding sufficiently. Camping too far away will leave the hunting grounds vulnerable to intruding hunters. In neither case, however, is there a mechanism for sustainable harvesting. As marmots are highly colonial and poor dispersers, it is easy to locate and harvest an entire colony. It is, unfortunately, unlikely that the colony can re-establish itself in any meaningful period of time.

Not surprisingly, many of the herders we interviewed articulated the problem of hunting regulation in terms of a spatial conflict, a competition between locals and 'outsiders' over access to wildlife habitat. The concentration of herding households around towns and natural water sources has had a compounding effect, leading to the under-use of more remote pastures. This trend has accelerated in recent years, as herders are simultaneously pushed out of the range by severe winters and major livestock losses, and pulled into urban settlements by the promise of alternative economic opportunities (Fernández-Giménez et al. 2006). Some herders remarked that wildlife species that avoid human activity, most notably gazelles, are concentrated in these outlying areas, far from 'occupied' pastures. Herders claim that these areas have become prime hunting sites for commercial hunters who can afford the time and expense required to access these remote areas, 'far from strangers' eyes', 'in areas that inspectors do not patrol regularly'. Relocating to uninhabited areas for a few weeks of the hunting season, unencumbered by livestock, these hunters stake claims on a first-come first-serve basis and conduct intensive harvests, sometimes using systematic and sophisticated techniques to harvest entire marmot colonies – juveniles as well as adults – without regard for future sustainability.

Ironically, this phenomenon may be intensified in designated protected areas, rendering them especially vulnerable to overexploitation by intensive, large-scale hunters. In 1992, the Mongolian Parliament announced its intention to place up to 30% of the country's land area in some form of protected status. The protected areas of eastern Mongolia were established in some of the region's most remote corners. Many of these protected areas are 'paper parks', which have little or no active management (Reading et al. 1999; Heffernan et al. 2005). The most strictly protected zones are often located along restricted border zones. Their sole inhabitants, military border guard units, are reportedly some of the most serious poachers. To herders who rely on temporary access to a broad, diverse landscape in order to satisfy needs for pasture, firewood, wildlife and other resources, the status of the reserve can be a source of ambivalence. Some herders feel that because a protected area is state-owned, like pasture, its use by nearby residents is justified. Others, particularly those excluded from the use of park resources in strictly protected zones, consider the protected area as a territory apart, alienated from the overlapping claims of use and exclusion that otherwise deter them from hunting in the rest of a *soum* or *aimag*. These herders reported that 'illegal hunting has increased in the protected area, where there is more wildlife', but were disinclined to challenge hunters who moved into the area to hunt: 'This is the problem of inspectors', said one. 'They are paid to do this job'.

In the years since Mongolia's political and economic transition, the harvest of wildlife has become an increasingly important dimension of herders' livelihood strategies on the eastern steppe. And yet, by official measure, most of this hunting is illegal. The Mongolian Law on Hunting, enacted in 1995 and amended in 2002, established a permitting system to control the wildlife harvest and to relay a certain percentage of permit sales back to state-funded conservation measures. By purchasing a license (pursuant to the Hunting Law), hunters agree to restrict their hunting activities to a prescribed season and a predetermined quota. In 2001, for example, marmot hunters were required to pay approximately US$0.60 per animal to hunt up to 15–25 marmots during a one- to two-week period. In reality, nearly all herder–hunters engage in some form of illegal hunting. Though most of the hunters we surveyed were well aware of the law, almost all hunted three to four times more than the permitted limit. Thirty-four percent of interviewed herder–hunters admitted to hunting out of season. Many did not even purchase a permit. Among the study sites, the percentage of hunting households holding permits ranged from 30%, down to only 9%. Wealthy hunters were more likely to hold permits (25–43%), but middle-ranking households exhibited the lowest proportion of licensed hunters (7–12%), suggesting that the decision to hunt without a permit is not a function of finances alone. 'Nobody speaks openly about it, but everyone knows that everyone hunts if they get a chance', admitted one survey participant.

Prospects for integrated rangeland management in Mongolia

The fact that herders expect that the conservation of wildlife in protected areas is a matter of state responsibility, and that community norms that otherwise govern natural resource use do not apply inside park boundaries, underscore the lack of integration of rangeland and wildlife management regimes. This phenomenon may be traced back, in part, to development experts' historical tendency to view pastoralists as 'irrational, wasteful and shortsighted' (Fratkin & Mearns 2003). Garrett Hardin's classic article on the 'tragedy of the commons' encouraged this view. Hardin posited that systems of communal land tenure are incapable of excluding resource users or regulating herd sizes, thus leading to overstocking, overgrazing, degradation and desertification (Hardin 1968). Between the 1960s and the 1980s, development assistance organizations like the World Bank adopted Hardin's view and promoted large-scale livestock development projects that advocated land privatization, sedentarization and intensification of livestock production in commercial ranches or grazing blocks (Fratkin 1997; Leach et al. 1997). These projects were implemented through a top-down approach and failed to generate autonomous local participation. New research has since revised the tragedy thesis, asserting that the tragedy of the commons was really a tragedy of open access. Overexploitation was not endemic to the commons, but resulted instead where formal or informal rules for the management of jointly used resources, such as extensive rangelands and mobile wildlife populations, were not effectively enforced (Bromley & Cernea 1989; Ostrom 1990). Accordingly, development organizations have since begun to emphasize pastoral institution building initiatives (Mearns 1997; Niamir-Fuller 1999).

Because Mongolia was politically isolated from the West until the early 1990s, Western development assistance arrived there relatively late in the evolution of these models. As a consequence, the country has been able to debate and even implement many of these models simultaneously. Our discussion of strategies for integrated rangeland and wildlife management in Mongolia therefore addresses three approaches: strengthening property rights and entitlements, developing community-based natural resource management and increasing the effectiveness of rangeland management through local government.

Strengthening property rights and entitlements

Unlike pastoralists in other Asian countries, Mongolian herders make up the majority of the country's population. Livestock products are a mainstay of the national economy. Despite this demographic and economic dominance, senior public officials have expressed ambivalence about the future of Mongolia's pastoral economy. For example, when speaking with the BBC in the wake of *dzuud* and drought losses in 2003, Mongolia's Prime Minister, Nambaryn Enkhbayar, predicted that nomadism would die out in the space of 10–15 years, as an inevitable consequence of competition and development (BBC 2003). During debates over amendments to Mongolia's Law on Land, this ambivalence has been even more pronounced. Some politicians have argued to retain the present system of state ownership of pasture land, not so much to preserve the practice of extensive livestock husbandry that depends on that system, but because they fear that Chinese citizens would buy up privatized Mongolian land (Agriteam Canada 1997).

Pastoralists and many rangeland managers, however, recognize that a tenure system that severely restricts mobility will serve the land and herders poorly and may result in accelerated degradation and increased household vulnerability to climatic stresses. While the lack of coordination and regulation of pasture use remains a challenge, privatizing pasture will not solve these problems. Instead, measures must be taken to increase the capacity of *soum* governments to coordinate use of state-owned pasture by regulating seasonal movements and stocking rates, as they are already authorized to do. Since large-scale climatic disasters will continue to affect Mongolian rangelands and livestock, *aimag* governments must also play a role in facilitating arrangements for reciprocal inter-*soum* and inter-*aimag* pasture use. Such arrangements are necessary to preserve pastoralists' flexibility and mobility while protecting the pasture rights of herders in areas unaffected by drought or *dzuud*.

Securing exclusive, collective management rights to key seasonal pastures for herder groups must also be a priority. New trends in land and resource use are increasing the need to formalize the herders' use rights to rangeland resources. For example, since the late 1990s, the pace of mining and energy exploration has accelerated dramatically in Mongolia, threatening the livelihoods of neighbouring pastoralists. Large quantities of water (both surface and underground) are needed for mine production. The influx of labour to mine sites also imperils wildlife, through loss of habitat and increased market and subsistence hunting to supply miners with food and firewood. As mining and energy exploration occupy more of Mongolia's territory, herders will be faced with a new source of competing claims to resources which were once taken for granted. If these rights are not clarified and strengthened, herders risk losing access to these resources without compensation for the loss of their livelihood, or the cost of relocation.

Non-state solutions: community-based natural resource management

Mongolia has benefited from a new emphasis on community-based natural resource management [CBNRM]. Experts have proposed a range of creative options for delegating rangeland management to local groups, including strengthening customary resource rights (through pasture land leases), implementing pastureland co-management schemes (Fernández-Giménez 2002), adjusting tax rates to incentivize lower stocking rates (Swift 1995), developing insurance and credit instruments and building market infrastructure to enable rapid destocking and restocking during natural disasters (Mearns 1997). Since the late 1990s, at least seven major donor- and government-sponsored projects have debuted in Mongolia, promoting various forms of community-based management and co-management of rangelands, water points and other resources. While some of the preliminary reports from these projects appear promising (Ykhanbai et al. 2004), as yet there has been no systematic attempt to compare the approaches and outcomes of these diverse efforts.

Meanwhile, conservation biologists have similarly shifted their approach to wildlife management. The 'sustainable' revolution in development policy drew attention to costs of wildlife conservation policies that give precedence to wildlife over pastoralists, evicting pastoralists from game preserves to reduce competition for forage between wildlife and livestock (Mearns 1997). Inspired by the CAMPFIRE programme in Zimbabwe, some conservationists have partnered with pastoralists to implement community-based approaches to wildlife management (Western et al. 1994; Murphree 2005; Mulder & Cop-polillo 2005). Pastoralists can be important allies in conservation work to the extent that they possess special ecological knowledge, or practice traditional or norm-based systems (such as limiting herd sizes, or regulating access to pastures or water sources) that can be adapted to wildlife management. And, as is often the case in Mongolia, pastoralists may also be the only people with a regular presence in remote wildlife habitat importance, making them both a major threat and indispensable collaborators. Because pastoralists need to resolve grazing conflict and secure sustainable livelihoods, there are strong incentives for cooperation where grazing areas and wildlife habitat overlap (Mearns 1997; Reading et al. 1999; Fernández-Giménez 2000; Siebert & Belsky 2002). Mechanisms for implementing CBNRM for wildlife typically include technical training for local leaders in rangeland and wildlife management; revenue-sharing agreements that distribute the proceeds from eco-tourism or trophy hunting among pastoralists residing near protected wildlife habitat; and direct compensation payments to livestock producers to finance the incre-mental costs of biodiversity conservation (Wells & Brandon 1992; Western et al. 1994; Wells 1995).

In Mongolia, there has been some recent progress on CBNRM pilot projects in forestry and rangeland management (Ykhanbai et al. 2004). In November 2005, the Mongolian Parliament approved amendments to the Law on Environmental Protection to allow the formation of community-based organizations (*nokhrolol*) from groups of at least 10–15 families, authorized to protect, utilize and possess certain natural resources. Development organizations are eager to experiment with community-based conservation under the new law, but there is substantial confusion about the precise rights and obligations of *nokhrolol* entities (Wingard & Zahler 2006). According to regulations issued by the Mongolian Ministry of Nature and the Environment in 2006, *nokhrolol* groups may contract with the government to assume certain management obligations from *soum* governments for particular resources. These obligations include calculating and monitoring pasture carrying capacity, stopping illegal activities, fencing hayfields, posting signs around protected areas (so long as visitor entrances are not obstructed) and setting up conservation funds derived from the sale of secondary resources. Neither the law nor these interpretive regulations, however, clearly authorize *nokhrolol* groups to exclude non-members from using the managed resources. Until this problem is resolved, it will be difficult to secure wildlife and rangeland management objectives through community herder groups or quasi-government pasture co-management teams, or any other CBNRM organization.

Even if the Mongolian Parliament vests the *nokhrolol* with the necessary legal entitlements, it is unclear whether herding communities have the organizational capacity to take advantage of an enabling legal environment. Herding communities tend to be small in size, comprised of households who have been neighbours for years. In theory, these factors should make it easier for groups to cooperate. But herders interviewed in 2006 expressed reluctance to join herder groups (including pasture management groups and herding cooperatives) because they did not trust other herders, because they perceived that the benefits of participation would be unequally distributed and because they were skeptical about the long-term viability of the groups (Fernández-Giménez et al. 2006). Other researchers have reported a similar lack of social capital – that is, relationships of trust and norms of reciprocity – within herding communities in Mongolia (Mearns 1996).

In addition to the problem of weak informal cooperation, the prospects for CBNRM are further limited by the spatial mismatch of herding encampments (clustered around towns and natural water sources) and wildlife habitat, described earlier in this chapter. The influence of customary campsite tenure and informal rights to nearby pasture imposes additional constraints on the spatial distribution of hunting activities, restricting access to wildlife resources around human settlements. But as our survey results suggest, this overlapping system of practical limitations and informal rights alone is not sufficient to ensure a sustainable wildlife harvest. It appears that organized, intensive commercial hunting, particularly in unpopulated, remote and often protected areas, can only be effectively controlled by state regulation.

The combination of legal impediments, limited social capital and spatial distribution of herders and wildlife suggests that CBNRM or co-management regimes will be most successful when they proceed in tandem with efforts to strengthen management by local government entities. Cooperation will depend on the ability of conservationists and wildlife managers to articulate their goals in common with the herders' need for a sustainable wildlife harvest, or reliable access to pasture. Involvement of herder–hunters in the design of *soum*-level hunting management strategies will be key to the success of any conservation strategy (Maroney 2005). Given the limited resources available for wildlife population surveys and enforcement patrols, cooperation with hunters could produce valuable data on hunting effort (number of kills per unit of time, an indirect indicator of wildlife population dynamics) and the cultivation of a network of informants or 'volunteer' rangers. In the next section, we examine potential roles for local government and alternative institutional arrangements to improve the prospects for rangeland and wildlife management in Mongolia.

Bringing the state back in: flexibility in local governance

The history of politics and natural resource in Mongolia, recounted earlier in this chapter, demonstrates a long-term trend of increasing state intervention in rangeland and wildlife management. This trend reached its peak with the organization of herding collectives and hunting brigades in the socialist era, thanks to rather unique geopolitical circumstances: a massive influx of Soviet aid and investment, and the suspension of informal trade between China and Mongolia. By comparison, Mongolia's central government is now starved for revenue, relying on a small operating budget derived mostly from the sale or use of the country's natural resources. The government is unable to fund basic enforcement to stop illegal hunting or out-of-season grazing. And it is unable to limit the flow of natural resources across the (4677 km) long, porous Sino-Mongolian border. It is simply not fiscally possible – even if it were desirable – to return to the days of strong, central control of natural resources.

Nevertheless, the experience of the socialist era (and the political eras that preceded it) continues to dominate legal and administrative structures, as well as popular expectations. The state retains ownership of pasture land, which is still divided among the same small administrative units set up in the *negdel* era. Most herders continue to follow *negdel*-era patterns of shorter seasonal movements. Wildlife continues to be harvested commercially for an international market. Local government entities bear the responsibility for implementing rangeland management and are almost entirely dependent on the central government to sustain their limited budgets. And hunters and herders continue to look to local government to moderate access to pasture and the harvest of wildlife, especially around protected areas.

This is especially true in the case of wildlife management. Wildlife management duties are delegated to untrained local authorities. Expert organizations like the Institute of Biology at the Mongolian Academy of Sciences are not empowered to set harvest quotas. In each *soum*, a single environmental inspector bears the responsibility of enforcing hunting laws. A handful of rangers are charged with patrolling the approximately 2 million hectares of protected areas in eastern Mongolia. Many of these rangers and inspectors lack vehicles, fuel or transport to carry out inspections. Customs officials are not trained to look for or identify illicit wildlife products at border checkpoints. The border guards responsible for policing illegal trade in the remote stretches of the Sino-Mongolian border must provision their own outposts by herding livestock or, not uncommonly, hunting gazelles (Wingard & Zahler 2006). Illegal hunting by government officials further undermines regulation. According to one former environmental protection inspector, local herder–hunters are well aware of such abuses of power and commonly justify their own illegal hunting activities by asking, 'Why can we not hunt when everybody, including high-up people, is hunting wildlife, even illegally?'

Following the transition from planned economy to free-market system, local governments seem to be struggling to find their role in controlling hunting, now that they are unable to plan, execute and profit directly from an organized wildlife harvest. The problem is typically understood as one of limited resources. Given the dispersed nature of the wildlife trade and the lack of quantitative data on its extent, Mongolian policymakers (like policymakers in many countries) tend to undervalue wildlife, particularly as compared to other commodities such as minerals (Oldfield 2003). And yet, the sale of hunting licenses – especially trophy-hunting licenses – generated US$1.9 million in 2003, the equivalent to almost two-thirds of the budget for the Ministry of Nature and Environment. Sale of saker falcons netted another US$2.2 million for the Ministry.

Mongolia's Law on Reinvestment of Natural Resource Use Fees mandates that 50% of revenues from hunting license sales must be reinvested in conservation and management. In 2004, the mandatory reinvestment would have totalled US$2.1 million; in fact, the Ministry invested less than one-quarter of that amount in wildlife management (Wingard & Zahler 2006). Officials tend to view wildlife management as a second-tier government service, to be provided for after priority budget items, like pensions or teacher salaries, have been paid. Either they do not view wildlife trade regulation as a revenue-generating investment, with enforcement driving up license sales, or they have calculated that the gains in license sales could not exceed the cost of patrols in remote areas.

Fiscal constraints on conservation efforts at the local level are not only a function of the small size of the *soum* budget, but also the lack of local control over the allocation of that budget. *Soum* operating budgets are funded primarily by large grants from the central government. In 2001, we spoke with several *soum* governors in the eastern steppe region who wished to exercise their discretion under the Law on Hunting to ban hunting for certain species or in certain areas of their *soum*. They were unable to do so, because the Ministry of Finance (which approves *soum* operating budgets) automatically offset the grants from the central government by the expected value of hunting license revenues in each *soum* (Scharf & Enkhbold 2002).

More recently, as wildlife populations have dwindled, the central government has itself begun imposing hunting bans. The harvest of 'very rare' species (including brown bear, snow leopard, Siberian moose, Saiga antelope and musk deer) has been banned since 1995. Commercial hunting of Mongolian gazelle was banned in 2001, and in 2000 red deer trophy and subsistence hunting was banned. Marmot hunting (both commercial and subsistence) was banned in 2005 for 2 years. Although these bans are seemingly 'costless' conservation measures, in effect, they have weakened the license system, by eliminating its major funding source, while inadequate enforcement and uncontrolled exploitation of wildlife resources continues.

How to reverse the downward spiral of Mongolia's wildlife management regime? There are many ways to tailor the licensing system to the needs of herder–hunters. As environmental regulators and conservation organizations seek more effective means to control hunting on the eastern steppe, it may be useful to consider common principles in the management of grazing and hunting. In both grazing management and hunting regulation, the major variables that can be manipulated to achieve management goals are (i) the number of animals grazed or hunted; (ii) the kind and class of animals used; (iii) the season of use; (iv) the duration and frequency of use and (v) the spatial distribution of use. Conventional grazing management (in the United States and elsewhere), like wildlife management, has focused mainly on controlling herd sizes, or the number of wildlife harvested. In mobile pastoral societies, however, it has proven extremely difficult to impose restrictions on stocking rates. This strategy, like the Mongolian Law on Hunting, is both ineffective and, from a herder's or hunter's perspective, irrational. By contrast, traditional regulation of grazing in Mongolia focused on the spatial and temporal distribution of grazing pressure. A first step toward more effective hunting management would be to work with *soum* governments or *nokhrolol* groups to map the spatial distribution of hunting pressure and to designate certain areas of a *soum* off-limits to hunters on a rotating basis, thus enabling local wildlife populations to recover. Inspectors currently responsible for patrolling entire *soums* could more effectively monitor these reserves.

Of course, efforts to improve enforcement of hunting controls in the countryside will not be effective so long as illegally-hunted wildlife products are traded and exported with impunity. At the parliamentary level, the Mongolian government has taken steps to control the open, illegal wildlife trade by endorsing a 'certificate of origin' system to allow illegally harvested pelts and carcasses to be distinguished from those hunted with a permit. A proliferation of agencies – police, customs officials, prosecutors, as well as environmental inspectors and ministry officials – must coordinate their efforts to crack down on the illegal wildlife trade at border crossings, and in raw materials markets and road inspection points near urban areas. In short, community-based natural resource management cannot succeed with market controls. Only by regulating the trade and export of wildlife resources will the Mongolian government be able to boost the hunting license sales needed to fund enforcement, wildlife surveys and sustainable hunting management by *soum* governments and *nokhrolol* groups.

Conclusion

The recent turn toward community-based natural resource management is encouraging. We are optimistic that this trend will enable herders to resolve grazing conflicts and secure access to key pasture resources, while at the same time reversing the heavy toll that unregulated hunting has taken on Mongolia's wildlife in recent years. As we have outlined above, the success of this integrated approach will depend not only on conservationists' and development experts' abilities to boost grassroots participation but also on their sensitivity to the history, politics, geography and sociology of herding and hunting in Mongolia. For example, to effectively advocate for more frequent ranger patrols, defined property rights over pasture resources, or legal empowerment of herding communities to manage wildlife populations, one must understand the fiscal relationships between central and local government, the spatial mismatch between wildlife populations and herders' encampments, and the interplay between formal rights and norms governing pasture usage. In the years since Mongolia began its transition to democracy and a market economy in the 1990s, economically distressed herders have become increasingly reliant on wildlife resources to supplement their economic and social welfare. As such, the management of Mongolia's wildlife has become not just a conservation issue but also a *development* issue, underscoring the need for an integrated management approach that promotes both rural livelihoods and biodiversity conservation.

References

Agriteam Canada. (1997) *Study of Extensive Livestock Production Systems*. Agriteam Canada Consulting, Calgary.

Amgalanbaatar, S., Reading, R.P., Lhagvasuren, B. & Batsukh, N. (2002) Argali sheep (*Ovis ammon*) trophy hunting in Mongolia. *Pirineos* 157, 129–150.

Avirmed, D. (1999) Hunting and Wild Animal Conservation in the History of the Mongols. [unpublished report]

Batbold, J. (2002) The problem of management of marmots in Mongolia. In: Armitage, K.B. & Rumiantsev, V.Y. (eds.) *Holarctic Marmots as a Factor of Biodiversity: Proceedings of the 3rd International Conference on Marmots, Cheboksary, Russia, 25–30 August 1997*. ABF Publishing House, Moscow.

Bawden, C.R. (1968) *The Modern History of Mongolia*. Kegan Paul International, London.

Bergholz, F.W. (1993) *The Partition of the Steppe: The Struggle of the Russians, Manchus, and the Zunghar Mongols for Empire in Central Asia, 1610–1758*. Peter Lang, New York.

British Broadcasting Corporation News/Asia Pacific. (2003) Mongolian nomadism 'to die out'. Friday, 24 October, 2003, available at http://news.bbc.co.uk/2/low/asia-pacific/3210457.stm (last accessed March 14, 2007).

Bromley, D. & Cernea, M. (1989) *The Management of Common Property Natural Resources: Some Conceptual and Operational Fallacies*. World Bank, Washington.

Fernández-Giménez, M.E. (1999) Sustaining the steppes: a geographical history of pastoral land use in Mongolia. *Geographical Review* 89, 315–342.

Fernández-Giménez, M.E. (2000) The role of Mongolian nomadic pastoralists' ecological knowledge in rangeland management. *Ecological Applications* 10, 1318–1326.

Fernández-Giménez, M.E. (2002) Spatial and social boundaries and the paradox of pastoral land tenure: a case study from post-socialist Mongolia. *Human Ecology* 30, 49–78.

Fernández-Giménez, M.E. & Batbuyan, B. (2004) Law and disorder: local implementation of Mongolia's land law. *Development and Change* 35, 141–165.

Fernández-Giménez, M.E., Batbuyan, B. & Oyungerel, J. (2006) Climate, Economy and Land Policy: Effects on Pastoral Mobility Patterns in Mongolia. Proceedings of the Kyoto Symposium Crossing Disciplinary Boundaries and Re-visioning Area Studies, Perspectives from Asia and Africa, November 9–13, 2006, University of Kyoto, Kyoto, Japan.

Fernández-Giménez, M.E. & Khishigbayar, J. (2005) Reflections on Mongolia's 2002 Land Law: Implications for Herders and Rangelands. Invited paper presented at the International Symposium on Mongolian Nomadic Society and Land Ownership, Center for Asian Legal Exchange, Nagoya University, Japan.

Fratkin, E. (1997) Pastoralism: governance and development issues. *Annual Review of Anthropology* 26, 235–261.

Fratkin, E. & Mearns, R. (2003) Sustainability and pastoral livelihoods: lessons from east African Maasai and Mongolia. *Human Organization* 62, 112–122.

Hardin, G. (1968) The tragedy of the commons. *Science* 162, 1234–1248.

Heffernan, D.E, Zahler, P., Merkel, J., Heffernan, C.A. & Jargalsaikhan, C. (2005) An assessment of the protected areas of the eastern steppe of Mongolia. *Mongolian Journal of Biological Sciences* 3, 25–29.

Humphrey, C. & Sneath, D. (1999) *The End of Nomadism? Society, State and the Environment in Inner Asia*. White Horse Press, Cambridge.

Kerven, C. (2006) *Review of the Literature on Pastoral Economics and Marketing: Central Asian, China, Mongolia, and Siberia*. World Initiative for Sustainable Pastoralism, International Union for Conservation of Nature and Natural Resources Eastern Africa Regional Office, Nairobi.

Lattimore, O. (1940, 1951 reprint) *Inner Asian Frontiers of China*. Beacon Press, Boston.

Leach, M., Mearns, R. & Scoones, I. (1997) Community-based sustainable development: consensus or conflict? *IDS Bulletin* 28(4), 1–3.

Maroney, R.L. (2005) Conservation of argali *Ovis ammon* in Western Mongolia and the Altai-Sayan. *Biological Conservation* 121, 231–241.

Mearns, R. (1996) Community, collective action and common grazing: the case of post-socialist Mongolia. *Journal of Development Studies* 32, 297–339.

Mearns, R. (1997) Balancing livestock production and environmental goals. *World Animal Review* 89, 24–33.

Millner-Gulland, E.J., Kholodova, M.V., Bekenov, A. et al. (2001) Dramatic declines in saiga antelope populations. *Oryx* 35, 340–345.

Mulder, M.B. & Coppolillo, P. (2005) *Conservation: Linking Ecology, Economics and Culture*. Princeton University Press, Princeton.

Murphree, M.W. (2005) Congruent objectives, competing interests, and strategic compromise: concept and process in the evolution of Zimbabwe's CAMPFIRE programme, 1984–1996. In: Brosius, J.P., Lowenhaupt Tsing, A. & Zerner, C. (eds.) *Communities and Conservation: Histories and Politics of Community-Based Natural Resource Management.* AltaMira Press, Walnut Creek, pp. 105–147.

National Statistical Office of Mongolia. (2006) *Mongolian Statistical Yearbook 2005.* National Statistical Office of Mongolia, Ulaanbaatar.

Niamir-Fuller, M. (ed.) (1999) *Managing Mobility in African Rangelands: The Legitimization of Transhumance.* Intermediate Technology Publications, London.

Oldfield, S. (ed.) (2003) *Trade in Wildlife: Regulation for Conservation,* Earthscan Publications, Ltd., London.

Ostrom, E. (1990) *Governing the Commons: The Evolution of Institutions for Collective Action.* Cambridge University Press, New York.

Reading, R.P., Johnstad, M., Batjargal, Z., Amgalanbaatar, S. & Mix, H. (1999) Expanding Mongolia's system of protected areas. *Natural Areas Journal* 19(3), 211–222.

Sanjdorj, M. (1980) *Manchu Chinese Colonial Rule in Northern Mongolia.* St. Martin's Press, New York.

Scharf, K. & Enkhbold, S. (2002) Hunting in Eastern Mongolia: The Challenge of Wildlife Management in a Post-Socialist Country. United Nations Development Programme-Global Environment Fund Eastern Steppe Biodiversity Project, Ulaanbaatar.

Scott, J. (1998) *Seeing Like a State: How Certain Schemes to Improve the Human Condition Have Failed.* Yale University Press, New Haven.

Siebert, S.F. & Belsky, J.M. (2002) Livelihood security and protected area management. *International Journal of Wilderness* 8, 38–42.

Swift, J. (1995) Dynamic ecological systems and the administration of pastoral development. In: Scoones, I. (ed.) *Living with Uncertainty: New Directions in Pastoral Development in Africa.* Intermediate Technology Publications, London.

Thornton, P.K., Kruska, R.L., Henninger, N. et al. (eds.) (2002) *Mapping Poverty and Livestock in the Developing World.* United Kingdom Department for International Development & Inter-Agency Group of Donors Supporting Research on Livestock Production and Health in the Developing World, Nairobi. http://www.ilri.cgiar.org/INFOSERV/WEBPUB/FULLDOCS/MAPPINGPLDW/media/index.htm

Wells, M. (1995) Social-economic strategies to sustainably use, conserve and share the benefits of biodiversity. In: United Nations Environment Programme (ed.) *Global Biodiversity Assessment.* Cambridge University Press, Cambridge.

Wells, M. & Brandon, K. (1992) *People and Parks: Linking Protected Area Management with Local Communities.* World Bank, World Wildlife Fund & United States Agency for International Development, Washington, D.C.

Western, D., Wright, R.M. & Strum, S. (eds.) (1994) *Natural Connections: Perspectives in Community-Based Conservation.* Island Press, Washington, D.C.

Wingard, J.R. & Zahler, P. (2006) *Silent Steppe: The Illegal Wildlife Trade Crisis in Mongolia*. World Bank, Washington, D.C.

Ykhanbai, H., Bulgan, E., Beket, U., Vernooy, R. & Graham, J. (2004) Reversing grassland degradation and improving herders' livelihoods in the Altai Mountains of Mongolia. *Mountain Research and Development* 24, 96–100.

$$\textbf{13}$$

Social and Economic Challenges for Conservation in East African Rangelands: Land use, Livelihoods and Wildlife Change in Maasailand

Katherine Homewood[1] *and D. Michael Thompson*[2]

[1]University College London, UK
[2]UK Environment Agency, UK

Introduction

Currently much conservation and development policy is predicated on the assumption that sustainable use of natural resources can underpin 'green development', delivering both significant revenue and increased equitability (Pearce & Moran 1994). A more sceptical view, however, suggests such interventions commonly allow better-placed individuals and sub-groups to control natural resources on which remote rural groups have hitherto depended, and from which the latter are progressively excluded (Escobar 1996; Castree 2003). This paper looks in detail at returns to local livelihoods from wildlife conservation in Maasailand, to better understand the changing patterns of control over these resources, and the decisions local people are making regarding economic activities and forms of land use that are either compatible with, or conversely conflict with, wildlife conservation.

Wild Rangelands: Conserving Wildlife While Maintaining Livestock in Semi-Arid Ecosystems,
1st edition. Edited by J.T. du Toit, R. Kock, and J.C. Deutsch.
© 2010 Blackwell Publishing

This paper first reviews the state of knowledge on how Maasai rangeland habitat and biodiversity have changed over the last decades, the main factors driving those changes and their relation to people's land-use decisions. It then sets out preliminary research findings on the returns provided by conservation in Maasai livelihoods in Kenyan and Tanzanian study sites.

These findings are used as a basis to explore the distribution of benefits from wildlife conservation and their impact on people's economic and land-use choices. The discussion considers how contrasting national and conservation-area policies affect the scale of conservation income, its distribution among local households, and the extent to which income works for local development while fostering conservation-compatible land-use choices.

Political and land-use policy context

Twenty-first century Maasailand is a particularly interesting context in which to investigate these questions of conservation and development. At the beginning of the twenty-first century Maasailand comprised $150,000-200,000 \text{ km}^2$ of arid and semi-arid rangelands straddling the Kenya–Tanzania border, including key highland, swamp and riverine habitats (Map 13.1). Fertile volcanic soils mean productive potential is comparatively very high. Maasailand is a focus of conservation interest with high savanna biodiversity and a large number of protected areas. The combination of spectacular savanna landscapes and wildlife with iconic Maasai pastoralists has long made this a major international tourism destination. Both Kenya and Tanzania rely heavily on tourism as a major source of revenue. In Kenya, despite fluctuations, tourism has consistently been one of the top earners of foreign currency; in Tanzania, tourism is estimated to account for 16% of gross domestic product (GDP) (CHL Consulting Group 2002). At the same time, in both Kenya and Tanzania, the rural (largely agro-pastoralist) populations of Maasailand lag far behind national standards in terms of literacy, health provision and food security (Coast 2002).

These populations thus include a high proportion of the very poor on whom international interest focuses, and in principle tourism revenue could underpin their sustainable development.

Maasailand therefore offers a powerful opportunity for investigating the possibility and conditions for wildlife tourism-based green development.

The variability of Maasailand makes rigorous analysis difficult. Encompassing immense biophysical variability (both spatial and temporal), Maasai

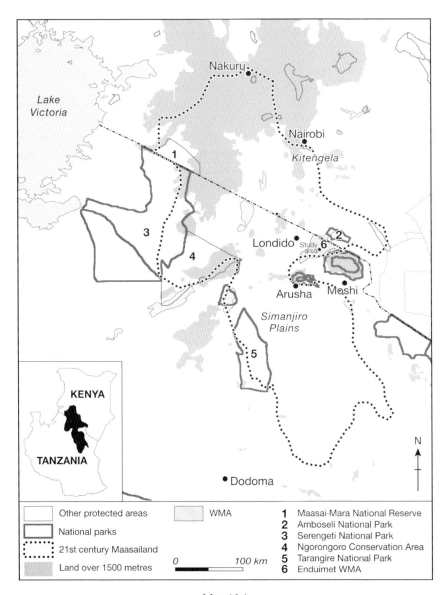

Map 13.1

rangelands are characterized by diversity and strong fluctuations in social, economic, demographic and political conditions, with often unpredictable shifts in the social as well as the natural environment. An analytical framework is provided by the combination of strong continuities across a wide range of conditions, overlaid by sharp boundaries and strong contrasts in macro-political and economic conditions. This means Maasai rangelands represent something of a natural experiment on the implications of different policies on conservation and development. The Kenya–Tanzania border sets up contrasts in terms of macro-economic and macro-political systems. Since Independence, Kenya has been strongly capitalist, emphasizing privatization and commercial enterprise. Tanzania was initially socialist, and despite economic liberalization since the mid 80s, land and natural resources remain under strong State control. In Kenya land tenure outside the core conservation areas has been privatized into individual small holdings, while in Tanzania land outside these core conservation areas is managed communally by village governments, though subject to State intervention (Ojalammi 2006). Part of Maasailand within each country is managed as Category I or II conservation estate, with no consumptive land use (Maasai–Mara National Reserve (MMNR), Serengeti National Park (SNP)), while other areas have experienced a range of land uses, from hunting (in Tanzania only) through mixed photographic tourism, livestock production and farming, mining and quarrying; to private land sales and property development in some sites (Homewood et al. 2009). These major economic and land use contrasts are superimposed on a landscape characterized by strong continuities on several major dimensions:

- **Ecological:** Maasailand is not uniform, but comparable agro-ecological and eco-climatic conditions and habitats repeat themselves across the whole area.
- **Ethnic:** More than 95% of the rural population is Maa-speaking, and self-defines as Maasai.
- **Micro-economic:** Though livelihoods are diversifying, the vast majority of rural people are primarily herders and farmers.

This superimposing of contrasting political and land-use policies on otherwise continuous landscapes creates a natural experiment making it possible to control for many of the confounding factors and determine the main factors driving quantitative change in habitat, wildlife populations and socio-economic indicators in the region.

At the same time, there is a wealth of historical, anthropological and development research data and analyses on which to build an understanding of institutional and other qualitative change (Rutten 1992; Spear & Waller 1993; Galaty 1999; Southgate & Hulme 2000; Woodhouse et al. 2000; Anderson 2002; Homewood et al. 2004; Hughes 2006). These make it possible to trace the site-specific historical particularities within wider patterns of changing management, control and use of natural resources.

Within the range of different types of land use found in Maasailand (category I and II conservation estate, consumptive wildlife management, photographic tourism, extensive or intensive livestock-rearing, small- or large-scale cultivation, mining or quarrying, intensive property development) some are more conservation-compatible than others. For example, wildlife and/or extensive livestock rearing involve the mobile use of unfenced rangelands, maintaining fire- and grazing-altered habitats conducive to a wide range of local savanna plant and animal species. Small-scale cultivation at low density can enrich the habitat mosaic and hence biodiversity, but at increasing densities clearance for cultivation, and intensive exploitation of key wetlands in these arid and semi-arid savannas, may impact adversely on conservation values reducing wildlife diversity and abundance. Finally, large-scale cultivation or urban settlement generally mean habitat conversion at a scale incompatible with conservation (Gichohi 2000; Kristjanson et al. 2002; Lamprey & Reid 2004). The very diversity of site-specific circumstances, combined with strong continuities and weight of background knowledge thus make Maasailand a fascinating and productive arena in which to study conservation and development.

Habitat and wildlife change in Maasailand

In both Kenya and Tanzania Maasailand, livestock on average accounts for over half of rural household income (Homewood et al. 2009), but livelihoods often also involve cultivation, whether commercial (some Mara households) or small-scale, as practised by approximately one-third of Mara, and two-thirds of Longido households (Homewood et al. 2009). Past collaborative research around the cross-border Serengeti–Mara Ecosystem (SME), has documented major differences in patterns of land use change between the Kenyan and Tanzanian sections of the SME. High spatial-resolution satellite imagery measured very significant loss of habitat around the Mara in Kenya

1975–95 (18% inner group ranch land converted from savanna to cropland; Homewood et al. 2001). By contrast, changes on the Tanzanian side were negligible, and habitat loss was balanced by regeneration elsewhere. Similarly, systematic aerial survey counts 1977–97 showed drastic decreases of 50–80% in all but two wild mammal species populations in Kenya SME (Ottichilo et al. 2000; Homewood et al. 2001). In the Tanzanian portion of the SME, wildlife species populations fluctuated from year to year with no overall trend. Livestock numbers fluctuated year to year in both the Kenyan and Tanzanian parts of the SME but show no overall trend (Homewood et al. 2001). The difference between Kenya and Tanzania in habitat change and in wildlife population trends runs counter to theoretical (Illius & O'Connor 1999; Vetter 2005), official government (United Republic of Tanzania 1997), media[1] and NGO[2] perceptions of major habitat loss due to small-scale agro-pastoral and particularly pastoralist impacts, and of drastic wildlife decline in both parts of the SME, rather than in the Kenyan part alone as suggested by long-term datasets. More recently, detailed analyses again confirm wildlife declines for Kenya (Western et al. 2006). Less well-substantiated analyses, based on snapshot surveys at the beginning and end of the 1990s (a decade of drought) posit some changes in Tanzania (Stoner et al. 2007). In SNP, 6 of 19 species surveyed showed a significant decline between 1991 and 2001 snapshot counts, while 13 species showed a significant increase or no significant change. In game reserves and game controlled areas (GCAs) surrounding the SNP, most species either showed no significant difference or the data were not adequate to show any significant difference. In contrast, 10 species showed a significant decline in Tarangire National Park, as against nine showing no significant difference or a significant increase. Unlike the Kenyan analyses, which draw on detailed long term datasets and can be taken as reliable trends, the Tanzanian data are based on changes between two points in time and cannot be taken as more than tentatively indicative in rangelands subject to extreme inter- and intra-annual variability. Nonetheless they do suggest, firstly, the possibility of significant declines in one-third to one-half of large mammal species surveyed (particularly around Tarangire as opposed to SME)

---

[1] Frequent Tanzanian media reports on eviction/displacement of cattle herders as a government environmental strategy, e.g.: http://www.thisday.co.tz/News/1446.html; http://www.thisday.co.tz/News/1842.html; http://www.thisday.co.tz/News/1565.html. Accessed January 2009.

[2] Serengeti National Park General Management plan 2006–16. Tanzania National Parks with Frankfurt Zoological Society; e.g. pp. 29–34. http://www.zgf.de/mitarbeiterbereich/pdf/Serengeti_GMP_%20July2006.pdf. Accessed January 2009.

and, secondly, the probability that such potential declines are currently neither as marked nor as all-pervasive as those seen in Kenya.

The main factor underpinning the sharp decline in Kenya's Mara wildlife has been linked to the spread of commercial cultivation across the Loita Plains, excluding wildlife from this key wet season resource (Homewood et al. 2001; Serneels & Lambin 2001). More recent work has focused on the equally severe impacts of small but intensive areas of irrigated cultivation extracting water from the Mara River, affecting the hydrology of the entire northern SME (Wolanski et al. 1999; Wolanski & Gereta 2001; Gereta et al. 2003; Mati et al. 2005). Quantitative and statistical analyses excluded climate, presence/density of livestock, rural population growth and uptake of smallholder cultivation as factors driving these strongly contrasting patterns of change in Kenya, versus fluctuation around average values in Tanzania. The major causes of this land use change have rather been shown as large-scale commercial investment in cultivation, most strongly associated with the high socio-economic status of households investing in this type of business, as well as proximity to markets and roads, rather than small-scale agro-pastoral development. Key local leaders – including Group Ranch chairman, treasurer or secretary, county councillor etc. – were 450 times more likely to be carrying out commercial cultivation (Thompson et al. 2002; Sachedina & Chenevix Trench 2009). Recently, smallholder cultivation spread around MMNR has emerged as a driver of landscape fragmentation and wildlife decline (Lamprey & Reid 2004).

Few households practised commercial cultivation around the Serengeti in Tanzania (Homewood et al. 2001), though this is changing fast in other areas of Tanzania Maasailand (e.g. Tarangire/Simanjiro: Sachedina 2008, Sachedina & Chenevix Trench 2009). This is not entirely surprising, but again flies in the face of government/NGO orthodoxy, which sees the spread of smallholder cultivation as being responsible for habitat loss and wildlife decline in Tanzania as much as in Kenya. Large-scale cultivation by commercial entrepreneurs, including local elites, absentee landowners and outside investors, is one of the main, and potentially the most important, drivers of land use change around the Mara and elsewhere in Kenya (Thompson et al. 2002), and is increasingly important in Tanzanian rangelands (Borner 1985; Igoe & Brockington 1999)[3].

Meanwhile conservation organizations, in an effort to promote wildlife-based conservation uses as a viable land-use policy, stress the potential

---

[3] http://www.sunbiofuels.co.uk/content/projects-tanzania. Accessed January 2009.

economic benefits of wildlife-based land use and propose conservation-based management strategies as economically and ecologically rational solutions to local development (IUCN 1988; ACC 2001; Stiles 2007).

To examine the viability of this approach and, particularly, the trends towards conversion of land to forms of use incompatible with conservation in several parts of Maasailand, it is worth investigating the value of wildlife income and its pattern of distribution amongst local households in the different policy areas found in this study, and the implications of this income for both outside investors and local household decisions on investment in conservation or alternative uses of land.

Income from tourism: wildlife viewing and hunting revenues

New work explores the levels of income from wildlife tourism (both game viewing and hunting) and the ways such income is distributed across households and communities in five specific sites across the Maasai rangelands: Mara, Kitengela and Amboseli in Kenya, Longido and Tarangire in Tanzania (Homewood et al. 2009). As well as data on revenues to the community or village government level, quantitative data on a standard set of core variables for 200–300 households in each site addresses household economy (income from livestock, cultivation, off-farm and wildlife-based activities), household characteristics (social/demographic/educational variables) and spatial data on agro-ecological zone, infrastructure and proximity to national parks. These data show the extent to which local livelihoods are diversifying away from pastoralism, the main patterns and correlates of such diversification (or its absence), and the impacts of wildlife conservation revenues on household economies and livelihoods. Household-level interviews and surveys show very considerable site-specific differences in returns from conservation (Thompson 2002; Thompson & Homewood 2002; Homewood et al. 2005) according to tourism potential, tenure regime, urban growth, etc. Also, within the overall category of wildlife- or conservation-related revenue, different income streams flow from game viewing, from hunting and from tourist expenditure. These are controlled and distributed through different channels and have different implications for returns to government, entrepreneurs, communities and the individual households within those communities. These data are set out elsewhere (Homewood et al. 2009). In the present paper, we summarize preliminary findings from two main and contrasting study sites: the Maasai Mara in Kenya, and Longido District in Tanzania.

Kenya: Maasai Mara

Resource tenure

Kenya Maasai rangelands were held communally as Trust land during the colonial period. Since Independence, all the group ranches adjacent to the MMNR (except Siana Group Ranch) have been privatized into individual private allocations of around 100 acres. The process of subdivision and titling has been extremely contentious, with opportunities for corruption and manipulation on every level (Galaty 1999; Thompson & Homewood 2002; Homewood et al. 2004; Homewood et al. 2009). Those landowners who have been able to secure title are free to choose to keep livestock, cultivate or set land aside for wildlife among other possibilities. They do not own the wildlife on their land, but they do control access and can negotiate independently with tour operators for game viewing deals. In Kenya, no big game hunting is allowed (c.f. Norton-Griffiths & Said, Chapter 14).

Returns from wildlife conservation in and around Maasai Mara National Reserve (MMNR)

The MMNR represents the northern extension of the Serengeti ecosystem. It is the second highest earning protected area in Kenya after Nakuru National Park (World Resources Institute 2007) with 296,000 visitors in 2005 generating some $20 million more than 1999 (see also Norton-Griffiths & Said, Chapter 14). From 1989 to 2001 (when the group ranches were sub-divided into individual plots), 19% of gate revenues were distributed to nine group ranch associations representing people resident around MMNR. Residents of the group ranches complained that they received little in the way of practical benefit from this (although some community projects were funded) and that the majority of funds from this source, as well as the remaining percentage that flowed to Narok County Council, have not been invested in the MMNR, but rather have remained within the County Council structure. Partly in response to this, from 1994 onwards, the local group ranch residents formed

Wildlife Associations (WAs), initially congruent with the Narok District Group Ranches located around MMNR. These WAs collected entrance fees and visitor bed-night fees on group ranches outside the MMNR and generated significant income (one WA generated some $950,000 in 1998, e.g. Thompson 2002; Thompson et al. 2009). Members of these WAs received income in the form of a dividend averaging some $100 paid annually. However, greater incomes were generated by the few residents appointed as directors to the WA compared to the wider membership (Thompson 2002, 2005; Thompson et al. 2009). Since subdivision into private plots, there has been a rapid disaggregation of former WAs, and a re-grouping into new WAs, along lines described later in this paper.

Greater conservation income levels are available to landowning households neighbouring the Mara, who can lease their land to tour operators for camping on private land around MMNR (Figure 13.1). There has been an ongoing contest over access to these sites. Those able to secure access to sites on which the lodges and luxury camp sites are located have been able to generate large personal incomes, and since land privatization, to cement their holdings over these income streams, and to exclude this formerly major source of income

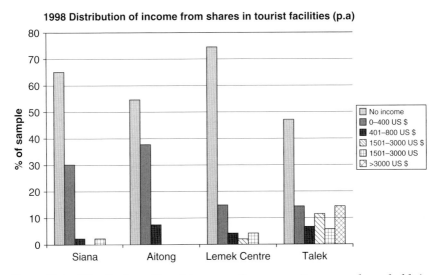

Figure 13.1   **Distribution of tourist revenue from campsites among households in different Mara study sites, 1998.**

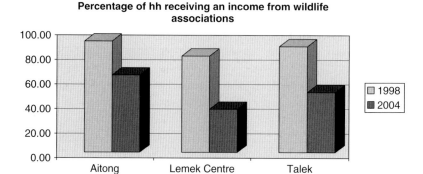

Figure 13.2 **Changing distribution of wildlife association income among house-holds in different Mara sites following land privatization.**

from the cooperatively run WAs (and from potentially more equitable forms of distribution through the annual WA's dividend).

In 1998–9, virtually all households sampled were members of WAs receiving a proportion of MMNR gate fees (Figure 13.2) and around 40% of 288 households surveyed around the Mara earned income from campsites or other tourist facilities (Figure 13.1). The distribution of that money among constituent households was anything but even but most households received some benefit from tourist revenues and some received very significant returns amounting to several thousand US dollars per annum (Figure 13.1). However, a follow-up survey in 2004 showed there had been a significant decline in tourism revenue since 9/11, and that returns had been concentrated into the hands of far fewer beneficiaries (Thompson et al. 2009). Around half of the households which in 1998 had access to WA revenue had lost this by 2004 (Figure 13.2). Similarly, the proportions of households with revenue from campsites and other tourist facilities declined in most areas (Thompson et al. 2009). This left around 50% of households earning some income from wildlife dividends or campsite shares in 2004, and an average of 5.4% of mean gross annual household income from these sources.

However, wildlife dividends and campsite shares only represent one part of the possible revenue potentially available from wildlife and tourism. There is also the possibility of earning income from off-farm work, which is based on tourism and conservation (e.g. as game warden, conservation agency employee or campsite labourer). Livelihoods analyses showed that taken together, employment in tourism-related enterprises, wildlife dividends and campsite shares mean that some 64% households receive wildlife-related income, and that on average such wildlife-related income accounts for 21% of total household annual income (Thompson et al. 2009). This puts wildlife-related earnings as second only to livestock income (which accounts for an average of 68% of mean annual household income). Analysis by wealth quintiles shows that (despite enormous variability in activities between households within quintiles), wildlife-related income makes up a fairly constant proportion of household income across the whole spectrum from best-off to poorest households in the Mara (Thompson et al. 2009). Cluster analyses identifying livelihood strategies (characterizing groups of households by the combinations and permutations of different income generating activities they engage in) show that wildlife-related income accounted for 20–30% of mean annual household income on average for four of the five livelihood strategies. Only poor agro-pastoralists, mainly based around Lemek in the agricultural belt remote from MMNR, received very little wildlife-related revenue (Thompson et al. 2009).

The picture of returns from wildlife-related sources around Mara is therefore complex. Income from dividends and shares has declined sharply and concentrated in significantly fewer hands between 1998 and 2004. Nonetheless, broader involvement in off-farm work arising from the wildlife conservation and tourism economy means wildlife-related income still forms a significant component of household income for two-thirds of Mara households. The different forms of wildlife-related income, taken all together, are reaching households across a broad spectrum of livelihood strategies and wealth ranks, second only to livestock and far outranking other forms of off-farm work (let alone cultivation) as a source of household income.

Nonetheless, the distributional issues around wildlife dividends and campsite shares drive a large part of local people's perceptions about the economic benefits of wildlife. The changes in distribution between 1998 and 2004 are the result of intense contestation at many different levels including:

1. the central State agency for conservation (Kenya Wildlife Service), and the county council which manages the MMNR;
2. the county council and WAs (originally congruent with group ranches, and at one time responsible for onward distribution of a part of the 19% MMNR revenues supposedly passed on by the county council for community projects);

3. the competing WAs, which are emerging as either breakaway exclusive subset associations of landowners with particularly high value tourist sites, able to deal with luxury end tour operators, or as new associations coalescing around political affiliation rather than spatial location (Thompson 2005; Thompson et al. 2009);
4. private conservancies and tourism concerns on formerly Maasai group ranch land (Figure 13.3a), now part-owned by outside investors (Figure 13.3b).

These local organizations and conservation agencies offer alternative structures for managing returns (Walpole & Leader-Williams 2001). Although two-thirds of households were earning some return from wildlife in 2004, and on average this was a significant contribution to household income, tourism revenues have fluctuated dramatically with perceptions of Kenya's relative security as a destination (with sharp drops in tourism visits following 9/11, the embassy bombings, and the 2007/8 post-election disturbances). Figures 13.1 and 13.2 show the proportion of households receiving an income from tourism-related sources falling from 1998 to 2004. This suggests that whenever tourism revenues fall, so do the proportions of households receiving wildlife-related returns, and so does the scale of those returns. Households which are excluded from wildlife revenue now express the intention to cultivate instead (Thompson 2005; Thompson et al. 2009). Local elites controlling the top-end campsites and dividends as well as the best jobs might be expected to be relatively protected against such fluctuations, but 2004 data suggest their returns have fallen significantly and disproportionately (Thompson et al. 2009). Land-use decisions made by those elites have a disproportionate impact, and it is not the case that elite households earning significant wildlife income revenues necessarily re-invest in conservation to safeguard their wildlife revenues. On the contrary, the same households are primarily responsible for investing in commercial cultivation. This is because the relative returns from wildlife, livestock and crops set up perverse incentives to convert arid and semi-arid rangelands to cultivation (Norton-Griffiths 2007; Norton-Griffiths & Said, Chapter 14), starting with key resources (e.g. Loita Plains, Mara River) where favourable agro-ecological conditions make cultivation a more profitable process (Norton-Griffiths 2007; Norton-Griffiths et al. 2008). Land-use decisions are further biased by the fact that crop and livestock revenues are relatively easy for a household to access and control, but wildlife revenue extremely difficult for most (Thompson & Homewood 2002). Only

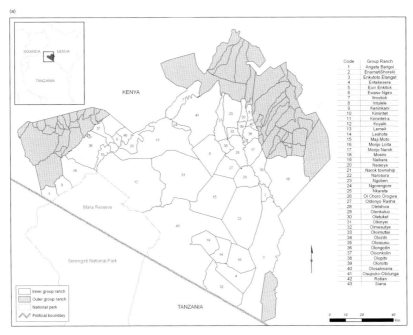

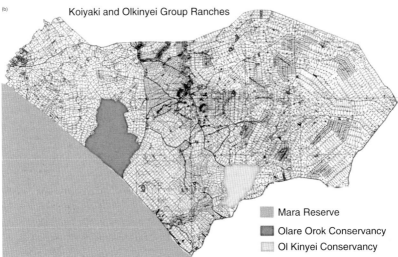

Figure 13.3   **Mara group ranches (a) before and (b) after subdivision and privatization.**

an estimated 5% of returns are captured by the landowner, as opposed to the rest of the tourism operator and support services (Norton-Griffiths 1998).

To summarize, in the Mara:

- Two-thirds of households earn some income from wildlife and on average, wildlife-related earnings account for 21% of household income. Households from all wealth ranks and from all livelihood strategies are involved in wildlife returns except poor agro-pastoralists who are mainly in the agricultural belt around Lemek and earning a significant part of their household income from cultivation.
- Kenya's policy of land privatization has meant that those who received a land allocation can potentially benefit from their secure private tenure in terms of receiving WA dividends and/or campsite shares.
- Wildlife dividends and campsite shares income is distributed very unevenly, with most captured by councillors, WA directors, and campsite owners. Disaggregated analysis shows a strong differentiation between households in these returns, and a decreasing proportion of households benefiting.
- Wildlife habitat and populations are declining. A major part of this decline was initially driven through land conversion for commercial cultivation (and, more recently, water capture for intensively irrigated commercial crops). Recent declines in the extent of large-scale cultivation have not been accompanied by any recovery of wildlife numbers.

The increasing differentiation in returns from WAs and campsite shares is consistent with concerns that the balance of access to and control of wildlife conservation-based revenues is shifting from local users to local and national elites, and ultimately to international investors (Figure 13.3b).

Tanzania: Longido District

Resource tenure

In Tanzania, in contrast to Kenya, most land is controlled by the State (Shivji 1998). Hunting is allowed under State license and it is important to distinguish between game viewing and hunting returns, which have until recently been controlled in very different ways. The protected areas adjacent to Longido

(Ngorongoro Conservation Area and SNP) are both owned and managed by the State. Funds do flow from them to adjacent District level government, but there is no system for dividends or even community-level projects for reserve adjacent dwellers. Hunting has major implications for the potential scale of wildlife revenues, but State and State-licensed entrepreneurs' control of land and of hunting and other tourism revenues affects the possibility of local households benefiting from such revenues.

Following the *Ujamaa* programme and creation of village land in the 1970s (Homewood & Rodgers 1991), the Land Act of 1992 was intended to extinguish customary rights altogether (Homewood et al. 2004). The new Village Land law which came into operation in Tanzanian Maasailand in 2001 has helped improve pastoralist land rights. However, pastoralist rangelands are defined in this land law as empty and therefore open to re-allocation by the government (Nelson et al. 2009). Throughout Tanzania, rural communities are voicing concern that the resources on which they depend are being set aside for conservation, or allocated to outside investors for agricultural development (Igoe & Brockington 1999), including, for example, biofuels[3], on the basis of centrally negotiated deals that yield no compensation or local benefit. An increasing number of villages, working with NGO support, have secured Certificates of Village Land as a basis for legally recognized village-based customary management of the rangelands. A growing number of villages have also worked with NGOs to create land-use plans as a way of safe-guarding village land rights. Village land law recognizes and provides for communal and customary forms of land tenure within and between villages. This has enabled Maasai villages to develop land-use plans that are largely based upon customary pastoralist communal range use practices, thereby precluding the need for sub-dividing their rangelands into smaller units. Ironically, sub-division has occurred in some villages due to villagers perceiving a direct threat from protected area expansion (A. Williams, Tanzania Natural Resources Forum, personal communication, 2007). In such cases, local people have sub-divided village land into plots. Although they continue to use these communally, individualization of the range has been adopted as a strategy to make it potentially harder for the government to annex their land into neighbouring protected areas. Finally, in some cases local village elites have misused the village land laws to allocate land to outsiders – particularly for farming – at the expense of local village members (Sachedina 2008).

At the time of writing, although individual *ujamaa* villages have had their lands surveyed and have nominal control over them through land-use plans, outside entrepreneurs can negotiate centrally with the State, and (minimally) with the village government, to lease high-potential land to cultivate, or, in GCAs, to hunt. Recently, the Tanzanian State moved towards adopting a policy of replacing GCAs with Wildlife Management Areas (WMAs) encouraged by donors and supported by NGOs (such as the African Wildlife Foundation, AWF). These are, in theory, community conservation areas where an *ujamaa* village sets aside part of its land for conservation and can derive wildlife revenue in return (from concessions for trekking, photographic tourism or hunting: URT 2005). However, State interests may align with those of outside entrepreneurs and of environmental NGOs in limiting villagers' access to land, to the detriment of local livelihoods and ultimately of conservation too, as local support for conservation wanes (Chapin 2004; Igoe 2007)[4].

Returns to communities from wildlife conservation in Longido District, Tanzania

Longido is arid, with hitherto relatively unremarked landscape and wildlife. It is ringed by Ngorongoro Conservation Area, Arusha and Mt. Kilimanjaro National Parks, and Amboseli National Reserve just across the border. As key resources around Amboseli and Kilimanjaro have been progressively converted to commercial cultivation (particularly Kimani and Namelok swamps and Loitokitok uplands) and as Amboseli elephant populations have increased, wildlife migratory routes have shifted to concentrate in Longido. International NGOs, particularly AWF, seek to protect such migratory routes and have collaborated with the Tanzanian State to establish Enduimet WMA, which initially set aside around 90% of the land of seven Longido villages for conservation.

Prior to this WMA's establishment, local deals with tour operators allowed game-viewing revenues to flow to such villages over the past decade

---

[4] 2007 Regulations for all non-consumptive wildlife use in all lands excluding National Parks and Ngorongoro Conservation Area. See the TNRF website for further details (<http://www.tnrf.org/node/6529>). Accessed January 2009.

(Nelson 2004; Nelson 2007). However, recently a serious conflict emerged between State and village interests (Nelson & Ole Makko 2005). Centrally-negotiated hunting licenses bring the State a larger income than do locally-negotiated game viewing/photographic safaris or village taxes. Deals between individual villages and tour operators were perceived to be constraining hunting companies' enterprises and, with them, the State's ability to capture revenue from tourist hunting in these areas. With the explicit support of the Division of Wildlife, Tanzanian hunting companies recently won a court ruling curtailing the right of particular villages to negotiate deals with game viewing tour operators. This was followed in November 2007 by new regulations criminalizing deals between tour operators and villages, and effectively channelling game viewing as well as hunting revenues through the State. At the same time, the Division of Wildlife has stalled over the move from GCAs to WMAs, in principle designed to shift control of revenue from either State or individual villages to collaborative multi-village groupings, such as is represented by Enduimet WMA. The lucrative hunting licences continue to be negotiated centrally, with the Division of Wildlife under no obligation to share revenue with the village[4](Kallonga et al. 2003). At the same time, village assemblies are challenging the WMA agreements that their village governments have signed.

Community-level benefits

Prior to the court ruling, some villages were making $15,000–$20,000 per year from tour operators (and considerably more in adjacent Loliondo District (Nelson 2004, 2007; Nelson et al. 2009). This revenue was managed by village governments, potentially funding significant improvements in local health, education and/or infrastructure. However, in a number of cases poor village governance meant little of this revenue benefited the village as a whole (Chenevix Trench et al. 2009). Even when benefits do flow to the community overall, individual households do not tend to perceive themselves as direct beneficiaries. It is not clear that WMA multi-village groupings would perform better on governance, even if the Division of Wildlife were to concede that they should receive a share of the game viewing (and/or hunting) revenue.

Table 13.1 **Longido: Household income from wildlife conservation sources.**

Village	No. households	No. households sampled	Percentage of households with wildlife/ conservation income	Mean (US$)	s.d.
Sinya	328	48	11	46.8	39.9
Elerai	240	31	3	31.5	
Olmolog	243	37	0	0	
Ngereyani	241	14	0	0	
Tingatinga	125	40	0	0	
Mairowa	188	59	0	0	
Total		229	3	0	

Returns to households from wildlife conservation in Longido District, Tanzania

The benefits at household level are even harder to see than at village level. Most of the 229 households surveyed in Longido fall well below the international poverty datum line of $1/person/day. Fewer than 3% households sampled earned wildlife income and less than 1% of household income is derived from wildlife tourism. Across all households the average return from this source was $1.2/household/year (Table 13.1). Our survey figures, collected before grazing and farming were affected by the creation of the WMA, show the extent of household dependence on livestock and crops and the implications of removing much of the land available for these activities (Homewood et al. 2009). Across Longido households overall, livestock account for just over 40% of income, off-farm work for around 35% and agriculture for just over 20%. Some of the off-farm work includes processing and sale of natural resources (wood, charcoal, thatch, plant medicines). Restrictions on the use of land, now designated as WMA and comprising the majority of village lands, for four out of six study villages, represent a major constraint on already pitifully poor livelihoods.

Wildlife conservation tourism in Maasailand: green development?

The different national and economic policy contexts of Kenya and Tanzania create some major contrasts between the potential for green development in the two parts of Maasailand.

- In Tanzania, state or parastatal ownership of the SNP and NCAA, and now central control of WMA revenues, means few wildlife-related returns are captured by local people. By contrast, in Kenya, county council management of the MMNR, and devolved management of tourism revenue, firstly to group ranch WAs and now to subset WAs and individual campsite owners, allows significant income to flow to local people.
- In Kenya, this income is quite widely distributed among households around the Mara, and has come to form a significant component of household income for all wealth ranks and livelihood strategies apart from poor agro-pastoralists particularly in the agricultural belt around Lemek.
- However, income from Kenyan WAs and campsites is extremely unevenly distributed. With land privatization and severe fluctuations in tourism over the last decade, dividends and campsite share income are concentrated in progressively fewer hands. Earlier work suggests that much is captured by a small socio-economic group with leadership positions and/or ownership of land where individual tourism facilities are found.
- Land subdivision in Kenya has removed the highest income-generating sites from WAs and placed them in individual ownership, removing the ability of such communal forms of management to redistribute wildlife-derived income to the wider population.

Maasailand seems an ideal scenario for green development, with sustainable use of natural resources through wildlife tourism creating revenue to underpin an improved standard of living and reduce poverty. In practice, this does not seem the case, either in terms of sustainable resource use or in terms of economic and infrastructure development. A review of land cover and wildlife population change data around the SME suggests that in Kenya, habitat loss, due primarily to commercial cultivation, is rapidly destroying the wildlife resource. By contrast, land use around the Tanzanian part of SME appears to have been sustainable at least until recently, with no significant loss of habitat or wildlife. This contrasts with the spread of large-scale cultivation elsewhere in Tanzanian Maasailand, such as around Tarangire National Park and across Simanjiro Plain.

As well as these differences in long-term trends over the last few decades, preliminary findings measuring conservation revenue to households illustrate very different tourism potential for the two contrasting cases of Mara in Kenya and Longido in Tanzania.

Wildlife tourism brings considerable revenue to the Mara, and many households there have benefited. However, wildlife returns to individual households and patterns of access to this revenue do not necessarily favour conservation. Returns from wildlife are significant, but limited compared to those potentially derived from commercial crops (Norton-Griffiths 2007; Norton-Griffiths et al. 2008), and are hotly contested between state, outside entrepreneurs, local elites and others in ways that increasingly cut ordinary residents out of the benefits. Excluded households have a strong incentive to cultivate, instead of choosing more conservation-compatible land use, while perverse incentives lead those households doing well from wildlife to invest their profits not in conservation but in cultivation. In Kenya, private land tenure and consequent control over revenues to land use enable individual landowners to freely make these choices.

In Tanzania, hunting adds another potential wildlife income stream, but State control restricts people's options and ability to benefit (whether from wildlife or alternative land uses). Together, with other macroeconomic and policy differences, this has meant less habitat change or wildlife decline around SME and Longido, if not around Tarangire/Simanjiro, but has also limited people's ability to benefit from wildlife returns. Currently, a negligible proportion of households have wildlife income; most of those which do earn trivial levels of revenue.

Alliances among State, outside entrepreneurs and environmental NGOs may further constrain and impoverish local users (Homewood et al. 2005; Nelson et al. 2009). The establishment of WMAs in Longido looks like an extension of conservation estate, driven in part from the outside, creating protected areas to compensate for destructive extraction elsewhere. At the same time, the Tanzanian State has moved to centralize, rather than devolve, control of revenues from non-consumptive as well as consumptive use of wildlife[4].This further restricts options for income from wildlife and will likely further degrade its importance in local livelihoods and land-use decisions.

So far, the WMA falls short of the lead NGO AWF's stated aims of enabling communities not only to benefit from wildlife resources, but to achieve fair and transparent sharing of those benefits and avoidance of costs (AWF 2005).

Conclusion

East African savannas, at first sight, have the conditions for synergies between conservation and development to work well, with sustainable use of natural resources underpinning development and poverty reduction. But the indications are that at the time of writing, ecological and social changes are not working well either for conservation or for development. Under Kenya's private land ownership system, both the relative returns to different land uses and the relative ease or difficulty of capturing and controlling such returns, alongside perverse incentives, mean that more and more landowners are excluded from incomes derived from wildlife and tend to opt for land uses that are not compatible with conservation. Under Tanzania's State-dominated system, alliances between entrepreneurs and State bypass local people, leaving negligible wildlife-based local income streams open to them, and thus undermining development based on wildlife. In Tanzania, ways of permitting local income generation from the wildlife resource are needed, which will require devolving control for revenue generation from State ownership to local village structures. Yet the patterns of income distribution that have evolved in Kenya show this will not be enough of itself. Conservation initiatives need to look at mechanisms such as the WA structures that evolved from 1994 to 2001 in Kenya, the distributional effects of wildlife income generation through these structures, and at the systems of land tenure and access which structure people's ability to pursue and benefit from conservation-compatible options, if incentives for local people to benefit from and to conserve wildlife are to be found.

References

African Conservation Centre (ACC). (2001) *Management Plan for Olchoro Oirua and Koiyaki-Lemek Wildlife Associations.* African Conservation Centre, Nairobi.

African Wildlife Foundation. (2005) *Kilimanjaro Heartland: A Summary of AWF's Engagement and Related GIS Data.* Belgium Directorate General for International Cooperation Reto-o-Reto meeting, International Livestock Research Institute, Nairobi.

Anderson, D. (2002) *Eroding the Commons: The Politics of Ecology in Baringo, Kenya, 1890s–1963*. James Currey Publishers, Oxford.

Borner, M. (1985) The increasing isolation of Tarangire National Park. *Oryx* 19, 91–96.

Castree, N. (2003) Bioprospecting: from theory to practice (and back again). *Transactions of the Institute of British Geographers* 28, 35–55.

Chapin, M. (2004) A challenge to conservationists. *World Watch Magazine* 17(6), 17–31.

Chenevix Trench, P., Kiruswa, S., Nelson, F. & Homewood, K. (2009) Still "people of cattle"? Livelihoods, diversification and community conservation in Longido District. In: Homewood, K., Kristjanson, P. & Chenevix Trench, P. (eds.) *Staying Maasai?: Livelihoods, Conservation and Development in East African Rangelands*. Springer, New York, pp. 217–256.

CHL Consulting Group. (2002) The United Republic of Tanzania Ministry of Natural Resources and Tourism, Tourism Master Plan: Strategy and Actions, Final Summary Update. CHL Consulting Group, Dublin. www.tzonline.org/pdf/tourismmasterplan.pdf.

Coast, E.E. (2002) Maasai socioeconomic conditions: a cross-border comparison. *Human Ecology* 30, 79–105.

Escobar, A. (1996) Constructing nature. In: Peet, R. & Watts, M. (eds.) *Liberation Ecologies*. Routledge, London, pp. 46–68.

Galaty, J. (1999) Grounding pastoralists: law, politics and dispossession in East Africa. *Nomadic Peoples* 3, 56–57.

Gereta, E., Wolanski, E.J. & Chiombola, E.A.T. (2003) *Assessment of the Environmental, Social and Economic Impacts on the Serengeti Ecosystem of the Developments in the Mara River Catchment in Kenya*. Tanzania National Parks and Frankfurt Zoological Society. Tanzania National Parks, Arusha.

Gichohi, H. (2000) Functional relationships between parks and agricultural areas in east Africa: the case of Nairobi National Park. In: Prins, H.H.T., Grootenhuis, J.G. & Dolan, T.T. (eds.) *Wildlife Conservation by Sustainable Use*. Kluwer Academic Publishers, Dordrecht, pp. 141–168.

Homewood, K., Coast, E. & Thompson, M. (2004) In-migration and exclusion in east African rangelands: access, tenure and conflict. *Africa* 74, 567–610.

Homewood, K., Kristjanson, P. & Chenevix Trench, P. (eds.) (2009) *Staying Maasai?: Livelihoods, Conservation and Development in East African Rangelands*. Springer, New York.

Homewood, K., Lambin, E.F., Coast, E. et al. (2001) Long-term changes in Serengeti-Mara wildebeest and land cover: pastoralism, population, or policies? *Proceedings of the National Academy of Sciences of the United States of America* 98, 12544–12549.

Homewood, K.M. & Rodgers, W.A. (1991) *Maasailand Ecology: Pastoralist Development and Wildlife Conservation in Ngorongoro, Tanzania.* Cambridge University Press, Cambridge.

Homewood, K., Thompson, M., Chenevix Trench, P., Kiruswa, S. & Coast, E. (2005) Community- and state-based natural resource management and local livelihoods in Maasailand. Gestione della risorse naturali su base communitaria e statale. Ambiente e sviluppo sostenibile in Africa australe. *Afriche e Orienti* Special issue 2, 84–101.

Hughes, L. (2006) *Moving the Maasai: A Colonial Misadventure.* Palgrave Macmillan, Hampshire.

Igoe, J. (2007) Human rights, conservation and the privatization of sovereignty in Africa: a discussion of recent changes in Tanzania. *Policy Matters* 15, 241–254.

Igoe, J. & Brockington, D. (1999) *Pastoral Land Tenure and Community Conservation: A Case Study from North-East Tanzania*, Pastoral Land Tenure Series, vol. 11. International Institute for Environment and Development, London.

Illius, A. & O'Connor, T. (1999) On the relevance of nonequilibrium concepts to arid and semiarid grazing systems. *Ecological Applications* 9, 798–813.

International Union for Conservation of Nature and Natural Resources. (1988) *Toward a Regional Conservation Strategy for the Serengeti.* IUCN, Nairobi.

Kallonga, E., Rodgers, A., Nelson, F., Ndoinyo, Y. & Nshala, R. (2003) Reforming environmental governance in Tanzania: natural resource management and the rural economy. Inaugural Tanzanian Biennial Development Forum, Dar es Salaam.

Kristjanson, P., Radeny, M., Nkedyanye, D. et al. (2002) *Valuing Alternative Land Use Options in the Kitengela Wildlife Dispersal Area of Kenya.* International Livestock Research Institute Impact Assessment Series, vol. 10. International Livestock Research Institute, Nairobi.

Lamprey, R.H. & Reid, R.S. (2004) Expansion of human settlement in Kenya's Maasai Mara: what future for pastoralism and wildlife? *Journal of Biogeography* 31, 997–1032.

Mati, B., Mutie, S., Horne, P., Mtato, F. & Gedain, H. (2005) Land use changes in the transboundary Mara basin. 8th International River Symposium, Brisbane.

Nelson, F. (2004) *The Evolution and Impacts of Community-Based Ecotourism in Northern Tanzania*, Drylands Issue Paper no. E131. International Institute for Environment and Development, London.

Nelson, F. (2007) *Emergent or Illusory? Community Wildlife Management in Tanzania*, Drylands Issue Paper 146. International Institute for Environment and Development, London.

Nelson, F., Gardner, B., Igoe, J. & Williams, A. (2009) Community-based conservation and Maasai livelihoods in Tanzania. In: Homewood, K., Kristjanson, P. &

Chenevix Trench, P. (eds.) *Staying Maasai?: Livelihoods, Conservation and Development in East African Rangelands*. Springer, New York, pp. 299–334.

Nelson, F. & Ole Makko, S. (2005) Communities, conservation, and conflicts in the Tanzanian Serengeti. In: Lyman, M.W. & Child, B. (eds.) *Natural Resources as Community Assets: Lessons from Two Continents*. Sand County Foundation & Aspen Institute, Monona & Washington, DC, pp. 121–145.

Norton-Griffiths, M. (1998) The economics of wildlife conservation policy in Kenya. In: Milner-Gulland, E.J. & Mace, R. (eds.) *Conservation of Biological Resources*. Blackwell Science, Oxford, pp. 279–293.

Norton-Griffiths, M. (2007) How many wildebeest do you need? *World Economics* 8, 41–64.

Norton-Griffiths, M., Said, M., Serneels, S. et al. (2008) Land use economics in the Mara area of the Serengeti ecosystem. In: Sinclair, A.R.E., Packer, C., Mduma, S.A.R. & Fryxell, J.M. (eds.) *Serengeti III: Human Impacts on Ecosystem Dynamics*. University of Chicago Press, Chicago, pp. 379–416.

Ojalammi, S. (2006) *Contested Lands: Land Disputes in Semi-Arid Parts of Northern Tanzania, Case Studies of the Loliondo and Sale Divisions in the Ngorongoro District*. Ph.D. Dissertation. Geography Department, University of Helsinki.

Ottichilo, W.K., de Leeuw, J., Skidmore, A.K., Prins, H.H.T. & Said, M.Y. (2000) Population trends of large non-migratory wild herbivores and livestock in the Masai Mara ecosystem, Kenya, between 1977 and 1997. *African Journal of Ecology* 38, 202–216.

Pearce, D. & Moran, D. (1994) *The Economic Value of Biodiversity*. Earthscan, London.

Rutten, M. (1992) *Selling Wealth to Buy Poverty: The Process of the Individualization of Landownership Among the Maasai Pastoralists of Kajiado District, Kenya, 1890–1990*. Verlag für Entwicklungspolitik, Saarbrücken.

Sachedina, H. (2008) *Wildlife is Our Oil: Conservation, Livelihoods and NGOs in the Tarangire Ecosystem*. Ph.D. Thesis. University of Oxford.

Sachedina, H. & Chenevix Trench, P. (2009) Cattle and crops, tourism and tanzanite: poverty, land-use change and conservation in Simanjiro District, Tanzania. In: Homewood, K., Kristjanson, P. & Chenevix Trench, P. (eds.) *Staying Maasai?: Livelihoods, Conservation and Development in East African Rangelands*. Springer, New York, pp. 263–298.

Serneels, S. & Lambin, E.F. (2001) Proximate causes of land-use changes in Narok District, Kenya: a spatial statistical model. *Agriculture, Ecosystems and Environment* 85, 65–81.

Shivji, I.G. (1998) *Not Yet Democracy: Reforming Land Tenure in Tanzania*. International Institute for Environment and Development & Land Rights Research and Resources Institute (LARRRI/HAKIARDHI), London & Dar es Salaam.

Southgate, C. & Hulme, D. (2000) Uncommon property: the scramble for wetland in southern Kenya. In: Woodhouse, P., Bernstein, H. & Hulme, D. (eds.) *African*

Enclosures? The Social Dynamics of Wetlands in Drylands. James Currey Publishers, Oxford, pp. 73–118.

Spear, T. & Waller, R. (eds.) (1993) *Being Maasai: Ethnicity and Identity in East Africa*. James Currey Publishers, Oxford.

Stiles, D. (2007) The Mara at the crossroads. *Swara* 30(3), 52–55.

Stoner, C., Caro, T., Mduma, S., Mlingwa, C., Sabuni, G. & Borner, M. (2007) Assessment of effectiveness of protection strategies in Tanzania based on a decade of survey data for large herbivores. *Conservation Biology* 21, 635–646.

Thompson, D.M. (2002) *Livestock, Cultivation and Tourism: Livelihood Choices and Conservation in Maasai Mara Buffer Zones*. Ph.D. Thesis. University of London.

Thompson, D.M., Serneels, S., Ole Kaelo, D. & Chenevix Trench, P. (2009) Maasai Mara - land privatization and wildlife decline: can conservation pay its way? In: Homewood, K., Kristjanson, P. & Chenevix Trench, P. (eds.) *Staying Maasai?: Livelihoods, Conservation and Development in East African Rangelands*. Springer, New York, pp. 77–114.

Thompson, M. (2005) *Valuing Land Use Options in the Maasai Mara, Interim Report*. International Livestock Research Institute, Nairobi.

Thompson, M. & Homewood, K. (2002) Elites, entrepreneurs and exclusion in Maasailand. *Human Ecology* 30, 107–138.

Thompson, M., Serneels, S. & Lambin, E.F. (2002) Land-use strategies in the Mara ecosystem (Kenya): a spatial analysis linking socio–economic data with landscape variables. In: Walsh, S.J. & Crews-Meyer, K.A. (eds.) *Remote Sensing and GIS Applications for Linking People, Place and Policy*. Kluwer Academic Publishers, Dordrecht, pp. 39–68.

United Republic of Tanzania. (1997) Livestock and Agriculture Policy, Section 4: Range Management. United Republic of Tanzania Policy Statements, Dar es Salaam, pp. 127–131.

United Republic of Tanzania. (2005) *Mkukuta: Tanzania's National Strategy for Growth and Reduction of Poverty*. United Republic of Tanzania Vice-President's Office, Dar-es-Salaam.

Vetter, S. (2005) Rangelands at equilibrium and non-equilibrium: recent developments in the debate. *Journal of Arid Environments* 62, 321–341.

Walpole, M.J. & Leader-Williams, N. (2001) Masai Mara tourism reveals partnership benefits. *Nature* 413, 771.

Western, D., Russell, S. & Mutu, K. (2006) *The Status of Wildlife in Kenya's Protected and Non-Protected Areas*. The First Stakeholders' Symposium of the Wildlife Policy and Legislation Review, Nairobi.

Wolanski, E. & Gereta, E. (2001) Water quantity and quality as the factor driving the Serengeti ecosystem. *Hydrobiologia* 458, 169–180.

Wolanski, E., Gereta, E., Borner, M. & Mduma, S. (1999) Water, migration and the Serengeti ecosystem. *American Scientist* 87, 526–533.

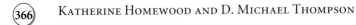
Woodhouse, P., Bernstein, H. & Hulme, D. (eds.) (2000) *African Enclosures?: The Social Dynamics of Wetlands in Drylands.* James Currey Publishers, Oxford.

World Resources Institute. (2007) *Nature's benefits in Kenya; an Atlas of Ecosystems and Human Wellbeing,* chapter 6 – Tourism, WRI, with DRSS, MENR, CBS, MPND, ILRI. WRI, Washington, D.C. & Nairobi.

14

The Future for Wildlife on Kenya's Rangelands: An Economic Perspective

Michael Norton-Griffiths[1] and Mohammed Y. Said[2]

[1]Independent Researcher, Kenya
[2]International Livestock Research Institute, Kenya

Introduction

Kenya's rangelands fall within 19 districts classified as arid and semi-arid lands (ASAL). These 19 districts cover 511,000 km^2 (88%) of Kenya's landmass and are populated primarily by some 4.5 million, mainly traditional, transhumant pastoralists along with some nine million head of livestock (Norton-Griffiths 1998; World Resources Institute et al. 2007). The backbone of Kenya's conservation effort is the 38,000 km^2 network National Parks, National Reserves, Country Council Reserves and other sanctuaries. The greater part (84%) of these Protected Areas is found in the ASAL Districts along with the great majority (>90%) of Kenya's wildlife. Uniquely to Kenya, some 70% or more of all large wildlife live, either permanently or seasonally, outside these formally protected areas.

Within the network of protected areas, the State, acting through the Kenya Wildlife Service (KWS) and its agents (local councils), has absolute regulatory powers over access, development and wildlife utilization. Within these protected areas, conservation is accordingly a matter of command and control.

However, outside the protected areas it is altogether a different matter. Here, command and control cannot be effective, for it is up to the individual

Wild Rangelands: Conserving Wildlife While Maintaining Livestock in Semi-Arid Ecosystems, 1st edition. Edited by J.T. du Toit, R. Kock, and J.C. Deutsch.
© 2010 Blackwell Publishing

landholders[1] to decide whether or not to maintain wildlife on their land. Outside the protected areas, incentives are required for landholders not only to conserve and husband the wildlife on their land but also to include habitat and wildlife conservation in their long term investment strategies for agricultural and livestock production.

Prior to 1977, the wildlife conservation policy in Kenya was very broad based and included complete preservation (within the protected areas); live capture for sale and export; ranching, cropping and culling; tanning, taxi-dermy and curios; and sport hunting within both long term and short term concession areas and hunting blocks. The consumptive wildlife industry may have been worth some $20 million in 1977 (personal calculations) of which a major component went directly to landholders as well as providing development funds at the District level (FAO 1978). Today the industry might be worth some $600 million annually, allowing for the discounted value of the dollar and the growth in the value of the industry throughout Africa.

In an abrupt policy change in 1977, all hunting was banned and this was followed in 1978 by a complete ban on all other consumptive utilization (live capture and sales, cropping, ranching, manufacture of trophies and curios). After this date, the sole use to which wildlife could be put was game viewing.

By the early 90s, there were the first indications that wildlife outside of the protected areas were in decline (Broten & Said 1995; Norton-Griffiths 1995, 1996; Sinclair 1995; de Leeuw et al. 1998; Ottichilo et al. 2000; Ottichilo et al. 2001) and in 1992, seeking to improve matters, the newly created KWS started to reinstate wildlife related benefits to landholders by permitting consumptive utilization (reduction cropping) on ranches infested with wildlife. Eventually, some 60 wildlife cropping, ranching and farming operations were licensed and game meat products could be sold on the open market. The KWS also encouraged neighbouring landholders to form licensed Wildlife Associations and Wildlife Forums to jointly manage their wildlife, much as neighbouring landholders do throughout southern Africa and Europe.

The new Community Wildlife Service of the KWS also started to provide tangible benefits to landholders by disbursing a proportion of gate receipts to communities living around the protected areas as Wildlife Development Funds

---

[1] The term 'landholder' is used here to include both landowners (with private or group tenure) and land users (usually with traditional usufruct rights).

for social investment (Berger 1993). They also began to assist landholders to negotiate more advantageous concession fees with tourism operators and to set up their own privately financed tourist operations, such as camp sites, tented camps and camel trekking.

Although returns from the cropping were low, typically around $0.5/hectare/year (Norton-Griffiths & Butt 2006), they were enough to encourage a benign attitude towards wildlife by landholders and especially to ignore some development options[2]. Specifically, it was not so much the earnings that were important but the increase in ranch profitability following the reduction in wildlife numbers (Norton-Griffiths et al. 2008; see Underlying economic forces). However, all cropping activities were again abruptly terminated in 2003.

Despite all these efforts, by the mid-90s it was becoming clear that some 50% of wildlife had been eliminated from the ASAL rangelands. The only good news was that loss rates seemed to be less within protected areas than outside (Norton-Griffiths 1998). However, recent analyses (Western et al. 2009) show that loss has continued unabated and is equally bad, both inside and outside protected areas, with some 70% of all large wildlife now eliminated from the protected areas and rangelands of Kenya – this despite the literally hundreds of millions of dollars spent on conservation efforts by international donors, conservation NGOs and the Government of Kenya. This is a terrible indictment of conservation policy and evidence of a massive failure by all concerned (Norton-Griffiths 2007).

Even in the mid-90s, the analysis of loss rates suggested a strong underlying economic component (Norton-Griffiths 1998). Losses, for example, were less on adjudicated land (group ranches) than on unadjudicated land; and were less where tourists visited than where they did not visit. Furthermore, where wildlife revenues went more clearly and transparently to landholders (group, communal or private) rather than to central government, wildlife was either holding its own or perhaps even increasing.

The recent analysis of Western et al. (2009) confirms these earlier analyses. Today, the only places where wildlife has held its own or actually increased are on large, private landholdings and private conservancies, or on Group Ranches savvy enough to strike profitable contracts with tour operators.

---

[2] In Laikipia District, over 990 km[2] of local community land (Ol Moran, Nagum and Sosian group ranches) were subdivided and fenced once the cropping programmes were terminated. The income from cropping, although low, was enough to make it worthwhile for the landowners to maintain their extensive, pastoral livestock production system.

System dynamics

Socio-economic changes on Kenya's rangelands

It is now clear that since the mid-1970s there have been major changes to the basic production systems on Kenya's rangelands (Figure 14.1). The human population growth of some 3.1% per annum has been accompanied by an 8.6% per annum growth in the area under cultivation, a trend seen in all ASAL Districts (Norton-Griffiths & Butt 2006; World Resources Institute et al. 2007). The livestock data are especially intriguing: while livestock populations vary appreciably year on year in response to rainfall, they appear to be stable in terms of overall trend (Broten & Said 1995; Norton-Griffiths 1998; Ottichilo et al. 2000). However, offtake in terms of sales at an auction is growing at some 4.4% per annum. Finally, wildlife everywhere are in steep decline (−3.2% per annum).

Taken together, these trends signal a fundamental switch in production strategies across the rangelands from an extensive, mainly pastoral production system, to a more intensive agro-pastoral system which excludes wildlife. While the growth in cultivated areas may well be a response to population growth, the

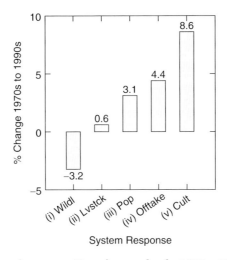

Figure 14.1   **System changes on Kenya's rangelands, 1970s–1990s: % per annum change in (i) wildlife numbers, (ii) livestock numbers, (iii) human population, (iv) livestock offtake (live sales) and (v) hectares under cultivation.**

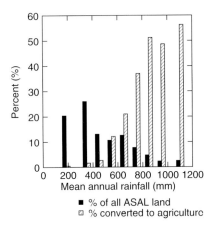

Figure 14.2   **Conversion of ASAL land to agriculture.**

gains in livestock, an offtake from the same size of herd, signals the adoption of more intensive methods of production geared to the cash economy.

Conversion of rangelands to cultivation

Figure 14.2, and Tables 14.1 and 14.2, demonstrate the extent to which the rangelands of the arid and semi-arid lands of Kenya are being converted to cultivation. Twelve per cent of the rangelands now support agriculture (Table 14.1 E) and it is in the areas of highest rainfall and therefore, of highest agricultural potential which have been preferentially converted (Figure 14.2). Of the 9% of rangelands outside the formally protected areas which receive more than 800 mm of rainfall (Table 14.2 C)[3], 52% now support agriculture (Table 14.2 E)[4]. In contrast,

---

[3] Rainfall data were obtained from the AWhere-ACT Kenya Database (Mudsprings Inc. 1999) on a 5.5 by 5.5 km grid. A topographically dependent climate surface, fitted to the point data and interpolated at a 1-km resolution using elevation, was overlaid with district boundaries to calculate the areas receiving different rainfalls.

[4] The Africover Project (Africover 2003) mapped all land uses, including both large-scale and small scale cultivation, across the whole land area of Kenya for the base year of 2000 using a combination of satellite imagery, aerial photography and extensive ground control. However, Africover does not map the actual hectares cultivated but rather the area within which certain types of cultivation occur. For example, a typical polygon in the Africover database might have the description '...isolated (in natural vegetation or other) rainfed herbaceous crop (field density 10–20% polygon area).....' along with an estimated area of the polygon. The rainfall surface (3. above) was overlaid onto the Africover surface to calculate the areas cultivated in each rainfall band within the 19 ASAL districts.

Table 14.1 Conversion of rangelands to cultivation.

A. Mean Annual Rainfall (mm)	≤200	300	400	500	600	700	800	900	≥1000	Totals
General										
B. Area (km²)	104,073	132,817	66,989	54,510	64,509	39,528	23,976	11,622	12,729	510,753
C. Area (km²) inside protected areas	1265	3428	4823	4340	11,123	7737	3006	1532	1085	38,339
D. Area (km²) outside protected areas	102,808	129,389	62,166	50,170	53,386	31,791	20,970	10,090	11,644	472,414
Agricultural rents outside the Protected Areas										
E. Area (km²) supporting agriculture	500	2099	1651	6105	11,216	11,767	10,742	4908	6556	55,544
F. % converted	<1%	2%	3%	12%	21%	37%	51%	49%	56%	12%
G. Mean agricultural rents ($/hectare/year)	$5.50	$11.95	$19.13	$30.60	$48.96	$78.34	$124.34	$200.54	$405.86	
H. Potential total agricultural rents ($ m)	$57 m	$155 m	$119 m	$154 m	$261 m	$249 m	$261 m	$202 m	$473 m	$1930 m
I. Actual agricultural rents ($ m) (% of potential)	$0 m (0%)	$3 m (2%)	$3 m (3%)	$19 m (12%)	$55 m (21%)	$92 m (37%)	$134 m (51%)	$98 m (49%)	$266 m (56%)	$670 m (35%)
Livestock rents outside the Protected Areas										
J. Mean livestock rents ($/hectare/year)	$1.05	$2.57	$4.42	$7.60	$13.07	$22.47	$38.63	$66.42	$149.76	
K. Total potential livestock rents ($ m)	$11 m	$33 m	$27 m	$38 m	$70 m	$71 m	$81 m	$67 m	$174 m	$573 m

Wildlife rents outside the Protected Areas

L. Area (km²) used by tourists	40	917	1401	3612	5120	3249	2373	1360	1927	19,999
M. Wildlife rents ($ m) @ $10.2/hectare/year	<$0 m	$5 m	$7 m	$18 m	$26 m	$16 m	$12 m	$7 m	$10 m	$100 m
N. Wildlife rents ($ m) @ $50/hectare/year	<$0 m	$1 m	$1 m	$4 m	$5 m	$3 m	$2 m	$1 m	$2 m	$20 m

A: Mean annual rainfalls (mm) (see footnote 3).

B: Areas (km²) in each rainfall band (see footnote 3).

C: Protected areas include national parks, national reserves and forest reserves.

D: On the land (D) outside the protected areas determined from the Africover data base (footnote 4).

E: % of the land outside the Protected Areas supporting cultivation (E as % of D).

G,J: Agricultural and livestock rents are defined as net returns to land ($/hectare/year) from agricultural or livestock production: mean rents as a function of rainfall are derived from equations 13.1 and 13.5 in Norton-Griffiths et al. (2008) (see Figure 14.4 and footnote 5).

H: Potential net returns to agriculture in millions of US $ ($ m) if all land outside the protected areas were to be converted to cultivation (D*100*G) .

I: Value of agricultural rents ($ m) on the land currently supporting cultivation (E*100*G).

K: Potential net returns to livestock ($ m) on the land outside the Protected Areas (D*100*J).

L: Wildlife tourists use only a very small proportion (~5%) of the rangelands (Norton-Griffiths & Butt 2006). Here, tourism use is broadly broken down by rainfall band.

M,N: Wildlife Rents (on the tourism land L) on the assumption of $10.2/hectare/year (average wildlife rents paid to landowners) and $50/hectare/year (highest rents paid to landowners) see Norton-Griffiths and Butt (2006) and Norton-Griffiths et al. (2008).

Table 14.2 **Changes to livestock and wildlife, 1970s–1990s (outside the protected areas).**

Stratum (on % supporting cultivation)	Low	Medium	High	Totals
A. Mean annual rainfall (mm)	≤500	>500 <800	>800	
B. Area (km^2) outside protected areas	294,363	135,347	42,704	472,414
C. % area outside protected areas	62%	29%	9%	
Agricultural, Livestock and Wildlife Rents outside the Protected Areas				
D. Area (km^2) supporting agriculture	4250	29,088	22,206	55,544
E. % converted	1%	22%	52%	12%
F. Actual agricultural rents ($ m) (% of potential)	$6 m (2%)	$166 m (25%)	$498 m (52%)	$670 m (35%)
G. Potential livestock rents ($ m)	$72 m	$179 m	$322 m	$573 m
H. Area (km^2) used by tourists	2358	11,981	5660	19,999
I: Wildlife rents @ $10.2/hectare/year ($ m)	$2 m	$12 m	$6 m	$20 m
J. Wildlife rents @ $50/hectare/year ($ m)	$12 m	$60 m	$28 m	$100 m
Change in Livestock 1970s–1990s				
K. % per annum change 1970s–1990s (ns)	(−1.34)	+2.21	(+0.97)	+0.60
p value of trend	0.19	0.02	0.39	0.40
L. % change over 20 years (1977–97) (ns)	(−21%)	+55%	(+25%)	(+13%)
M. % of all livestock 1990s	50%	32%	18%	100%
Change in Wildlife 1970s–1990s				
N. % per annum change 1970s–1990s (ns)	−4.17	(−1.31)	−3.86	−3.24
p value of trend	0.00	0.21	0.00	0.00
O. % change over 20 years (1977–97) (ns)	−54%	(−24%)	−58%	−55%
P. % of all wildlife 1990s	25%	44%	31%	100%

A–J, Summarized from Table 14.1 within the three strata low, medium and high; $ m millions of US $; K–P, KREMU/DRSRS census data from 1970s through to the 1990s (footnote 5); ns, trend or % change not statistically significant.

only 1% of the lowest rainfall areas (<500 mm annual rainfall) have been converted.

It is clear from Tables 14.1 and 14.2 that agricultural conversion is cascading down the rainfall gradient and out across the rangelands of Kenya.

Differential impacts of agricultural conversion on livestock and wildlife outside protected areas

This conversion of rangelands to agriculture has had quite different effects on livestock when compared with wildlife[5]. With livestock, the majority (50%, Table 14.2 M) are still to be found in the drier (<500 mm of rainfall) rangeland areas and relatively fewer in the wetter areas of more agricultural potential. Across the entirety of the rangelands there is no significant trend in livestock numbers over the last 20 years or so (Table 14.3 K). There is, however, some indication of a redistribution of livestock with numbers increasing significantly in the medium rainfall stratum (>500–<800 mm) and perhaps decreasing in the areas of highest agricultural potential (but the trend is not statistically significant), a trend not seen in a more detailed study around the Maasai Mara National Reserve (Norton-Griffiths et al. 2008).

In contrast, the chronic loss of wildlife across the rangelands of – 3.24% per annum is clear to see (Table 14.2 N). While loss rates are very high (−4.17% per annum) in the areas of highest agricultural potential and greatest agricultural conversion, wildlife are perhaps holding their own in the medium rainfall band with lower rates of agricultural conversion (the negative trend is not statistically significant). In contrast, loss rates are again high (−3.8% per annum) in the most arid areas where the marginal impacts of wildlife on livestock production are greatest (Norton-Griffiths & Butt 2006). In general terms, while livestock is being absorbed along the expanding agricultural frontier into a developing agro-pastoral land use matrix, wildlife is being displaced and eliminated (Figure 14.3; Norton-Griffiths et al. 2008). Indeed, a

---

[5] Data Source: Aerial surveys of livestock and wildlife carried out by the Kenya Rangeland Ecological Monitoring Unit (KREMU), later renamed the Department of Remote Sensing and Resource Surveys (DRSRS) from 1977 onwards, reported at District level for each wildlife and livestock species (GOK 1995a,b). Analysis follows Norton-Griffiths (1998), simple OLS equation of the form ln(numbers) = constant + b * year, where year is a decimal variable with January 1997 = 0.083 etc, and with dummy variables (within each stratum) for each District and for each livestock or wildlife species. In Table 14.2, %pa = instantaneous rate of change, p values are two tailed.

Table 14.3   **Dynamics of land values on the Kitengela in northern Kajiado district, Kenya 1997–2007.**

Value of land in Rongai township, 2007 ($/hectare)	$300,000–$400,000
Value of land at Orly Air Park ($/hectare)	
2007	$8800
1997	$2999
Drop in land values between Rongai and Orly Air Park, 25 km (%/km)	−13.7%/km
Growth in land values at Orly Air Park 1997–2007 (%/year)	11.4%/year
Average return on 10 year Treasury Bills (%/year)	10%–11%/year
Orly Air Park: rainfall 550 mm: net returns to land (%), 2002 prices	
Combined agricultural and livestock production @ $35.98/hectare/year	0.41%
Potential from wildlife production @ $10/hectare/year	0.11%

Sources: Land Values in Rongai (2007), land vendors and purchasers, quantity surveyors and solicitors: at Orly Air Park, Aero Club of East Africa (purchasers of the land in 1997 and 2007): Yield of Treasury Bonds from National Bank of Kenya website: returns to agricultural, livestock and wildlife production (Norton-Griffiths & Butt 2006; Norton-Griffiths et al. 2008).

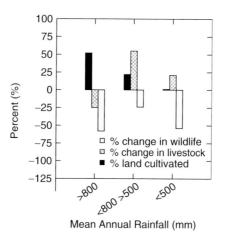

Figure 14.3   **Changes in cultivation, livestock and wildlife in the ASAL: 1970s–1990s.**

simple elasticity analysis suggests that for every 1% increase in land supporting cultivation there is a 0.85% decrease in wildlife densities.

The rapid evolution of property rights (see Evolution of property rights below) throughout the rangeland areas, from large land parcels under group or communal tenure into small land parcels under private tenure, is accompanied by an expansion of more permanent settlements and by the general development of agricultural infrastructure, such as fencing. Wildlife are displaced and eliminated along this expanding settlement frontier (Lamprey & Reid 2004; Norton-Griffiths et al. 2008) and by the associated land fragmentation (Norton-Griffiths 1998).

Underlying economic forces

This fundamental change on Kenya's rangelands from an extensive pastoral production system to a more intensive agricultural and agro-pastoral production system, with the attendant elimination of wildlife, can be most simply explained in terms of the differential net returns[6] to agricultural, livestock and wildlife production along the rainfall gradient (Figure 14.4).

The potential agricultural rents (Table 14.1 H) to be captured from Kenya's rangelands can be found by applying the relationship graphed in Figure 14.4 to each rainfall band (Table 14.1 A) and multiplying it by the area potentially available for cultivation outside of the protected areas (Table 14.1 D). Next, the areas actually supporting agriculture (Table 14.1 E) can be used to find the proportion of these rents that have been captured. Totally $670 million (35%) of the $1930 million of potential agricultural rents have been captured (Tables 14.1 I, 14.2 F), the greatest proportion being in the areas of higher rainfall and therefore, of higher agricultural potential.

A similar procedure using the relationship between rainfall and livestock production graphed in Figure 14.4 shows that the total livestock rents available from the rangelands are significantly lower and amount to around $573 million (Tables 14.1 K, 14.2 G). Since there is little evidence of a displacement or elimination of livestock by the conversion of land to cultivation, the captured

---

[6] These net returns to agricultural, livestock and wildlife production in Kenya (Norton-Griffiths & Butt 2006; Norton-Griffiths et al. 2008) are strictly financial and represent the net cash returns to landholders from these activities. They are cash-in-hand once all the direct costs of production and marketing have been met. They are accordingly quite different to revenues which represent gross returns before deducting the expenses of production and marketing.

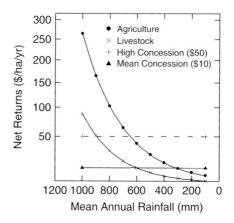

Figure 14.4 **Differential returns ($/hectare/year) to agricultural, livestock and wildlife production in the ASAL districts.**

agricultural rents of $670 million represent an absolute net gain, year on year, to the pastoral economy.

Potential wildlife rents can be calculated in a similar way from the areas of the rangelands actually used by tourists (Table 14.1 L, following Table 5.4 in Norton-Griffiths & Butt 2006). The most important of the wildlife utilization activities from the viewpoint of conservation are the fees paid to landholders for setting land aside and keeping it undeveloped so that tourists can view, undisturbed, the wildlife that live there. These concession and access fees paid to landholders by the tourism cartels average $10.2/hectare/year but are sometimes as high as $50/hectare/year (Figure 14.4). These give potential wildlife rents (to landholders) of $20–$100 million a year, vastly less than the rents from either livestock or agricultural production.

The reciprocal costs between these different production systems – agriculture, livestock and wildlife – are as important as are their differential returns in shaping the decisions by landholders as to which production system to favour. Livestock seem to be seamlessly integrated along the expanding agricultural frontier to the extent that converting rangeland to cultivation creates net gains, year on year, of some $670 million to the pastoral economy.

In contrast, wildlife can have a devastating impact on agricultural production and can literally wipe out a year's production in a single visit (Action Aid 2006). This undoubtedly explains why large wildlife have been completely eliminated from the high potential agricultural areas of Kenya and why they are being eliminated from the expanding agricultural frontier in the rangelands (Mizutani et al. 2005).

Wildlife have equally pernicious effects on livestock production (Norton-Griffiths & Butt 2006; Norton-Griffiths et al. 2008). While wildlife adds, perhaps, only 6% to the total operating costs of a livestock operation, this can represent anywhere up to 50% of the net operating profits. In other words, eliminating wildlife can effectively double the operating profits of livestock production. Given that wildlife benefits are restricted to 5% of the livestock range, it is not difficult to understand why wildlife are being eliminated.

Evolution of property rights

This pervasive expansion of cultivation across Kenya's rangelands is accompanied by a rapid evolution of property rights from large parcels of land under group or communal tenure to small parcels of land under private tenure. Perhaps the most striking example of this is on the high potential agricultural land around the Maasai Mara Game Reserve where the original 42 group ranches of an average size of 35,000 hectares have been transformed into some 30,000 private land holdings of around 50 hectares each (Map 1 in Norton-Griffiths et al. 2008). The Kitengela area of Kajiado District, abutting Nairobi, affords another striking example (Table 14.3; Kristjanson et al. 2002; Reid et al. 2007).

This process of land subdivision is driven by two main sets of incentives, some defensive and others opportunistic in nature. As defensive behaviour, subdivision secures and strengthens an individual's property rights against in-migration; against land alienation by political or economic elites (Homewood et al. 2004; Mwangi 2007); and against land alienation by conservation NGOs wishing to enlarge the area of conservation estate or even impose conservation easements onto productive land. In contrast, as opportunistic behaviour subdivision allows individuals to assume personal control of their social and economic future, in terms of capturing the economic benefits of agricultural, livestock and wildlife production directly at the household level rather than through communal institutions such as group ranch committees (Thompson & Homewood 2002; BurnSilver & Mwangi 2007); and capturing, again at the household level, the striking growth in land values throughout Kenya's rangelands.

These incentives manifest themselves in different ways. In Kenyan Maasailand, for example, the impetus to develop cultivation is driven primarily by the marked differential returns to agriculture with respect to both livestock and wildlife (Figure 14.4) and the motivation to capture these benefits, individually, at the household level. In contrast, in Tanzanian Maasailand, the main incentive is to simply demonstrate that the land is in 'productive use' and thereby strengthen the communal property rights against land alienation.

This subdivision of rangelands may not always lead to the elimination of wildlife. For example, on the now subdivided group ranches around the Mara, especially on the former Koiaki and Olkinyei group ranches, neighbouring private landholders are pooling their access and user rights to wildlife and forming Wildlife Conservancies or Wildlife Associations in which each landowner's share is proportional to the size of their landholding (Stiles 2007). These associations now negotiate directly with an individual tour operator and in return for exclusive access (and for leaving their land undeveloped and unused for either settlements or livestock) are obtaining significantly higher rents (at the household level) than before when everything was negotiated for, and channelled through, the group ranch committees (Thompson & Homewood 2002). Furthermore, these rents are now captured directly by the individual private landholders who make up the association and provide them with the working capital needed to develop farms elsewhere.

Nonetheless, this rapid evolution of private property rights on rangelands has two important large-scale economic impacts which do not bode well for wildlife. First, land values rise sharply with subdivision (the new owner now has something of value to sell), making it easier to raise capital for land development but in turn making the land more attractive to outside investors. Second, subdivision and private ownership imposes on the landholder a change from extensive to more intensive methods of production which in turn leads to raised productivity and an increase in the value of production (but see Kimani & Pickard 1998; Boone et al. 2005; Mwangi 2006; Thornton et al. 2006).

It is the process of urbanization that is creating and driving these new and powerful incentives to subdivide land throughout Kenya's rangelands, specifically the influence of burgeoning urban markets and periurban sprawl. The city of Nairobi, for example, now purchases annually $1.3–$2.4 billion of agricultural and livestock produce from the surrounding agricultural lands and rangelands[7], worth around $400–$500 million in farmgate prices. While this massive transfer of wealth from urban to rural hinterland contributes substantially towards the elimination of rural poverty, a secondary effect is to inflate the value of land in response to the growing demand for both quantity and quality of produce, which in turn, creates further incentives to invest in land subdivision, conversion and production (Tiffen 2006).

Urban sprawl creates in turn an insatiable demand for land for domestic and commercial use, especially in periurban areas. An excellent example of this is shown where northern Kajiado district abuts onto Nairobi Province, on the Kitengela, which surrounds the Nairobi National Park and over the entire Athi-Kaputiei plains (Reid et al. 2007). Thirty years ago most of this land was in group ranches and the main land uses were nomadic pastoralism and sport hunting of wildlife, although even then the first middle-class pioneers from Nairobi were purchasing land from the Maasai for domestic use. Today the land is wholly subdivided into individually owned land parcels; domestic, agricultural (horticulture and woodlots) and light-industrial development is widespread; pastoralism (and wildlife utilization) is effectively a thing of the past and most household income is from off farm sources (Kristjanson et al. 2002; Reid et al. 2007).

But the most dramatic change is to the value of this periurban land. Only 30 years ago the township of Rongai was a shabby slum settlement on the outskirts of Nairobi; today, land in Rongai sells at between $300,000–$400,000 per hectare depending upon location (Table 14.3). There is, naturally, a steep gradient of decreasing land values away from the Nairobi provincial boundary and out across the Kitengela, and by 25 km away land sells today at around $9000 a hectare (Table 14.3), a loss of some 14% in value for every kilometre away from Rongai.

Land values, 25 km out on the Kitengela have appreciated at over 11% per annum over the last 10 years (Table 14.3) which compares well against average ten-year returns from Treasury Bills. In comparison, the net returns to land from agricultural and livestock production, or from wildlife production, of 0.41% per annum and 0.11% per annum respectively, are simply derisory.

---

[7] A rough calculation based on 5 million people with a median daily food budget of between $1 and $1.3.

Today, the value of land in periurban areas like the Kitengela and the Athi-Kaputiei Plains has become completely dislinked from its agro-ecological potential and reflects solely its future value for periurban and urban development, where wildlife has no future.

There is no doubt that the political process of subdividing group ranches has been riven by disputes, with political and economic elites often winning out in terms of land allocation over less powerful individuals, and with some individuals sadly squandering their allocated land through highly injudicious land sales (Mwangi 2006; BurnSilver & Mwangi 2007; Mwangi 2007; Norton-Griffiths et al. 2008). But it is equally clear that these experiences of the past have not gone unnoticed and that the more recent subdivisions among group ranch members (for example, group ranches in the Mara area) have been to some extent more equitable. Be that as it may, however flawed the subdivision process may have been, the individual title holders are at least protected against the provisions in the recent Draft National Land Policy (of May 2007) to revoke all group ranch titles and revert the land back to some as yet unspecified form of communal tenure under the control of a central land board rather than local government. In this situation, half a loaf is clearly better than no bread at all.

Policy environment

Risk analysis of wildlife

Wildlife are still most at risk from further agricultural expansion on the 9% of rangelands receiving more than 800 mm of annual rainfall. While 52% of these rangelands already support agriculture (Table 14.2 E), resulting in the loss of some 54% of the wildlife that used to be there (Table 14.2 O), they still hold 31% of all the remaining wildlife. Given the huge discrepancies between the returns to agricultural and livestock production versus wildlife production (Figure 14.4), these wildlife must be considered to be at a very high risk of elimination.

Another 44% of wildlife are found on the rangelands receiving between 500 and 800 mm of rainfall. Even though there remain significant agricultural rents to be captured, agricultural expansion is undoubtedly slower in these areas and will probably not be so extensive. Wildlife here seem to be holding their own so should be considered to be at a medium-to-low risk of elimination.

The remaining 25% of wildlife are found in the areas receiving less than 500 mm annual rainfall. Even though agricultural expansion is extremely unlikely in these areas in the foreseeable future (only some 1% currently supports cultivation), the high rate of loss (−4.17%/annum) shows these wildlife also to be at high risk.

Why are returns to wildlife so low?

Wildlife returns to pastoral landholders are low and uncompetitive through a combination of Policy Failures, Institutional Failures and Market Failures.

Policy Failures originate from the continuing ban on all consumptive utilization of large wildlife which restricts the opportunities for pastoral landholders to generate revenues from their wildlife resources. The impact of this policy failure is to largely disenfranchise 95% of the pastoral rangelands from any income generating opportunities from wildlife: tourist wildlife viewing (and its associated income generating opportunities) is restricted to a mere 20,000 km^2 (5% of the total) in only 8 out of the 19 ASAL districts where wildlife is found (Norton-Griffiths & Butt 2006; Table 14.1 L). Other policy failures concern the continuing investment of wildlife ownership and user rights entirely in the State and the denial of compensation to landholders for the costs of raising wildlife[8].

Together, these three policy failures ensure that for most landholders wildlife remain a cost while yielding meagre benefits. Perversely, current policy actively prevents wildlife from becoming fully marketable goods making them ever yet more uncompetitive against agricultural and livestock production. For the

---

[8] The sheer inconsistency in all this is completely astonishing since some consumptive utilization of wildlife is still permitted but with very restricted benefit streams. The companies ranching crocodiles, ostrich and butterflies create local benefits primarily through employment opportunities. In contrast, bird shooting (either pest control on rice schemes, or game birds on ranchland) creates significant revenues, between $10,000 and $20,000 a year, for some group ranches. Returns from bird shooting could be significantly higher if the landholders were more skilled in negotiating contracts with the shooting operators (see footnote 10). Furthermore, the State accepts wildlife from the Private Sector to restock protected areas – but without making any payment, and provides wildlife (typically rhinoceros) to the Private Sector, again without accepting any payment – even though it is fully recognized that the Private Sector makes profits from this same wildlife through tourism activities.

great majority of landholders, the sensible option is to disinvest completely in the wildlife resource.

Institutional Failures are found in the KWS, among the conservation NGOs and in local institutions. First, the KWS acts as a regulatory and enforcement service rather than an enabling institution; it lacks technical expertise in wildlife production and management and endlessly vacillates in applying regulations. The impact of this is to reduce incentives on the part of pastoral landholders to invest in and encourage wildlife.

Second, many conservation NGOs are often too focused on single issues which rarely concern the economics of producing wildlife; they are largely unaware of the importance of market forces in determining land use and production decisions by pastoral landholders and they are far too reticent in challenging government over policy issues. The impact of this is inappropriate investments on the part of the NGO community and too little support for the development of free and unencumbered markets for wildlife goods and services.

Finally, the leadership of many communal institutions, such as group ranch committees, pander to locally powerful elites and fail to keep the interests of their ordinary members in mind when entering into development or tourism contracts and when disbursing revenues from such contracts. The impact of this has been to fuel demands for subdivision so that economic benefits can be captured directly at the household level.

Market Failures for the provision of wildlife goods and services stem primarily from the tourism cartels which divert the major portion of all wildlife generated revenues away from the producers of wildlife – the pastoral landholders – to the service side of the industry (agents and the providers of transport and accommodation). In general terms, landholders (which here include private landholders, the KWS and County Councils) see perhaps 5% at most of the total revenues generated by wildlife. The tourism cartels also maintain barriers that prevent landholders from becoming more directly involved in the tourism business (e.g. transport, accommodation) and thus capturing for themselves more of the potential revenues[9]; and finally they pass

---

[9] With the exception of guiding (to which there are now severe barriers in the form of 'standards'), landholders find it difficult to engage in other income generating opportunities. Few have the capital or management capacity to enter the transport or accommodation sectors (unless heavily subsidized).

onto the landholders a disproportionate amount of the business risk involved in tourism[10].

Policy responses

To most landholders, wildlife has become a liability and it is their best economic interests to disinvest in the resource and eliminate it. If wildlife are to survive in any meaningful way on Kenya's rangelands then the overall policy objective must be to transform wildlife from a liability into an asset by creating economic incentives for landholders to manage, conserve and invest in wildlife.

Three 'policy bundles' are called for. The first is an economic bundle to improve the revenues that landholders receive from wildlife; next is a property rights bundle to settle issues of ownership and user rights to wildlife; and finally an institutional bundle to create the required enabling environment within which what is essentially private sector conservation can flourish.

The **Economic Bundle** would aim to improve wildlife generated revenues for landholders. In all rangeland areas there could be wider and more equitable sharing of protected area revenues with neighbouring landholders; regulated access rights to harvest natural products (medicinal plants, honey, etc.); enhanced payments for ecosystem services (perhaps from the NGO and donor communities) for maintaining critical corridors or dispersal areas; fairer and more transparent compensation schemes for loss of life and damage to property; and expansion of wildlife tourism into new areas (but without harming economically the areas where they currently go).

In the current wildlife tourism areas, there is a need to improve the negotiating skills of landholders with the tourism cartels so that they can capture a larger share of wildlife revenues[11]: specifically to negotiate contracts

---

[10] An operator will typically pay a relatively small amount as a concession or access fee but will load up the bed night fee. When business is slack, both the landholder's and operator's revenue falls – but the landholder, unlike the operator, cannot reduce his costs. Such arrangements should be replaced with a fixed lease – as with agricultural leases. After all, in one case an operator is renting land to grow wheat and in the other he is renting land to grow wildebeest – so why should the terms of business be any different? To be fair, some operators have now seen the light and are paying something approaching the true opportunity costs of the land (Stiles 2007; Norton-Griffiths et al. 2008).

[11] This is seen even more clearly with bird shooting where landholders typically receive less than 2% of the daily fee paid by the sportsmen to the operators. In Europe, sportsmen willingly pay well over £1000 per gun day on premium shoots – the potential in Kenya is great, but landholders must learn negotiating skills.

which provide for concession and access fees that match the agricultural and/or livestock potential of the land (i.e. to the true opportunity cost of leaving the land undeveloped), which lower the barriers to involvement in the tourism industry, and which equitably share business risks. The capacity of landholders to establish and manage tourism ventures as profitable individual firms must also be strengthened.

But most importantly it is the clear and immediate need to relax the current restrictions on income generating opportunities from wildlife and open up again the full range of utilization and value added activities to landholders. These include live sales of wildlife between landholders (restocking or destocking ranches), between landholders and the State (restocking protected areas), and between the State and landholders (restocking ranches); wildlife ranching (antelope, birds, reptiles, insects) for local or overseas markets, in either live sales or in wildlife products; culling of locally abundant populations; value added activities of tanning and sales of skins, and production of trophies and curios; and to include mammals in the current sport hunting of birds and fish (Barnett & Patterson 2006; Lindsey et al. 2006; Lindsey et al. 2007).

The **Property Rights Bundle** would aim to settle issues of ownership and user rights to wildlife[12]. Important measures are to recognize wildlife management and production as a legal form of land use; devolve user rights to wildlife to the landholders on whose land the wildlife are found; devolve ownership of wildlife to the landholders on whose land the wildlife are found; facilitate neighbouring landholders to pool their access, user and ownership rights; and strengthen the legal foundation to foster the formation and registration of local wildlife resource use institutions, including wildlife forums, private conservancies and wildlife associations.

Finally, the **Institutional Bundle** would aim to create the required enabling environment within which private sector conservation can flourish.

The KWS requires major reform. Specifically, the KWS must be removed from political interference and control and be made answerable and responsible solely to its Board of Directors; the Board of Directors itself, while maintaining a 'majority' of government appointees, should have greater representation from the landholders who have to live with wildlife on their

---

[12] In Botswana, wildlife is state owned but private land owners have been given custodial rights to use it (Fauna Conservation Act 1982); in Zimbabwe, the Wildlife Act of 1975 gave private farmers the right to utilize and derive the full benefit of their wildlife resources; and in Namibia, the Nature Conservation Ordinance of 1967 privatized the ownership of wildlife on privately owned land.

land; and the KWS must transform itself from a regulatory and enforcement institution into one that encourages and supports in every way possible the conservation, management and utilization of wildlife on private land outside the protected areas.

Private landholders, wildlife forums, conservancies and associations must be recognized as the custodians of wildlife outside the protected areas; economic incentives, rather than regulation or enforcement, should form the basis of an enduring partnership with the private sector in meeting national conservation goals; and the management of all protected areas (National Parks) and County Council reserves should gradually be brought under Private Sector management – much like the Mara Conservancy is today.

The conservation NGOs should openly support the development of a free and unencumbered market for wildlife goods and services, both non-consumptive and consumptive alike. Specific interventions could include wildlife and habitat management; develop and strengthen local wildlife resource use institutions such as wildlife forums, private conservancies and wildlife associations; and enhance the commercial capabilities of private land-holders, forums, conservancies and associations to develop and manage wildlife utilization ventures, both non-consumptive (tourism) and consumptive, as profitable individual firms.

Finally, local institutions such as Group Ranch Committees are gradually being displaced by the newer institutions of wildlife forums, associations and conservancies. These are more firmly embedded in commercial practices and are less open to corruption. It is these new institutions that should be encouraged and supported, leaving the older ones to wither away on the vine.

The Wildlife (Conservation and Management) Bill (2009)

In September 2006, the Government of Kenya instituted a national consultative review of wildlife policy which was to lead to a new Wildlife Act. A National Steering Committee was established along with a policy drafting team and views were sought from one and all throughout the country in a series of 2 national and 22 regional seminars.

The consultative process resulted in a Draft Wildlife Policy (dated April 17 2007) that went some of the way towards promoting wildlife conservation as an economic form of land use in Kenya. The Policy tackled the serious wildlife governance issues that currently persist and created some genuine incentives for pastoral landowners to conserve and invest in wildlife, including a more relaxed policy towards consumptive utilization.

But the new Wildlife (Conservation and Management) Bill (Government of Kenya 2009) now largely ignores, and even undermines, the spirit of the earlier Draft Wildlife Policy. It imposes a top-down autocratic approach to wildlife conservation and management which does nothing to address the catastrophic decline of wildlife in Kenya and which effectively disenfranchises communities and landowners from wildlife management decisions (Norton-Griffiths 2007).

The worst aspect of the new Wildlife Bill is that it offers no real incentives to pastoral landholders and communities to conserve and invest in wildlife. On the contrary, it proposes such crippling compliance costs and regulations that all but the most wealthy community groups and landholders will be inhibited from creating viable wildlife conservation areas and will have to abandon their current activities. For example, in a seemingly innocuous move the Act bundles 'wildlife tourism' under 'wildlife user rights' which communities and landowners must apply for from the Ministry of Forestry and Wildlife. However, the application (along with detailed financial plans and management plans and wildlife monitoring schemes) must be made through interlocking levels of local, sub-regional and regional wildlife committees before either the Ministry or the KWS will consider an application for approval.

Furthermore, the Bill places such tight restrictions on the implementation of any consumptive utilization (only game ranching and cropping are contemplated) that they are unlikely ever to be sanctioned, while banning outright sport hunting, bird shooting and all fishing, whether artisanal, commercial or sport. In a final irony, the Bill creates the unconstitutional power to annex private land for conservation in that conservation easements can be imposed on land against a landowner's wishes.

Kenya's wildlife is poorly served by this Bill, which broadly fails to address the serious problems facing wildlife conservation.

Conclusion

Kenya's rangelands are undergoing a fundamental transformation in land use from nomadic, transhumant pastoralism to a more sedentary agropastoralism. This process of transformation is cascading down the rainfall gradient and is proceeding faster and is more advanced, in the areas of higher agricultural potential.

This transformation is being driven by macro- and micro- economic forces within the Kenyan economy which scarcely existed even 25 years ago. At the macro-economic scale both domestic (primarily urban) and international markets are expanding, there are real gains in producer prices, ever increasing opportunities for off-farm jobs and investment, and a wider availability and choice of goods and services. At the micro-economic level are improved market and transport networks, improved information networks about market conditions and improved access to financial services. All of these create incentives for pastoral landholders to increase returns to land by investing in land development and production.

Even greater incentives are created by the clear differentials between the returns to agricultural, livestock and wildlife production, which are so great, that the benefits from agricultural production overwhelm those from either livestock or wildlife, even in the areas of highest use by wildlife tourists.

Over half of the most productive rangelands in Kenya, which used to hold the great majority of wildlife, are now supporting agricultural production with an associated rapid evolution of property rights from large land parcels under communal or group ownership to small land parcels under private ownership. While livestock seem to be generally absorbed within the expanding agricultural and settlement frontier, wildlife are being displaced and eliminated.

Current conservation policy in Kenya, as evidenced by the new Wildlife (Conservation and Management) Bill (Government of Kenya 2009), demonstrates little awareness by the Government of Kenya of either the magnitude of these problems or of any real commitment to redress them.

References[13]

Action Aid. (2006) *Wildlife conservation concerns in Kenya*. Action Aid International, Nairobi.

Africover. (2003) www.africover.org for full details of methodologies, applications and downloads.

Barnett, R. & Patterson, C. (2006) *Sport Hunting in the Southern African Development Community (SADC) Region: An Overview*. TRAFFIC East/Southern Africa, Johannesburg.

---

[13] Papers authored (or co-authored) by Norton-Griffiths may be downloaded from the website mng5.com. The Draft National Land Policy (May 2007) and the Wildlife Act may also be found there.

Berger, D. (1993) *Wildlife Extension: Participatory Conservation by the Maasai of Kenya*, ACTS Environment Policy Series, vol. 4. African Centre for Technology Studies, Nairobi.

Boone, R.B., Burnsilver, S.B., Thornton, P.K., Worden, J.S. & Galvin, K.A. (2005) Quantifying declines in livestock due to land subdivision. *Rangeland Ecology and Management* 58, 523–532.

Broten, M.D. & Said, M. (1995) Population trends of ungulates in and around Kenya's Maasai Mara Reserve. In: Sinclair, A.R.E. & Arcese, P. (eds.) *Serengeti II: Dynamics, Management, and Conservation of an Ecosystem*. University of Chicago Press, Chicago, pp. 169–193.

BurnSilver, S. & Mwangi, E.N. (2007) *Beyond Group Ranch Subdivision: Collective Action for Livestock Mobility, Ecological Viability, and Livelihoods*, CAPRi Working Paper No. 66. International Food Policy Research Institute, Washington, D.C.

Food and Agricultural Organization of the United Nations. (1978) Wildlife Management Kenya: Plans for Rural Incomes from Wildlife in Kajiado District, AG/KEN/71/525 Technical Report 1. Food and Agricultural Organization of the United Nations, Rome.

Government of Kenya. (1995a) *Data Summary Report for the Kenyan Rangelands 1977–1994*. Ministry of Planning and National Development (Department of Resource Surveys and Remote Sensing), Nairobi.

Government of Kenya. (1995b) *National Rangelands Report: Summary of Population Estimates for Wildlife and Livestock*. Ministry of Planning and National Development (Department of Resource Surveys and Remote Sensing), Nairobi.

Government of Kenya. (2009) *Wildlife (Conservation and Management) Bill (2009)*. Government Printers, Nairobi.

Homewood, K., Coast, E. & Thompson, D.M. (2004) In-migrants and exclusion in east African rangelands: access, tenure and conflict. *Africa* 74, 567–610.

Kimani, K. & Pickard, J. (1998) Recent trends and implications of group ranch subdivision and fragmentation in Kajiado district, Kenya. *The Geographical Journal* 164, 202–213.

Kristjanson, P.M., Radeny, D., Nkedianye, R. et al. (2002) *Valuing Alternative Land-Use Options in the Kitengela Nairobi, Kenya*, International Livestock Research Institute (ILRI) Impact Assessment Series, vol. 10. International Livestock Research Institute & African Conservation Centre, Nairobi.

Lamprey, R.H. & Reid, R.S. (2004) Expansion of human settlement in Kenya's Maasai Mara: What future for pastoralism and wildlife? *Journal of Biogeography* 31, 997–1032.

de Leeuw, J., Prins, H.H.T., Njuguna, E.C., Said, M.Y. & de By, R.A. (1998) *Interpretation of DRSRS Animal Counts (1977–1997) in the Rangeland Districts of Kenya*. International Institute for Aerospace Survey and Earth Sciences, Enschede.

Lindsey, P.A., Alexander, R., Frank, L.G., Mathieson, A. & Romanach, S.S. (2006) Potential for trophy hunting to create incentives for wildlife conservation in Africa where alternative wildlife-based land uses may not be viable. *Animal Conservation* 9, 283–298.

Lindsey, P.A., Roulet, P.A. & Romanach, S.S. (2007) Economic and conservation significance of the trophy hunting industry in sub-Saharan Africa. *Biological Conservation* 134, 455–469.

Mizutani, F., Muthiani, E., Kristjanson, P. & Recke, H. (2005) Impact and value of wildlife in pastoral livestock production systems in Kenya: possibilities for healthy ecosystem conservation and livestock development for the poor. In: Osofsky, S.A. (ed.) Cleaveland, S., Karesh, W.B., Kock, M.D., Nyhus, P.J., Starr, L. & Yang, A. (assoc. eds.) *Conservation and Development Interventions at the Wildlife/Livestock Interface: Implications for Wildlife, Livestock and Human Health*. Occasional Paper of the IUCN Species Survival Commission No. 30. International Union for Conservation of Nature and Natural Resources (IUCN), Gland.

Mudsprings Inc. (1999) Download meta data from www.mudsprings.com.

Mwangi, E.N. (2006) *Subdividing the Commons: The Politics of Property Rights Transformation in Kenya's Maasailand*, CAPRi Working Paper No. 46. International Food Policy Research Institute, Washington, D.C.

Mwangi, E.N. (2007) The puzzle of group ranch subdivision in Kenya's Maasailand. *Development and Change* 38, 889–910.

Norton-Griffiths, M. (1995) Economic incentives to develop the rangelands of the Serengeti: implications for wildlife conservation. In: Sinclair, A.R.E. & Arcese, P. (eds.) *Serengeti II: Dynamics, Management, and Conservation of an Ecosystem*. University of Chicago Press, Chicago, pp. 588–604.

Norton-Griffiths, M. (1996) Property rights and the marginal wildebeest: an economic analysis of wildlife conservation options in Kenya. *Biodiversity and Conservation* 5, 1557–1577.

Norton-Griffiths, M. (1998) The economics of wildlife conservation policy in Kenya. In: Milner-Gulland, E.J. & Mace, R. (eds.) *Conservation of Biological Resources*. Blackwell Science, Oxford, pp. 279–293.

Norton-Griffiths, M. (2007) How many wildebeest do you need? *World Economics* 8, 41–64.

Norton-Griffiths, M. & Butt, B. (2006) *The economics of land use change Loitokitok Division, Kajiado District, Kenya*, Lucid Working Paper 34 International Livestock Research Institute, Nairobi. www.lucideastafrica.org.

Norton-Griffiths, M., Said, M.Y., Serneels, S. et al. (2008) Land use economics in the Mara area of the Serengeti ecosystem. In: Sinclair, A.R.E., Packer, C., Mduma, S.A.R. & Fryxell, J.M. (eds.) *Serengeti III: Human Impacts on Ecosystem Dynamics*. University of Chicago Press, Chicago, pp. 379–416.

Ottichilo, W.K., Grunblatt, J.M., Said, M.Y. & Wargute, P. (2000) Wildlife and livestock population trends in the Kenya rangeland. In: Prins, H.H.T., Grootenhuis, J.G. & Dolan, T.T. (eds.) *Wildlife Conservation by Sustainable Use.* Kluwer Academic Publishers, Boston, pp. 203–218.

Ottichilo, W.K., de Leeuw, J., Skidmore, A.K., Prins, H.H.T. & Said, M.Y. (2001) Population trends of large non-migratory wild herbivores and livestock in the Maasai Mara ecosystem, Kenya 1977–1997. *African Journal of Ecology* 38, 202–216.

Reid, R.S., Gichohi, H., Said, M.Y. et al. (2007) Fragmentation of a peri-urban savanna in the Athi-Kaputiei plains, Kenya. In: Galvin, K.A., Reid, R.S., Behnke, R.H. & Hobbs, N.T. (eds.) *Fragmentation of Semi-Arid and Arid Landscapes: Consequences for Human and Natural Systems.* Springer, Dordrecht, pp. 195–224.

Sinclair, A.R.E. (1995) Serengeti past and present. In: Sinclair, A.R.E. & Arcese, P. (eds.) *Serengeti II: Dynamics, Management, and Conservation of an Ecosystem.* University of Chicago Press, Chicago, pp. 3–30.

Stiles, D. (2007) The Mara at the crossroads. *SWARA* 30(3), 52–55.

Thompson, M. & Homewood, K. (2002) Entrepreneurs, elites and exclusion in Maasailand: trends in wildlife conservation and pastoral development. *Human Ecology* 30, 107–138.

Thornton, P.K., BurnSilver, S.B., Boone, R.B. & Galvin, K.A. (2006) Modelling the impacts of group ranch subdivision on agro-pastoral households in Kajiado, Kenya. *Agricultural Systems* 87, 331–356.

Tiffen, M. (2006) Urbanization: impacts on the evolution of 'mixed farming' systems in sub-Saharan Africa. *Experimental Agriculture* 42, 259–287.

Western, D., Russell, S. & Cuthill, I. (2009) The status of wildlife in protected areas compared to non-protected areas of Kenya. *PLoS ONE* 4(7), e6140. doi:10.1371/journal.pone.0006140.

World Resources Institute; Department of Resource Surveys and Remote Sensing, Ministry of Environment and Natural Resources, Kenya; Central Bureau of Statistics, Ministry of Planning and National Development, Kenya & International Livestock Research Institute (2007) *Nature's Benefits in Kenya, An Atlas of Ecosystems and Human Well-Being.* World Resources Institute, Washington, D.C., Nairobi.

15

Synthesis: Local and Global Solutions to the Challenge of Keeping Rangelands Wild

James C. Deutsch

Wildlife Conservation Society, NY, USA

Will our children, and our children's children, inherit 'wild' rangelands – expansive ecosystems used by people to raise livestock which also retain most of their native wildlife species, habitat structure and function? Will these iconic wildlife landscapes continue to sustain unique human cultures and provide reliable livelihoods for some of the world's most marginalized people (Redford et al. 2008)? Will they still contribute local and global environmental services: animal protein, leather, fibre, water, tourism and carbon storage? Or will these ecosystems continue to degrade and simplify, losing what is left of their native biodiversity and habitat structure? Will they become wastelands of unpalatable bushes and barren compacted soils or sites for the industrial production of crops or livestock? What will happen to the biological and human communities that depend on them? And what can we, as academics, conservationists and range-managers, do to influence the future of rangelands?

The chapters in this book, some focusing on specific rangelands around the world and some seeking to draw general lessons about rangeland function and management, document the decline of rangeland habitat and wildlife and the challenges to keeping them wild. They paint a generally bleak picture but also offer tentative solutions and causes for hope. In this conclusion, I revisit the question 'Why conserve wild rangelands?', review some solutions to local

Wild Rangelands: Conserving Wildlife While Maintaining Livestock in Semi-Arid Ecosystems, 1st edition. Edited by J.T. du Toit, R. Kock, and J.C. Deutsch.
© 2010 Blackwell Publishing

and global challenges put forward by contributors to this book and pose some questions that future studies could fruitfully address.

From the birth of the national park and conservation movement in the late nineteenth century through the first three quarters of the twentieth, the spectacular landscapes, vast ungulate herds and complex predator guilds of the world's open landscapes were valued as much as forests. Indeed, many of the first national parks [Yellowstone, Virunga (formerly Albert), Kruger] protected grass- or shrub-land as well as forests. With the invention of the term 'biodiversity' in the 1980s and the imperative to slow the rate of species extinction, the global conservation movement shifted focus to tropical rain-forests where more than half of the world's terrestrial species are thought to live (Wilson 1986). More recently, the contribution of tropical deforestation to carbon emissions and global warming has intensified this shift. Except in a few parts of the world (e.g. eastern and southern Africa) where wildlife-based tourism is economically important, conservation of open lands has languished, with these vast and valuable areas as often as not considered wastelands.

This volume and the symposium on which it is based are a small effort to redress this imbalance. The contributors were asked to analyse the conservation challenges facing wild rangelands – places where domestic livestock still coexist with wild ungulates and predators. Why? Why care about the conservation of open landscapes, and why consider areas where wildlife coexist with livestock rather than focusing on national parks and other protected areas which exclude people and livestock?

First, because the globe that we humans steward would be vastly impoverished, aesthetically and materially, should the last wild ecosystems of grass, bush, ungulates and predators be lost. Having evolved in the savanna, we value it, perhaps above all other landscape types. I grew up in America's Northeast, and like many Americans I cherish memories of my family's first trip 'out west' to Yellowstone. Across southern Africa, the most rapidly growing source of income is ecotourism based on foreign visitors, much of it focused on savanna wildlife. Nor is such appreciation limited to the descendents of Europeans – rangelands sustain unique and rich cultures across Africa and Asia. We will have done our children and grandchildren a great disservice if we bequest them a world with no migrations of pronghorn, wildebeest, white-eared kob or Mongolian gazelle, no hunting prides of lions or packs of African wild dogs.

Open ecosystems provide us with material benefits too. This is most obvious where their destruction has led to catastrophe, as in the American dustbowl of the 1930s. Grass- and bush-lands function as water catchments, support harvests of wild meat especially valuable in hard times and provide communities and nations with vital foreign exchange from tourism and hunting. There is also evidence for their importance in maintaining globally significant carbon stores in soil and peat (Conant et al. 2001).

Could conservation efforts focused solely on national parks and protected areas preserve the aesthetic and material benefits of grass- and bush-lands? The clear answer, reflected in the global conservation movement's shift over the past two decades to a 'landscape approach', is 'no'. Wildlife migrations and seasonal movements, watershed management and carbon storage all operate on spatial scales much too large to be contained within protected areas, and this applies even more to relatively arid lands, with their greater seasonality, than to forests. Disease transmission between wildlife, livestock and people, legal and illegal hunting, alien invasives and human-wildlife conflict all link wildlife in protected areas with the people, livestock and lands that surround them, making 'island conservation' a futile endeavour. Thus, protecting the global value of grass- and bush-lands requires that we conserve not only protected areas but much larger rangelands – vast areas used by people to produce livestock which also support wildlife and generate environmental goods and services.

Yet, wild rangelands have fared miserably over the past century. In North America and most of southern Africa, livestock and free-ranging wildlife have been rigorously isolated from each other to prevent disease transmission between domestic and wild bovids. Those native ungulates that remain have been mostly confined within fenced and managed parks and ranches, halting migrations and drastically altering landscapes, ecosystems and human cultures. In Australia and South America, wild rangelands have been permanently altered by extinctions, introduced plant and animal species and overgrazing by livestock. In Africa's Sahel and much of Asia, coexistence between livestock and wildlife continues in principle, but in practice over-hunting has decimated most wildlife populations and much of the range has been degraded, overgrown with alien invasive plants or desertified.

Four of the chapters in this book focus on challenges and solutions in some of the last places where significant populations of native wildlife still coexist with traditional pastoralism: in East Africa, Tanzania (Chapter 13 by Homewood & Thompson) and Kenya (Chapter 14 by Norton-Griffiths & Said and Chapter 13), and in Central Asia, Mongolia (Chapter 12 by Scharf and co-authors) and Trans-Himalaya (Chapter 11 by Mishra and co-authors).

In each of these places, the traditional cultural and political systems that pastoralists developed over centuries to regulate hunting and range use by livestock decayed during the twentieth century as a result of rapid political and social change. Nowhere have modern regulatory regimes filled the void, and the predictable result has been over-hunting, over-grazing and (in East Africa) accelerating conversion to agriculture. Yet in each of these places, large areas of rangeland remain relatively intact with most native species still present, if in reduced numbers, and habitat structure relatively unchanged. Each of these places thus represents a globally important opportunity to preserve or rehabilitate wild rangelands.

In two of these countries, Tanzania and Mongolia, periods of centralized control under socialist governments (and before that British administration in Tanganyika) have given way to less regulated capitalist systems. Traditional control systems were forcibly abolished under socialism but replaced by centralized efforts at management. Since 1989, with a shift to capitalism in both countries, central control has languished, creating open access and accelerating over-hunting and over-grazing. Interestingly, the authors considering each of these countries advocate the same solution: empowering local communities to reinstate traditional controls on both hunting and range-management (stocking and seasonal use). It seems likely, however, that in order to succeed this strategy would require substantial capacity building and oversight from government or other authorities.

In contrast, capitalist Kenya never attempted to replace traditional range management with central control. Land ownership was privatized shortly after independence, relying on economic incentives for landowners to manage their rangelands wisely. Yet this subdivision of the range ended large seasonal movement of livestock (for a time – people are now negotiating reciprocal rights to forage with their neighbours), and Kenya also outlawed all hunting, reducing the value of wildlife to local people. Over-grazing and over-hunting predictably followed. East Africa is also barred from exporting beef to developed countries because of foot and mouth disease, and few benefits from tourism accrue to local people. With so few financial benefits accruing to local people from traditional range management, Norton-Griffiths and Said show that most people are better off financially in converting all but the most arid rangelands (<250 mm rainfall/year) to agriculture – which is exactly what they are doing. They advocate bringing back safari-hunting to increase incentives to pastoralists to maintain rangelands and wildlife populations.

Like Mongolia, Trans-Himalayas (including areas of China, India and neighbouring countries) remains sparsely populated and undeveloped economically. Traditional pastoralists herding a variety of species still coexist with native ungulates and carnivores, both within the 8% of the land designated as protected areas and right across the sub-region. Mishra and co-authors argue that what is needed is large-scale participative landscape planning and zoning to identify the areas most important for wildlife and reconcile these needs with those of local people.

Compared to the rangelands of East Africa or Central Asia, those of Australia, South America and North America are even less 'wild', as species have been lost; wildlife is mostly restricted to protected areas or domesticated herds, and the range has been substantially altered by intense grazing and introduced species. Nevertheless, authors considering these areas argue for preserving the biodiversity and environmental services that remain and shifting the goal of management regimes to promoting conservation.

Garnett and co-authors (Chapter 8) consider the challenges to maintaining or reconstructing wild rangelands in tropical (North and Central) Australia. Australia lost most of its native large mammal species shortly after the arrival of humans 50,000–20,000 years ago. Moreover, plant and animal introductions over the past hundred years have changed the ecosystem and overgrazing has led to increased woody cover. The authors argue that considerable value, nevertheless, remains in biodiversity and the provision of ecosystem services. They advocate elevating conservation of this biodiversity and ecosystem service provision to become the main goal of range management across Australia's tropical rangelands and practising active management. This would be a significant shift from the current policy where conservation is the chief objective on only an ecologically unrepresentative 6% of the rangeland set aside as protected area.

Baldi and co-authors (Chapter 10) discuss conservation of Patagonia's rangelands, especially the only large herbivore, the guanaco. Only 1% of Patagonia is set aside in protected areas; severe over-stocking with sheep during the last century has resulted in permanent degradation of the range including increased bushiness, and the guanaco population has been reduced by 90% to remnant sub-populations. The authors advocate careful study of the impact of shearing on wild guanacos, protection of remaining sub-populations and the removal of sheep fences to allow migration between these. They do not, however, suggest additional economic or political incentives to encourage conservation – perhaps creative incentive regimes of the type suggested by Victurine and Curtin (Chapter 7 and below) need to be considered.

In the final geographically focused chapter (Chapter 9), Fleischner considers the history, conservation challenges and solutions for wildlife management on the vast rangelands of America's West. Until the 1970s, he reports, rangelands were managed solely in the interest of ranchers, to maximize cattle production. The result has been drastic ecological change generally detrimental to wildlife but impacting specific plant species and communities, birds and mammals differently. Fleischner advocates two management changes to benefit wildlife: the exclusion of cattle from riparian (riverine) habitat which is particularly fragile and vital to large numbers of wildlife species and (admittedly much more complicated) efforts to protect biological soil crusts which stabilize soil, retain moisture and fix nitrogen. Implementing these changes though, will require balancing the national interest in biodiversity preservation with the local interest in livestock production.

While the six chapters considering geographically focused challenges to wild rangelands suggest a variety of solutions (reinstating traditional range-management controls, improving economic incentives for maintaining rangeland and preserving wildlife, prioritizing conservation rather than subsidizing economically inefficient commercial pastoralism and protecting key habitats), the other six chapters attempt to draw general lessons from challenges that face wild rangelands around the world.

One of the most challenging obstacles to maintaining wild rangelands – where that means wildlife and livestock co-mingling in a single ecosystem – is disease transmission between wildlife and livestock. Indeed, this is the main issue that has resulted in the exclusion of wildlife from most rangelands in North America and southern Africa. Kock and co-authors (Chapter 5) suggest various solutions, including 'compartmentalization', or isolating animals headed for the human food-chain, 'commodity based trade', or processing disease-free food products for export even where diseases such as foot and mouth have not been eliminated and targeted control measures, such as vaccination. If people such as East Africa and Central Asia's traditional pastoralists are to be incentivized to maintain wild rangelands, combinations of measures such as these will need to be implemented swiftly.

Zimmermann and co-authors (Chapter 6) consider the related but vexing challenge of human–predator conflict in rangelands. Wherever livestock and predators coexist – wolves in North America, wild dogs in Africa, dingoes in Australia, jaguars in South America, tigers in Asia, lynx in

Scandinavia – conflict ensues. The problem is exacerbated where people do not accompany their livestock on the range and where wild prey has been depleted. Solutions vary and must be based on knowledge of the biology and behaviour of the predators, the attitudes and behaviour of the people, common sense, trial and error and good science. What Zimmermann and co-authors do not mention is that no solution is likely to be perfect and that the coexistence of livestock and predators will necessarily require that those who value predator existence must be willing to contribute financially over the long term, ideally for predator tolerance rather than compensation.

The two chapters by Archer (Chapter 4) and Walker (Chapter 2) consider general patterns of rangeland ecological evolution common across the globe and implications for management. From Australia, to Patagonia, to the United States of America, to the Sahel and East Africa, a common challenge facing rangeland managers (and both wild and domestic grazers) is an increase in woody vegetation cover and decrease in palatable grass accompanied by compaction of the soil. Archer surveys the various possible causes contributing to this trend, including overgrazing by livestock and fertilization by the increased atmospheric concentrations of carbon dioxide (CO_2) caused mainly by burning of fossil fuels since the Industrial Revolution. Walker presents a summary of complex modelling and empirical work demonstrating that systems such as rangelands, often have paired sets of equilibrium states. Thus, for example, rangelands that have been overgrazed by domestic stock may shift from a grassland-dominated equilibrium (with continuous grass cover, permeable soil, sparse woody cover and frequent fires) to a bushland-dominated equilibrium (with evenly spaced woody shrubs separated by bare earth across which fire cannot easily spread). Though caused by overgrazing, this new equilibrium may persist even after grazers are removed from the system. Walker and his colleagues advocate a 'resilience approach', in which rangeland managers seek to increase the resilience of the system to perturbation in order to keep it in the preferred state.

Considering practical approaches to wild rangeland conservation, du Toit (Chapter 3) argues that selecting the appropriate spatial and temporal scales and institutional level for intervention is crucial to success. For example, only recently have Serengeti conservationists focused on water usage far outside the landscape in the Mara River catchment, though the risk of the Mara running dry during the dry season may be a much greater threat to the Serengeti migration than illegal, but sustainable, hunting. An example of managers operating at the wrong temporal scale is the unwise prevention of fire in an effort to increase stability in the short term – often resulting in long-term rangeland changes that are difficult to undo. Du Toit also advises that, in a world of limited resources, conservationists must choose their fights and they would be wise to focus on places where the opportunity costs to society of conservation are low (such as arid habitats unsuitable for farming) and where the cost-effectiveness of intervention is expected to be high (generally where human infrastructure is sparse but not non-existent).

In perhaps the most practically useful chapter in the book, Victurine and Curtin (Chapter 7) address the challenge of creating economic incentives for pastoralists and ranchers to keep rangelands wild. They review examples of three types of legal and financial mechanisms: the provision of direct payment incentives for maintaining land as rangeland tolerating the presence of wildlife, the purchase of conservation easements restricting landowners' freedom to convert land to other uses, and the creation of markets for rangeland-friendly and wildlife-friendly products. Successful examples of direct payments include payments from tourism operators in Tarangire, Tanzania, to Masai pastoralists to maintain wet season dispersal ranges for elephants and wildebeest, and a novel and cost-effective auction of land-management contracts among ranchers in Victoria, Australia. The American West provides examples of both innovative conservation easements (including 'grass-banking' or the trading of certain everyday usage rites for the right to use set-aside common forage during droughts) and new markets for sustainably raised beef. Victurine and Curtin point out that conservationists wishing to devise such solutions face two challenges – of designing an intervention appropriate to the ecological, sociological and legal context, and of finding a conservation-minded stakeholder willing to provide the payments, purchase the easements or buy the products. But, in a world in which conservation benefits often accrue to the global community or concerned people in the developed world while their costs are often borne by poor people in developing countries, designing such transactions is the first step towards negotiating successful deals and finding sustainable solutions.

To what extent do the ideas and advice offered in the general chapters in this volume address the challenges posed by the geographically specific chapters? The problem of conversion from pastoral to agricultural land and a low tolerance for wildlife in East Africa seems directly amenable to the direct payment options offered by Victurine and Curtin, especially given the economic importance of wildlife to Kenya and Tanzania's economies. For Trans-Himalaya, Mishra and co-authors seem to have taken on board du Toit's lessons about finding the appropriate spatial scale in their recommendation that the conservation focus shift from local protected area management to large-scale land-use planning across the plateau. Perhaps creating innovative markets for guanaco-friendly wool could go some way towards incentivizing Argentinean ranchers to manage habitat for guanacos. The far more ecologically challenging task of restoring Australia and North America's wild rangelands may benefit from Walker's resilience approach – not surprisingly, since that is just what the Resilience Group had in mind. And few rangelands anywhere are likely to co-mingle wild and domestic ungulates in the future unless alternative solutions to disease transmission between wild and domestic ungulates can be devised as outlined by Kock and co-authors.

The global perspective of this volume presents a paradox: in developing countries, especially in eastern and southern Africa and central Asia, native wildlife still co-mingle with livestock and rangelands remain relatively 'wild'. Yet here the poverty of both rangeland residents and the nations in which they live means that effective conservation of rangelands will occur only if creative economic incentives are devised to reward rangeland residents for stewarding the range and its wildlife. The sort of economic analysis presented by Norton-Griffiths and Said and the creative financial mechanisms reviewed by Victurine and Curtin are thus crucial. Funding for these mechanisms will need to come from wealthier stakeholders, such as international donors or tourists. In contrast, the general public of many developed countries (e.g. the United States and Australia) are coming to value biodiversity and the provision of ecosystem services above the meagre returns from managing marginal lands for livestock, but here the range has already been radically converted; so much greater technical and financial input will be required to return it to anything approaching 'wild'.

Future symposia or volumes should extend this search for ways forward to important areas both geographical and topical lying beyond the scope of this volume. For example, the arctic and sub-arctic rangelands of North America, Scandinavia and Asia represent some of the largest and most intact rangeland habitats and wildlife assemblages of the world. Zimmermann and co-authors (Chapter 6) touch on conflicts between newly reintroduced or recovered predators (such as wolves and lynx) and unaccompanied herds in these areas, but an analysis of other challenges facing these rangelands could provide an important comparison to the tropical and temperate examples in this book. The Llanos of Venezuela, Pantanal of Brazil, Beni of Bolivia and grasslands of West Africa and the Sahel have experienced dramatic reductions in wildlife numbers but still harbour globally important wildlife species and provide valuable environmental services. A global approach to rangeland management and restoration needs to include them too. To Victurine and Curtin's menu of financial incentives for rangeland conservation should be added the efforts to quantify the economic value of ecological services offered by rangelands in order to persuade national governments or donor agencies to purchase these, as well as requirements that companies, nations or development agencies offset the residual biodiversity impact on rangelands of their projects.

Over the past few years, environmental discussion globally has come to be dominated by climate change, but very little of this discussion has focused on rangelands. Climate models suggest that many rangelands, especially arid ones, will be disproportionately impacted by climate change (Pachauri & Reisinger 2007). For example, much of southern Africa is likely to become substantially drier (Pachauri & Reisinger 2007). Future studies need to document the likely impact of climate change on rangelands and discuss adaptive strategies for promoting persistence of rangeland habitats and wildlife. Climate change mitigation and rangelands is also a neglected topic. Current climate change agreements include credit for re-forestation, and future agreements are likely also to include incentives to reduce rates of tropical deforestation. Research is just beginning, however, on the impact on global carbon flux of different range-management regimes, especially those altering the quantity of carbon stored in soils. Should these carbon fluxes prove globally significant, discouraging CO_2 release and promoting storage in rangelands should be included in global market systems and overseas development aid incentives.

References

Conant, R.T., Paustian, K. & Elliott, E.T. (2001) Grassland management and conversion into grassland: effects on soil carbon. *Ecological Applications* 11, 343–355.

Pachauri, R.K. & Reisinger, A. (eds.) (2007) *Report of the Intergovernmental Panel on Climate Change* Core Writing Team, Geneva: Intergovernmental Panel on Climate Change.

Redford, K.H., Levy, M.A., Sanderson, E.W. & de Sherbinin, A. (2008) What is the role for conservation organizations in poverty alleviation in the world's wild places? *Oryx* 42, 516–528.

Wilson, E.O. (1986) *Biodiversity*. National Academy Press, Washington, D.C.

Index

Tables in **bold** and figures in *italic*